UNION CARBIDE CORP.
So. Charleston, WV

MAR 6 1997

LIBRARY

ACS SYMPOSIUM SERIES **618**

Enzymatic Degradation of Insoluble Carbohydrates

John N. Saddler, EDITOR
University of British Columbia

Michael H. Penner, EDITOR
Oregon State University

Developed from a symposium sponsored
by the Division of Agricultural and Food Chemistry
at the 207th National Meeting
of the American Chemical Society,
San Diego, California,
March 13–17, 1994

American Chemical Society, Washington, DC 1995

Library of Congress Cataloging-in-Publication Data

Enzymatic degradation of insoluble carbohydrates / John N. Saddler, editor, Michael H. Penner, editor.

p. cm.—(ACS symposium series, ISSN 0097–6156; 618)

"Developed from a symposium sponsored by the Division of Agricultural and Food Chemistry at the 207th National Meeting of the American Chemical Society, San Diego, California, March 13–17, 1994."

Includes bibliographical references and indexes.

ISBN 0–8412–3341–1

1. Polysaccharides—Biotechnology—Congresses.
2. Polysaccharides—Metabolism—Congresses. 3. Glucosidases—Biotechnology—Congresses. 4. Cellulase—Biotechnology—Congresses.

I. Saddler, John N., 1953– . II. Penner, Michael Henry. III. American Chemical Society. Division of Agricultural and Food Chemistry. IV. American Chemical Society. Meeting (207th: 1994: San Diego, Calif.) V. Series.

TP248.65.P64E59 1995
661.8'02—dc20   95–50257
                                                      CIP

This book is printed on acid-free, recycled paper.

Copyright © 1995

American Chemical Society

All Rights Reserved. The appearance of the code at the bottom of the first page of each chapter in this volume indicates the copyright owner's consent that reprographic copies of the chapter may be made for personal or internal use or for the personal or internal use of specific clients. This consent is given on the condition, however, that the copier pay the stated per-copy fee through the Copyright Clearance Center, Inc., 222 Rosewood Drive, Danvers, MA 01923, for copying beyond that permitted by Sections 107 or 108 of the U.S. Copyright Law. This consent does not extend to copying or transmission by any means—graphic or electronic—for any other purpose, such as for general distribution, for advertising or promotional purposes, for creating a new collective work, for resale, or for information storage and retrieval systems. The copying fee for each chapter is indicated in the code at the bottom of the first page of the chapter.

The citation of trade names and/or names of manufacturers in this publication is not to be construed as an endorsement or as approval by ACS of the commercial products or services referenced herein; nor should the mere reference herein to any drawing, specification, chemical process, or other data be regarded as a license or as a conveyance of any right or permission to the holder, reader, or any other person or corporation, to manufacture, reproduce, use, or sell any patented invention or copyrighted work that may in any way be related thereto. Registered names, trademarks, etc., used in this publication, even without specific indication thereof, are not to be considered unprotected by law.

PRINTED IN THE UNITED STATES OF AMERICA

1995 Advisory Board

ACS Symposium Series

Robert J. Alaimo
Procter & Gamble Pharmaceuticals

Mark Arnold
University of Iowa

David Baker
University of Tennessee

Arindam Bose
Pfizer Central Research

Robert F. Brady, Jr.
Naval Research Laboratory

Mary E. Castellion
ChemEdit Company

Margaret A. Cavanaugh
National Science Foundation

Arthur B. Ellis
University of Wisconsin at Madison

Gunda I. Georg
University of Kansas

Madeleine M. Joullie
University of Pennsylvania

Lawrence P. Klemann
Nabisco Foods Group

Douglas R. Lloyd
The University of Texas at Austin

Cynthia A. Maryanoff
R. W. Johnson Pharmaceutical Research Institute

Roger A. Minear
University of Illinois at Urbana–Champaign

Omkaram Nalamasu
AT&T Bell Laboratories

Vincent Pecoraro
University of Michigan

George W. Roberts
North Carolina State University

John R. Shapley
University of Illinois at Urbana–Champaign

Douglas A. Smith
Concurrent Technologies Corporation

L. Somasundaram
DuPont

Michael D. Taylor
Parke-Davis Pharmaceutical Research

William C. Walker
DuPont

Peter Willett
University of Sheffield (England)

Foreword

THE ACS SYMPOSIUM SERIES was first published in 1974 to provide a mechanism for publishing symposia quickly in book form. The purpose of this series is to publish comprehensive books developed from symposia, which are usually "snapshots in time" of the current research being done on a topic, plus some review material on the topic. For this reason, it is necessary that the papers be published as quickly as possible.

Before a symposium-based book is put under contract, the proposed table of contents is reviewed for appropriateness to the topic and for comprehensiveness of the collection. Some papers are excluded at this point, and others are added to round out the scope of the volume. In addition, a draft of each paper is peer-reviewed prior to final acceptance or rejection. This anonymous review process is supervised by the organizer(s) of the symposium, who become the editor(s) of the book. The authors then revise their papers according to the recommendations of both the reviewers and the editors, prepare camera-ready copy, and submit the final papers to the editors, who check that all necessary revisions have been made.

As a rule, only original research papers and original review papers are included in the volumes. Verbatim reproductions of previously published papers are not accepted.

Contents

Preface ix

1. Comparison of Enzymes Catalyzing the Hydrolysis
   of Insoluble Polysaccharides 1
   David B. Wilson, Mike Spezio, Diana Irwin, Andrew Karplus,
   and Jeff Taylor

2. Molecular Deformations and Lattice Energies of Models
   of Solid Saccharides 13
   A. D. French, M. K. Dowd, S. K. Cousins, R. M. Brown,
   and D. P. Miller

3. Glycosidases, Large and Small: T4 Lysozyme and
   Escherichia coli β-Galactosidase 38
   R. H. Jacobson, R. Kuroki, L. H. Weaver, X-J. Zhang,
   and B. W. Matthews

4. Barley Seed α-Glucosidases: Their Characteristics and Roles
   in Starch Degradation 51
   C. A. Henson and Z. Sun

5. Three-Dimensional Structure of an Endochitinase
   from Barley 59
   Jon D. Robertus and John Hart

6. Existence of Separately Controlled Plastic and Conserved
   Phases in Glycosylase Catalysis 66
   Edward J. Hehre

7. A Mutant α-Amylase with Enhanced Activity Specific
   for Short Substrates 79
   Ikuo Matsui, Kazuhoko Ishikawa, Eriko Matsui,
   Sachio Miyairi, Sakuzo Fukui, and Koichi Honda

8. Synergistic Interaction of Cellulases from *Trichoderma reesei*
   During Cellulose Degradation 90
   Bernd Nidetzky, Walter Steiner, and Marc Claeyssens

9. **Synergism Between Purified Bacterial and Fungal Cellulases** 113
   John O. Baker, William S. Adney, Steven R. Thomas,
   Rafael A. Nieves, Yat-Chen Chou, Todd B. Vinzant,
   Melvin P. Tucker, Robert A. Laymon,
   and Michael E. Himmel

10. **Cellulose-Binding Domains: Classification and Properties** 142
    Peter Tomme, R. Antony J. Warren, Robert C. Miller, Jr.,
    Douglas G. Kilburn, and Neil R. Gilkes

11. **Identification of Two Tryptophan Residues in Endoglucanase III from *Trichoderma reesei* Essential for Cellulose Binding and Catalytic Activity** 164
    Ricardo Macarrón, Bernard Henrissat, Jozef van Beeuman,
    Juan Manuel Dominguez, and Marc Claeyssens

12. ***Cellulomonas fimi* Cellobiohydrolases** 174
    Hua Shen, Andreas Meinke, Peter Tomme,
    Howard G. Damude, Emily Kwan, Douglas G. Kilburn,
    Robert C. Miller, Jr., R. Antony J. Warren,
    and Neil R. Gilkes

13. **Thermostable β-Glucosidases** 197
    Badal C. Saha, Shelby N. Freer, and Rodney J. Bothast

14. **Initial Approaches to Artificial Cellulase Systems for Conversion of Biomass to Ethanol** 208
    Steven R. Thomas, Robert A. Laymon, Yat-Chen Chou,
    Melvin P. Tucker, Todd B. Vinzant, William S. Adney,
    John O. Baker, Rafael A. Nieves, J. R. Mielenz, and
    Michael E. Himmel

15. **Enhanced Enzyme Activities on Hydrated Lignocellulosic Substrates** 237
    K. L. Kohlmann, A. Sarikaya, P. J. Westgate, J. Weil,
    A. Velayudhan, R. Hendrickson, and M. R. Ladisch

16. **Comparison of Protein Contents of Cellulase Preparations in a Worldwide Round-Robin Assay** 256
    William S. Adney, A. Mohagheghi, Steven R. Thomas,
    and Michael E. Himmel

17. Economic Fundamentals of Ethanol Production from Lignocellulosic Biomass 272
    Charles E. Wyman

18. Cellulose Degradation by Ruminal Microbes: Physiological and Hydrolytic Diversity Among Ruminal Cellulolytic Bacteria 291
    Paul J. Weimer and Christine L. Odt

19. Induction of Xylanase and Cellulase in *Schizophyllum commune* 305
    Dietmar Haltrich, Brigitte Sebesta, and Walter Steiner

20. Simultaneous Production of Xylanase and Mannanase by Several Hemicellulolytic Fungi 319
    G. M. Gübitz and Walter Steiner

21. Xylanase Delignification in Traditional and Chlorine-Free Bleaching Sequences in Hardwood Kraft Pulps 332
    Nelson Durán, Adriane M. F. Milagres, Elisa Esposito, and Marcela Haun

22. Paper Biopulping of Agricultural Wastes by *Lentinus edodes* 339
    G. Giovannozzi-Sermanni, A. D'Annibale, N. Vitale, C. Perani, A. Porri, and V. Minelli

23. Possible Roles of Xylan-Derived Chromophores in Xylanase Prebleaching of Softwood Kraft Pulp 352
    Ken K. Y. Wong, Patricia Clarke, and Sandra L. Nelson

Author Index 363

Affiliation Index 364

Subject Index 364

Preface

THE USE OF HYDROLYTIC ENZYMES to break down insoluble carbohydrates into potentially valuable products continues to be the focus of intense research activity. In areas such as the enzymatic hydrolysis of cellulose, the initial application of fermenting the released glucose to ethanol has broadened as we learn more about enzymes to include detergent, textile, and food applications. However, fundamental problems such as difficulties in obtaining good agreement in the protein content of a similar enzyme preparation (*see* Chapter 16) have yet to be fully resolved.

This book is intended to provide a general review of enzyme and substrate factors influencing the hydrolysis of insoluble carbohydrates, discuss several glucosidases and their mechanism of catalysis, and provide a more in-depth description of cellulase and hemicellulase enzymes and their possible applications. This was an ambitious goal, and we are grateful to the authors who strove to meet this objective. We hope that this text will provide an overview of specific and general issues that still limit our understanding of the mechanisms and applications of hydrolytic enzymes acting on insoluble polysaccharides.

Acknowledgments

We are grateful to the Agricultural and Food Chemistry and the Cellulose, Paper, and Textile divisions of the American Chemical Society for supplying the forum for this work. We also thank the International Energy Agency for providing the resources and the focus for much of the carbohydrate-to-fermentable sugar work that is profiled in the symposia. In particular, we thank the authors and reviewers for their time and expertise that should ensure that this book is a useful contribution to the continually expanding area of enzyme modification of polysaccharides.

JOHN N. SADDLER
Faculty of Forestry
University of British Columbia
Vancouver V6T 1Z4
Canada

MICHAEL H. PENNER
Department of Food Science
 and Technology
Oregon State University
Corvallis, OR 97331–6602

September 27, 1995

Chapter 1

Comparison of Enzymes Catalyzing the Hydrolysis of Insoluble Polysaccharides

David B. Wilson, Mike Spezio, Diana Irwin, Andrew Karplus, and Jeff Taylor

Department of Biochemistry, Molecular and Cell Biology, Cornell University, 458 Biotechnology Building, Ithaca, NY 14853

Amylases, cellulases, chitinases, lysozymes and xylanases are all insoluble polysaccharide hydrolases. Many of these enzymes possess a separate substrate binding domain that is pointed to the catalytic domain by a hinge peptide. All of them contain multiple binding sites for sugar monomers (from four to eight) in their active sites and either two or three carboxyl side chains that function in catalysis. There are three amylase families, nine cellulase families, two chitinase families, five lysozyme families and two xylanase families. The enzymes in some families invert the conformation of the glycoside oxygen during hydrolysis while the enzymes in the other families retain the conformation. The structure of an inverting endocellulase, *Thermomonospora fusca* E2, has been determined by X-ray crystallography at 1.18 Å resolution. It has an open active site cleft, while CBHII which is an exocellulase with the same polypeptide chain fold has its active site in a tunnel. Several E2 mutants have been isolated and characterized to help determine its catalytic mechanism.

Insoluble polysaccharides function either as storage polymers or as the major structural components of bacteria, crustacea, insects, fungi and plants. Because of their specific functions and the abundance of these polymers, many organisms produce enzymes which can degrade them. Most of these enzymes only hydrolyze one type of polysaccharide but some hydrolyze more than one. This either results from a single active site that can bind polymers with related structures such as cellulose and xylan (1) or chitin and bacterial cell walls (2) or because of the presence of two or more different catalytic domains in one protein (3).
A given type of enzyme can play quite different roles in different organisms. Thus plants produce cellulases to modify their cell walls to allow growth (4) or to ripen fruits (5) while plant pathogens produce cellulases to allow entry through plant cell walls (6). Many bacteria and fungi produce cellulases to utilize cellulose as a carbon source and they are essential for recycling plant cell walls in nature. Few animals make cellulases; so that ruminants, shipworms and termites utilize the cellulases produced by symbiotic bacteria and fungi to digest cellulose (7-9). Chitinases are produced by crustacea, fungi and insects to allow the organisms to

0097–6156/95/0618–0001$12.00/0
© 1995 American Chemical Society

grow; since chitin is a major structural material in these organisms (10). Chitinases also are produced by bacteria that utilize chitin as a carbon source. Interestingly, chitinases are produced by many plants as part of a defense mechanism against fungal pathogens (11).
There are several properties that are shared by many insoluble polysaccharide hydrolases. One is the presence of a separate domain that binds tightly to the insoluble polymer substrate. These domains have been found in amylases (12), cellulases (13), chitinases (14) and xylanases (15). Surprisingly all of the xylanase binding domains that have been studied so far bind cellulose while some of them bind xylan (16) and some do not (15). A *Streptomyces* chitinase was also found to contain a cellulose binding domain (17). Binding domains are C-terminal to the catalytic domain in some hydrolases and N-terminal in others. For the six *Thermomonospora fusca* cellulases we have studied the binding domain is C-terminal in three (E1, E2, and E4) and N-terminal in the others (E3, E5 and E6) (18). In all of the hydrolases containing binding domains it appears that the binding domain is joined to the catalytic domain by a flexible peptide called a hinge or linker region (19). In many cases these peptides are easily recognised because they are rich in proline, and either threonine, serine or glutamine and often have a repeating sequence such as x-Pro. The presence of a cellulose binding domain increases the activity of the cellulase on crystalline substrates such as Avicel, cotton or filter paper but usually does not effect the activity of the enzyme on soluble substrates like carboxymethyl cellulose or oligodextrins (20).
We have constructed two mutants that alter the hinge region of *T. fusca* endocellulase E2. The amount of insoluble and total reducing sugar produced from filter paper by the E2 catalytic domain, a mutant lacking the linker peptide and a mutant with a linker that is twice as long as the native linker were measured. The catalytic domain has about half of the activity of the native enzyme and produces a higher percentage of insoluble reducing sugar (48% versus 31%). Interestingly, addition of pure binding domain does not stimulate the catalytic domain, showing that the binding domain has to be joined to the catalytic domain to effect activity. The enzyme lacking the hinge peptide was more active than the E2 catalytic domain but less active than E2. The double linker enzyme is slightly more active than the native enzyme. These results support a model in which the enzyme is bound to its substrate by the binding domain and the hinge region allows the catalytic domain to catalyze a number of cleavages due to its flexibility.
Domains with amino acid sequence similarity to fibronectin type III domains are present in a number of bacterial hydrolases including an amylase, several cellulases and several chitinases. However the function of these domains is not known and most hydrolases do not contain them (10). All insoluble polysaccharide hydrolyases appear to have active sites that contain multiple binding sites for sugar monomers with a range of from four to as many as eight subsites. In addition these enzymes appear to utilize catalytic mechanisms that involve two carboxyl side chains with different pKs, as was first proposed for egg white lysozyme (21). One of these carboxyl groups is thought to be protonated and donates a proton to the leaving sugar while the other is thought to be ionized and either stabilizes the carbonium ion present in the transition state or forms a covalent linkage with the rest of the substrate. In some enzymes, a bound water may replace one of the carboxyl groups in this mechanism.
The strongest evidence for this mechanism comes from studies in which the three dimensional structure of enzyme-inhibitor complexes have been determined by x-ray crystallography and the carboxyl side chains identified in the structure have been changed to other residues by site directed mutagenesis. Such studies have been carried out on both hen egg white (22) and T4 lysozyme (23), the exocellulase *T. reesei* CBHII (24), *T. fusca* endoglucanase E2 (25), an amylase (26) and a xylanase (27). In each structure, carboxyl side chains were identified

that were correctly positioned to function in catalysis and mutants lacking these residues had greatly reduced activity. Additional evidence for the presence of essential carboxyl groups comes from experiments in which hydrolases have been reacted with carbodiimides which usually modify carboxyl groups and which inhibit most of the hydrolases that have been studied (28). Another common inhibitor of most of these enzymes is Hg^{++}, which normally reacts with sulfhydrol groups but since most of the insoluble polysaccaride hydrolases lack free sulfhydrol groups, Hg^{++} must be inhibiting in some other fashion, possibly by binding to the active site carboxyl groups.

For each class of enzyme, many different genes from a number of different organisms have been cloned and sequenced. These genes can be grouped into a number of families based on the amino acid sequences of their catalytic domains (19,29). It is very likely that all members of a given family have a similar three-dimensional structure and it is also likely that members of different families have different structures, but this is less certain since some proteins with very little sequence identity have similar three-dimensional structures. The number of families that have been proposed for each class of these enzymes is shown in Table I. Most cellulolytic organisms contain cellulases from different families and where they contain enzymes from the same family the enzymes are often quite divergent. Usually one is an endocellulase and one is an exocellulase. This is true for *T. reesei* where CBHI and Endo I both belong to family C and for *T. fusca* where E2, an endocellulase, and E3, an exocellulase, belong to family B while E1, an endocellulase, and E4, an exocellulase, belong to family E (30).

One of two stereochemically different mechanisms is utilized by each polysaccharide hydrolase. Some enzymes catalyze hydrolysis with inversion of the glycoside linkage, while the others catalyze hydrolysis with retention of the stereochemistry of the glycoside linkage. It appears that enzymes that give inversion catalyze a direct attack of water in an Sn_2 reaction on the glycoside linkage as shown in Figure 1a. Enzymes that give retention appear to catalyze a two step process in which an active site caboxyl group attacks the glycoside linkage with inversion, usually forming an enzyme bound intermediate (Fig. 1b). Then the intermediate is attacked by water, again inverting the conformation, with the overall result of two inversions being retention of the original stereochemistry. Most retaining enzymes can catalyze transglycosylation by reacting the enzyme bound intermediate with a sugar molecule rather than water (31). However, transglycosylation only occurs when there is a high concentration of an acceptor species to complete with water. Enzymes that invert do not catalyze transglycosylation. Studies have shown that all members of a given family appear to utilize the same mechanism (32). Both inventory and retaining enzymes have been demonstrated for amylases, with β amylases causing inversion and α amylases giving retention. Cellulases in family A and family C give retention while cellulases in family B and E give inversion. In the case of xylanases, the two major families both give retention of the β linkage (32). This is also true for lysozymes while the stereochemistry of chitinases has not been reported. However it has been reported that a bacterial chitinase can catalyze transglycosylation which indicates that it probably gives retention (33).

An important property of cellulases is their ability to give synergism in hydrolysis of crystalline cellulose. No cellulase acting alone has a specific activity on filter paper that is greater than one μmole/min/μ mole of enzyme while the best synergistic mixture of three cellulases has an activity of nearly 16 μ moles/min/μ mole of enzyme (20). We define synergism as the ratio of the activity of a mixture of enzymes divided by the sum of the activities of the enzymes in the mixture activity acting alone. Synergism is seen in most mixtures containing an endocellulase and an exocellulase and in some mixtures containing two different

Table I

Hydrolase Families (29,34)

Enzyme	Families
Amylase	13,14,15
Cellulase	5,6,7,8,9,12,44,45,48
Chitanase	18,19
Lysozyme	21,22,23,24,25
Xylanase	10,11

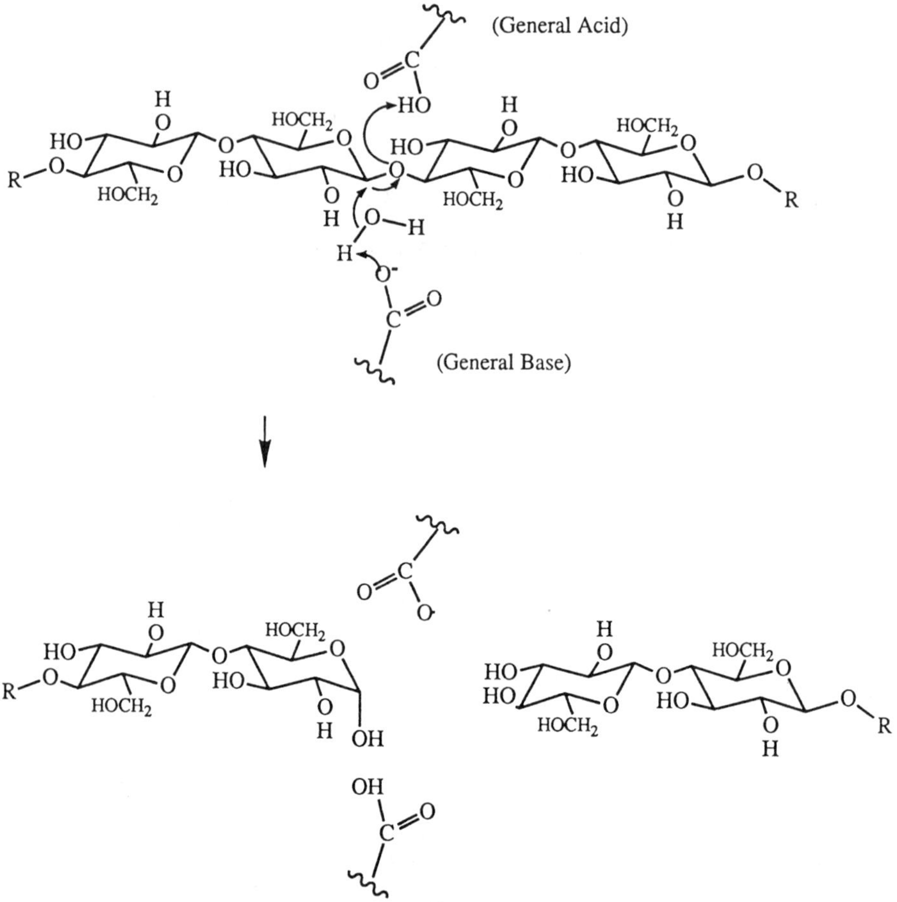

Figure 1a. Mechanism of hydrolysis of cellulose with inversion.

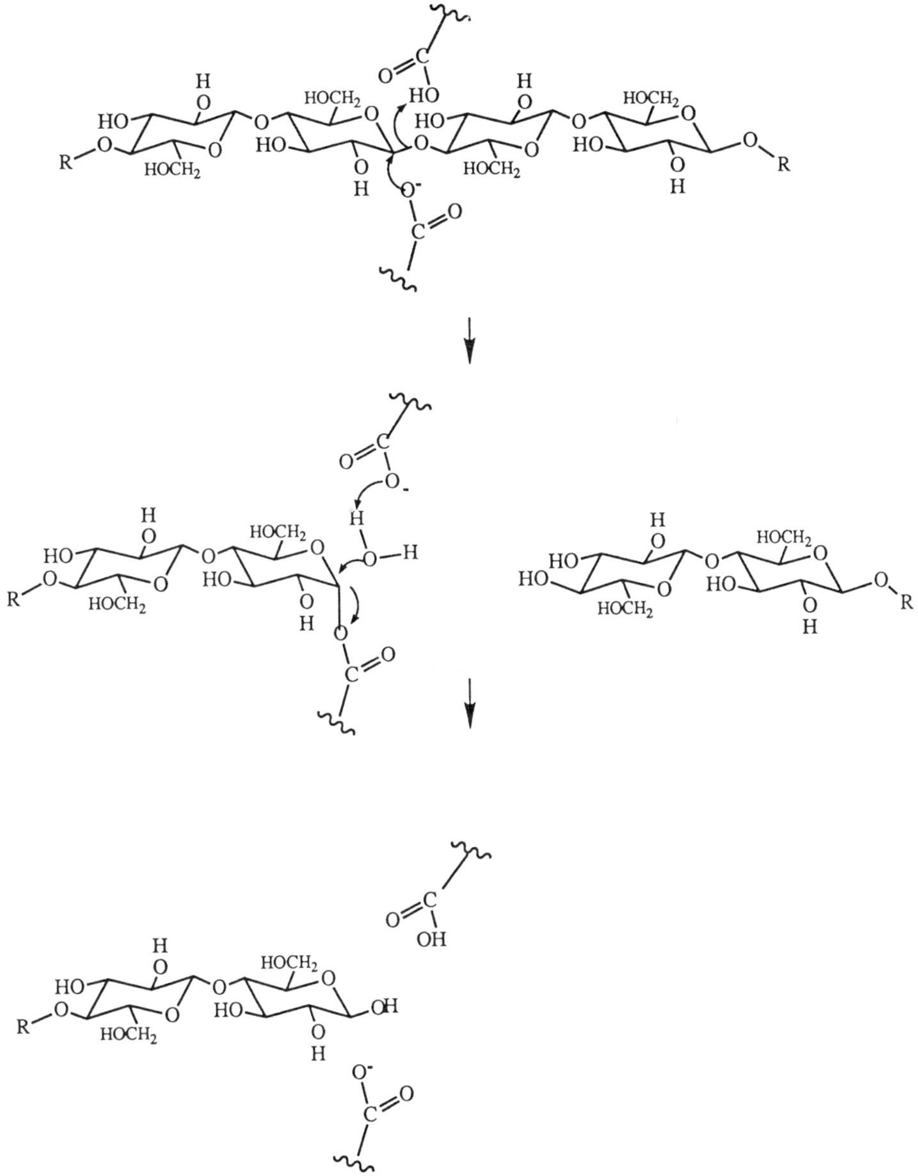

Figure 1b. Mechanism of hydrolysis of cellulose with retention.

exocellulases. It is very rare to see synergism between endocellulases. The optimum mixture contains two different exocellulases plus an endocellulase with the endocellulase making up only 20% of the total enzyme.

A major goal of our research on *T. fusca* cellulases is to engineer a cellulase mixture that has a higher activity in degrading insoluble cellulose then *T. reesei* crude cellulase. An important step forward toward this goal was the determination of the three-dimensional structure of the catalytic domain of endoglucanase E2 at 1.18Å resolution which was carried out by Mike Spezio, in collaboration with Dr. Andrew Karplus at Cornell. We first tried to crystallize the native enzyme but as I believe is always the case for proteins that contain hinge peptides we could not get crystals of the multidomain enzyme; so we then tried to crystallize the isolated catalytic domain which contains 286 amino acids and has a molecular weight of 30,000. Crystals were obtained from ammonium sulfate and by seeding we can reproducibly obtain crystals that diffract to 1.0Å (25). The structure was solved by multiple isomorphous replacement using four derivatives at 2.6 Å and then refined first to 1.8 Å and then to 1.18 Å. The R value for the refined structure including all reflections from 66 to 1.18 Å is 23.2% while the RMS deviation from ideality in the bond lengths is 0.008 Å and in the bond angles is 1.4°. The structure contains 230 water molecules and two bound sulfate ions. An example of the quality of the electron density that we obtain at our highest resolution (1.0Å) is shown in Figure 2 which shows the electron density map for Trp 41. At this level of resolution side chains that adopt two different conformations can be detected and about 15 of the 286 side chains show alternate conformations.

The overall model of E2cd that we have determined is shown in Figure 3 and it is a globular protein with a pronounced cleft at the top. The active site is in the cleft as was shown by the fact that when the structure of a cocrystal of E2cd and the competitive inhibitor, cellobiose, was determined, the cellobiose bound into the cleft. This is shown in Figure 4 which is a model of this structure. The basic chain fold of E2cd is an unusual parallel β barrel which is similar to the fold determined for the exocellulase, CBHII, from *Trichoderma reesei* (24) (Figure 5). The major difference between these two enzymes is that the open active site cleft in E2 is a tunnel in CBHII. The peptide backbones of the two enzymes are shown in Figure 6, where the thick line is the E2 α carbon trace while the thin line is the α carbon trace for CBHII. The closed cleft in CBHII results from an extra 20 amino acid loop on the left which is missing in E2 and from the alternate position of another loop in CBHII which completes the roof of the tunnel.

Some of the residues in the active site of E2 are shown in Figure 7. We believe that Asp[117] and Asp[265] probably provide the catalytically important carboxyl groups. Asp[265] is likely to be ionized as it appears to form a salt bridge with Arg[221] while Asp[117] is likely to be protonated as it is close to Asp[156]. We have isolated mutants in which Asp[265] was changed to either a Phe or a Val residue and find that they have greatly reduced activity. Work with other family B cellulases including CBHII have shown that the residue equivalent to Asp[117] is an essential residue, as changes in that residue reduce the activity to 1% or less (24). We have also isolated a mutant in which Glu[263] was changed to Gly. Glu[263] appears to participation substrate binding but the mutant has wild type activity on all cellulosic substrates. It does show lower activity on cellotetraose and that appears to be caused by a higher Km value. The fact that the Glu[263] mutant has normal activity provides indirect evidence that the low activity found for the Asp[117] and Asp[265] mutants does not result from the inability of E2 to fold properly due to the loss of a carboxyl group in the active site. We are trying to use the information about the structure of E2 to first understand how it catalyzes the hydrolysis of crystalline cellulose and then to design more active endocellulases.

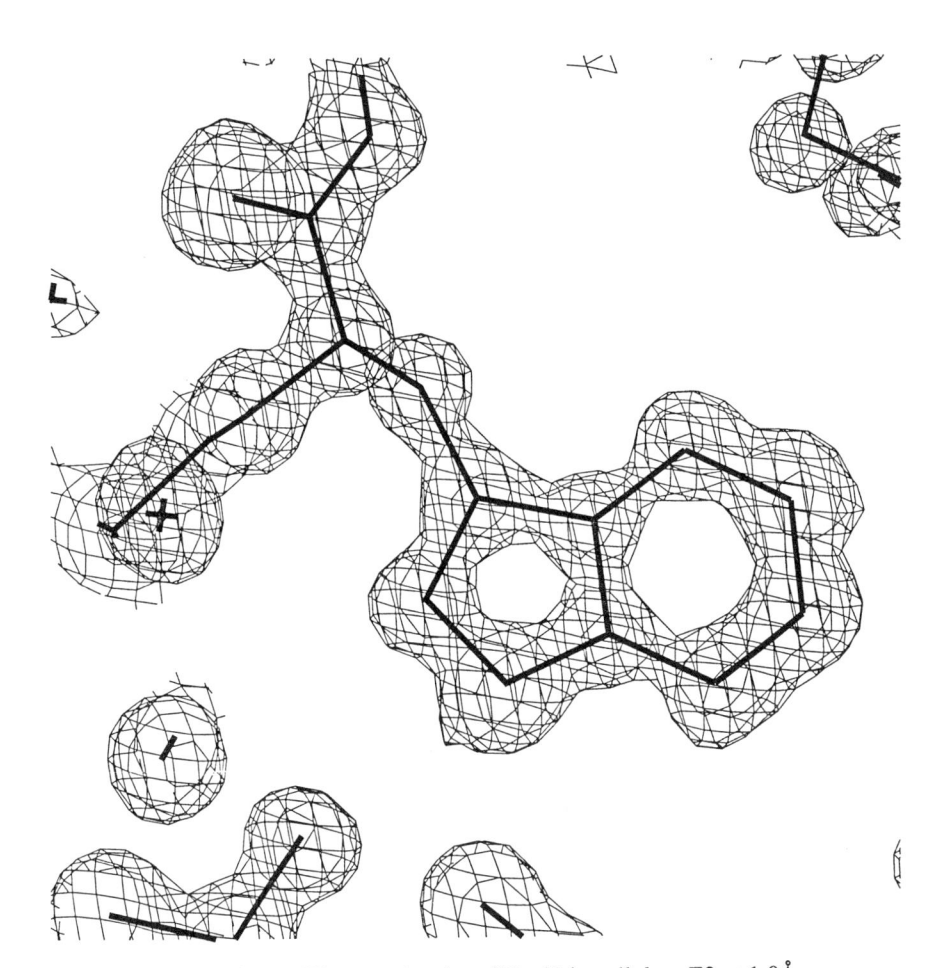

Figure 2. Electron density of Trp41 in cellulase E2 at 1.0Å resolution.

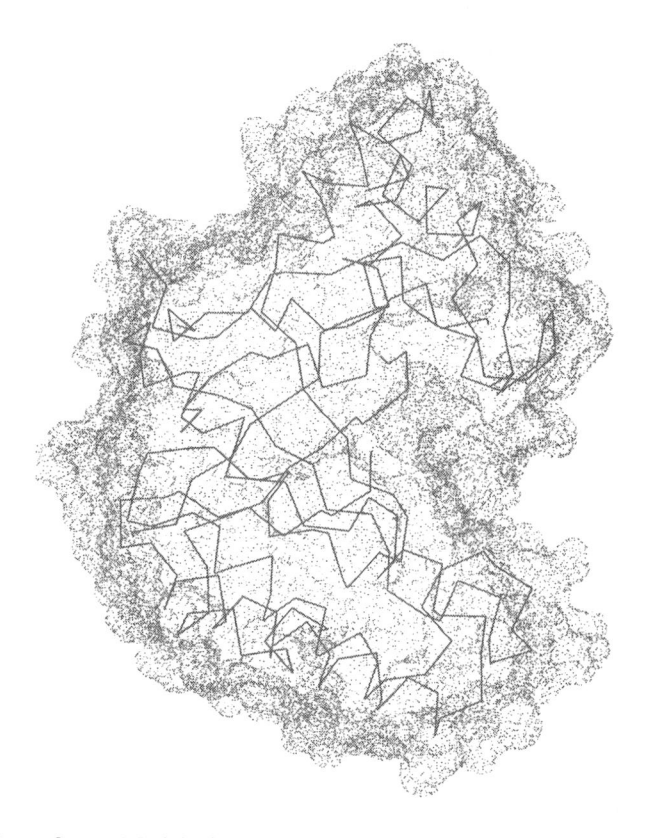

Figure 3. Model of E2 showing the peptide chain and accessible surface.

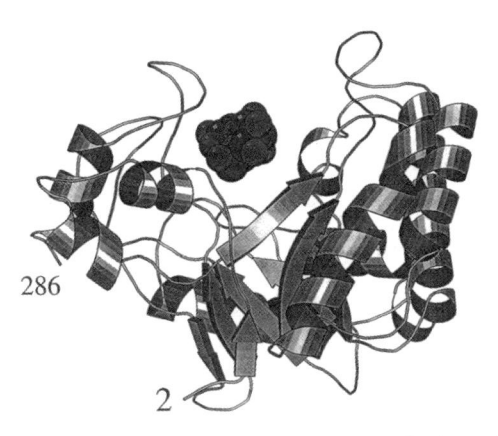

Figure 4. Ribon diagram of the E2 structure with bound cellobiose (side view).

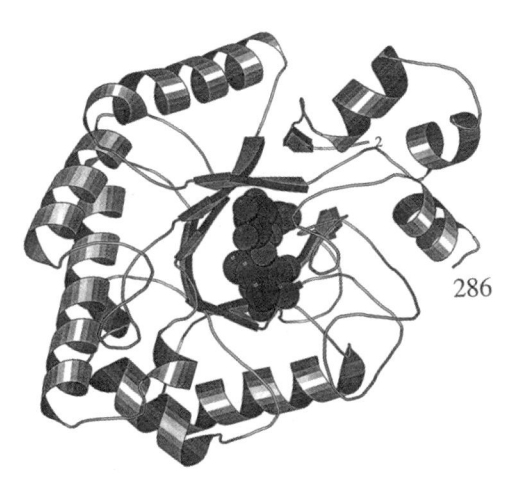

Figure 5. Ribon diagram of the E2 structure with bound cellobiose (top view).

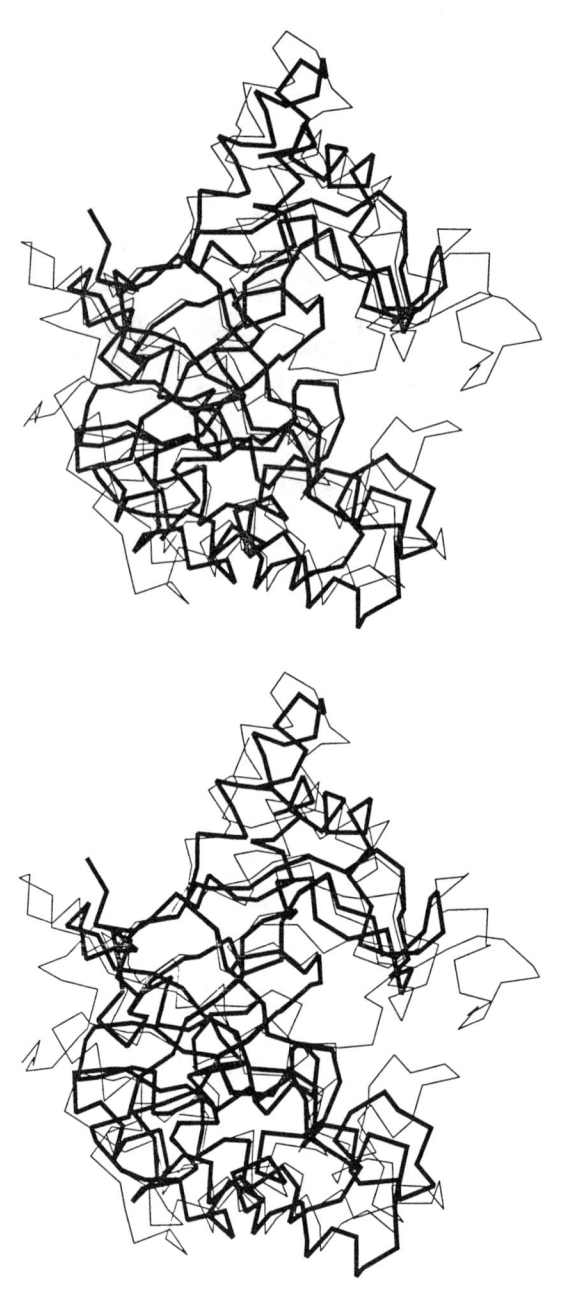

Figure 6. Comparison of the peptide chains of E2 (thick line) and CBHII (thin line).

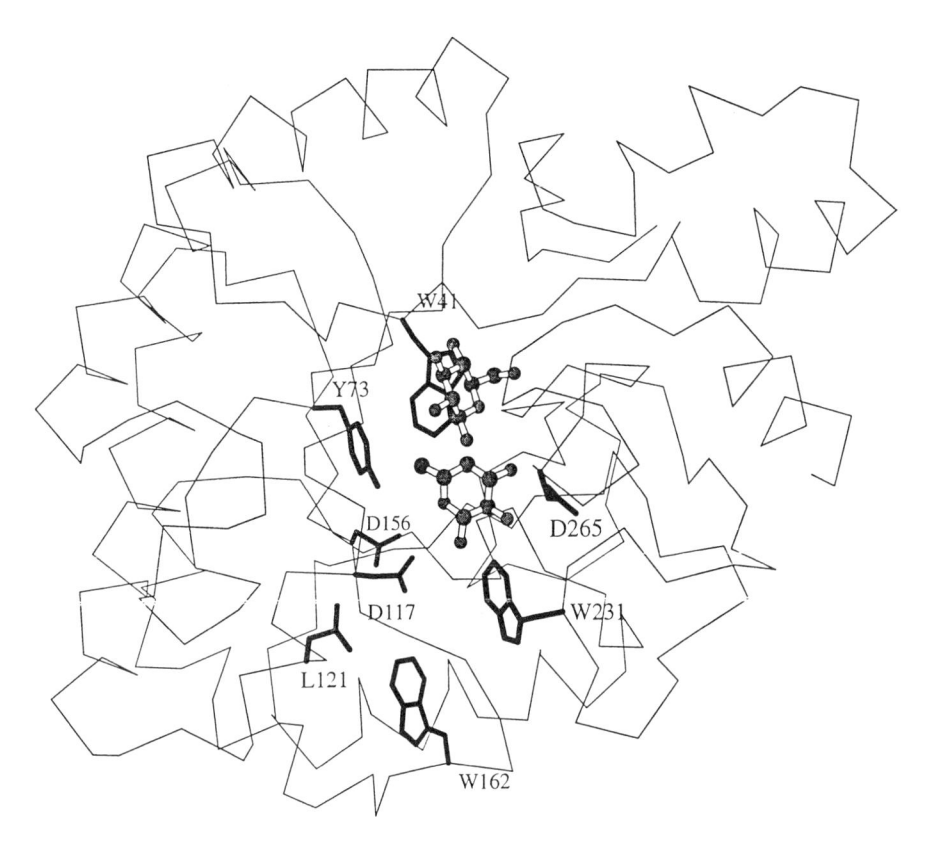

Figure 7. Potential catalytic residues in the active site of E2.

Literature Cited

1. Shoemaker, S.; Watt, K.; Tsitousky, G. and Cox, R. *Biotechnology* **1993**, *1*, pp.687-690.
2. Stintzi, A.; Hertz, T.; Prasad, V.; Weidemann-Merdinoglu, S.; Kauffmann, S.; Geoffrey, P. and Fritig, B. *Biochimie* **1993**, *75*, pp.687-706.
3. Gibbs, M.D.; Saul, D.J.; Luethi, E. and Berquist, P.L. *Appl. Environ. Microbiol.* **1992**, *58*, pp.3864-3867.
4. Truedsen, T.A. and Windaele, R. *J. of Plant Physiol.* **1991**, *139*, pp.129-134.
5. Buse, E. and Laties, G. *Plant Physiol.* **1993**, *102*, pp.417-423.
6. He, S.Y.; Lindberg, M.; Chaterjee, E.K. and Collmer, M. *Proc. Natl. Acad. Sci. USA* **1991**, *88*, pp.1079-1083.
7. Weimer, P.J. *Crit. Rev. Biotechnol.* **1992**, *12*, pp.189-223.
8. Waterbury, J.B.; Calloway, C.B. and Turner, R.D. *Science* **1983**, *221*, pp.1401-1403.
9. Martin, M.M. *Philos. Trans. R. Soc. Lond B. Biol. Sci.* **1991**, *333*, pp.281-288.
10. Flach, J.; Pilet, P.-E. and Jolles, P. *Experientia* **1992**, *48*, pp.702-716.
11. Mauch, F.; Hadwinger, L.A. and Boller, T. *Pl. Physiol.* **1988**, *87*, pp.325-333.
12. Coutinho, P.M. and Reilly, P.J. *Protein Engin.* **1994**, *7*, pp.291-299.

13. Vantilbeurgh, H., Tomme, P., Claoeyssens, M., Bikhabhai, R., and Petterson, G. Limited proteolysis of the cellbiohydrolase I from *Trichoderma reesei*: separation of functional domains. FEBS Lett. 204, 223-227 (1986).
14. Lucas, J.; Henshen, A.; Lottspeich, F.; Voegeli, U. and Boller, T. *FEBS Lett.***1985**, *193*, pp.208-210.
15. Hall, J.; Hazelwood, G.P.; Huskisson, N.S.; Durrant A.J. and Gilbert, H.J. *Mol. Microbiol.* **1989**, *3*, pp.1211-1219.
16. Irwin, D.; Jung, D.D. and Wilson, D.B. *Appl. Environ. Microbiol.* **1994**, *60*, pp.763-770.
17. Fujii, T. and Miyashita, K. *J. Gen. Microbiol.* **xxxx**, *139*, pp.677-686.
18. Wilson, D.B. *Crit. Rev. Biotechnol.* **1992**, *12*, pp.45-63.
19. Gilkes, N.R.; Henrissat, B.; Kilburn, D.G.; Miller, R.C. Jr. and Warren, R.A.J. *Microbiol. Rev.* **1991**, *55*, pp.303-315.
20. Irwin, D.C.; Spezio, M.; Walker, L.P. and Wilson, D.B. *Biotechnol. and Bioengin.* **1993**, *42*, pp.1002-1013.
21. Jolles, P. and Jolles, J. *Mol. Cell. Biochem.* **1984**, *63*, pp.165-189.
22. Malcolm, B.A.; Rosenberg, S.; Corey, M.J.; Allan, J.S.; Baetselier, A. and Kirsh, J.F. *Proc. Natl. Acad. Sci USA* **1989**, *86*, pp.133-137.
23. Anand, N.N.; Stephen, E.R. and Narang, S.A. *Biochem. Biophys. Res. Commun.* **1988**, *153*, pp.862-880.
24. Rouviner, J.; Bergfors, T.; Teeri, T.; Knowles, J.; and Jones, A. *Science* **1990**, *249*, pp.380-385.
25. Spezio, M.; Wilson, D.B. and P. A. Karplus. *Biochemistry* **1993**, *32*, pp.9906-9916.
26. Matsuura, Y.; Kusonoki, M.; Harada, N. and Kakudo M. *J. Biochem.* **1984**, *95*, pp.697-702.
27. Ko, E.P.; Akatsuka, H.; Moriyama, H.; Shinmyo, A.; Hata, Y.; Katsobe, Y.; Urabe, L.; Okada, H. *Biochem. J.* **1992**, *288*, pp.117-121.
28. McGinnis, K.; Wilson, D.B. *Biochemistry* **1993**, *32*, pp.8151-8156.
29. Henrissat, B. and Bairoch, A. *Biochem. J.* **1993**, *293*, pp.781-788.
30. Jung, E.D.; Lao, G.; Irwin, D.; Barr, B.K.; Benjamin, A. and Wilson, D.B. *Appl. and Env. Microbiol.* **1993**, *59*, pp.3032-3043.
31. Matsui, L.; Yoneda, S.; Ishikawa, K.; Miyaira, S.; Fukui, S.; Umeyama, H., and Honda, K. *Biochemistry* **1994**, *33*, pp.451-458.
32. Gebler, J.; Gilkes, N.; Claeyssens, M.; Wilson, D.B.; Beguin, P.; Wakarchuk, W.; Kilburn, D.; Miller, Jr. R., Warren ; R.A. and Withers, S.G. *J. Biol. Chemistry* **1992**, *267*, pp.12559-12561.
33. Nanjo, F.; Sakai; Ishikawa, M.; Isobe, K. and Usui, T. *Agric Biol. Chem.* **1989**, *53*, pp.2189-2195.

RECEIVED August 18, 1995

Chapter 2

Molecular Deformations and Lattice Energies of Models of Solid Saccharides

A. D. French[1], M. K. Dowd[1], S. K. Cousins[2], R. M. Brown[2], and D. P. Miller[3]

[1]Southern Regional Research Center, Agricultural Research Service, U.S. Department of Agriculture, P.O. Box 19687, New Orleans, LA 70179
[2]Department of Botany, University of Texas, Austin, TX 78713
[3]P.O. Box 423, Waveland, MS 39576

A context for studying molecular deformations in crystals was established by reviewing conformational studies of isolated mono- and di-saccharides. In new work, models of miniature crystals of cellulose allomorphs, cellobiose, maltosyl di- and tri-saccharides and panose were based on published crystal structures. Their lattice energies were computed after complete energy minimization with MM3(92). Comparisons of the conformations of molecules in the model crystals with the models of isolated individual molecules indicated the extents of distortion resulting from the crystal fields. Ring puckerings, linkage conformations and side group orientations can each be deformed by the crystal field by as much as three kcal/mol. The sum of such distortions is typically about five kcal/mol of glucose residues, compared to the molecule modeled in isolation. These increases in intramolecular energy are stabilized by lattice energies ranging from about 15 to almost 50 kcal/mol of monomer residues, with oligomers and polymers having smaller lattice energies per residue than monosaccharides. Contrary to conventional thought, the van der Waals force is usually a stronger component of the lattice energy than the intermolecular hydrogen bonding, especially for cellulose models.

Although experimental techniques continue to improve, computerized molecular modeling of carbohydrates will be increasingly useful in studies of enzyme interactions with polysaccharides. In the present work, we study interactions of carbohydrates with their neighbors in the well-defined environments offered by carbohydrate crystal structures. We also show how modeling can be used both to further understand experiment and to probe structures of insoluble polysaccharides for which experimental evidence is not definitive.

Three main methods exist for modeling structures and energies. Of these, ab initio quantum mechanics calculations at sufficiently high levels of theory take large amounts of computer time. Therefore, they cannot model all the potential variants of monosaccharides or the large assemblies of molecules described in the present paper. Available semi-empirical quantum mechanical models, although not prohibitively time-consuming, are not sufficiently accurate for such work (1-3). In view of these limitations on the quantum mechanical methods, we use molecular mechanics (4), in particular the general purpose program (MM3) (5-7). Numerous studies of carbohydrates (8-10) and other organic molecules have been published with MM3 and its predecessor, MM2 (in several official and unofficial variants) (11-13). MM3 is less well known for modeling proteins, partly because versions of MM2 were limited to 255 atoms or less, far fewer than in proteins. Now, larger systems can be studied, and MM3 is parameterized for peptides (14).

Our first step in learning about the deformations and interaction energies in carbohydrate crystals is the construction of potential energy surfaces for isolated model molecules. The minimum energies, considering ranges of exocyclic group orientations, are plotted against the important variables of molecular conformation, e.g., the puckering parameters of monosaccharide rings or the linkage torsion angles of disaccharides. (The latter plots are the familiar ϕ, ψ, energy maps used by Ramachandran more than 30 years ago (15).) Next, the puckering parameters or ϕ and ψ values of relevant (in a fairly broad sense) molecules in crystal structures are plotted on these surfaces. For each observed structure, the corresponding energy above the global minimum on these contoured surfaces is then attributed (in principle) to the energy of deformation caused by the crystal field.

One type of deformation is a displacement from a local minimum on the energy hypersurface, while a second type is the entrapment in a local minimum higher in energy than the global minimum. Because polymers are connections of monomeric units, our definition of molecular deformation includes interactions arising from adjacent residues of the same molecule. We expect all of these types of distortion and their combinations to be found in crystalline carbohydrates. Of course, energies above the modeled global minimum could also arise from errors in the modeling program or from inadequate strategies for its application to molecules that are as complex as carbohydrates. Additional increases in energy are brought about by specific intramolecular interactions in molecules that are "relevant" but not exactly the modeled molecule. Therefore, especially when molecules have energies higher than usual on the conformational surfaces, we have also used models that incorporate neighboring molecules located according to the published crystal structure. We call these models miniature crystals, or minicrystals. If the modeling program is nearly correct, the higher-energy conformation will be kept during energy minimization of the minicrystals.

One test of the quality of the model is the extent of movement of the individual atoms from their experimentally determined positions during the energy minimization. Further evidence of the validity is provided by the lattice energy of the model. This is calculated by removing the central molecule to an infinite distance and subtracting the energy of the intact minicrystal from the total energy of the separated structures (stabilizing lattice energies are positive). Obtaining an experimental value is more difficult. Experimental heats of combustion for carbohydrates can be converted to heats of formation for the solid compound. MM3 and some other programs permit calculation of the heat of formation of a vapor phase molecule. Our "experimental" lattice energy is the difference between the MM3 heat of formation for a vapor phase molecule and the experimental heat of formation for the solid. Although the resulting "experimental" lattice energy depends on MM3 with its tabulated enthalpies for individual bonds, that part of MM3 is apparently quite accurate. Calculated heats of formation of 41 alcohols and ethers in the vapor phase had a standard deviation of 0.38 kcal/mol when compared to experiment (6). Studies of molecules in lattices are not new. Williams has used such calculations for many years (16) for the study of nonbonded parameters, and similar models were used in a study of peptides (14). We have found miniature crystal models to be useful additions to structural study even when single crystal studies have been performed on very small molecules such as erythritol that are disordered in the crystal (17).

In the present paper, we first review our studies of mono- and disaccharide energy surfaces. We then describe new studies of miniature crystals of cellobiose and cellulose allomorphs, using a new method of comparing the relative stabilities of the various allomorphs of cellulose. We also study four molecules that represent fragments of starch: β-Maltose monohydrate, β-methyl maltoside, β-methyl maltotrioside and panose (α-D-glucopyranosyl-(1→6)-α-D-glucopyranosyl-(1→4)-α-D-glucopyranose). Panose is a model for the (1→6) branch point of the amylopectin polysaccharide in starch, as well as the (1→4)-linked amylose chain. The maltooligomers are also models for amylose and the amylopectin backbone. Three of these maltooligomers are hydrated so the role of water can be considered, too.

Methods

The procedures for making energy surfaces for furanose and pyranose monosaccharides are detailed in a study of the ketosugar psicose (18). Ring conformations are varied by raising and lowering non-adjacent ring atoms in increments of 0.1 Å, while holding the three other ring atoms in a plane. For each of the 38 characteristic pyranose geometries (two chairs, six boats, six skews, 12 half-chairs and 12 envelopes), all 729 combinations of staggered primary and secondary alcohol groups were optimized. The 15 to 23 unique combinations of side group orientations favored at the 38 characteristic geometries were used at 4913 points in ring-puckering space for each pyranosyl molecule.

Disaccharide modeling strategies were described in a study of cellobiose (19). Increments of the linkage torsion angles were 20° or less. For the energy surface of cellobiose presented here, 16 combinations of exocyclic group orientations were used as initial structures. We used 24 initial structures for the maltose surface.

Miniature crystals were constructed using the published atomic coordinates, unit cell dimensions, and space group from the original crystal structure reports. As described previously (20), the miniature crystals of the various molecules were constructed with CHEM-X (developed and distributed by Chemical Design, Ltd.). Unlike other work that retains space group symmetry and unit cell dimensions (14), our studies place no restriction on the atomic movement during energy minimization. The cellobiose and panose models contained 1215 and 1866 atoms, respectively; the other models had fewer than 750 atoms. Cellulose models were based on groups of seven cellotetraose molecules.

Minimizations terminated when the energy change per five iterations was less than 0.00008 kcal/mol times the number of atoms in the structure. As described before, however, it is also necessary to observe the atomic movement. If the average atomic movement per iteration was more than 0.00011 Å when minimization terminated, one (or more) hydroxyl hydrogen atom(s) on the surface was probably fluctuating. (This artifact of MM3 minimizations is thoroughly described in reference 20.) Fluctuating groups were oriented manually to be in a different energy well and the minimization was resumed. Energies were calculated with dielectric constants (ϵ) in the range of three to four, as stated below. This attenuates the electrostatic interactions, compared with the vacuum phase (MM3 uses $\epsilon=1.5$ for the vacuum phase), and yields closer comparisons with conformations determined by crystallographic experiments.

MM3(92) ran on various computers, including Digital VAX and DECstations and on IBM-PC compatibles, and CHEM-X was used on both PC-compatible and VAX computers. Root mean square (RMS) atomic movements were computed with CHEM-X. RMS energies were computed with the SLICE utility provided with SURFER, the program used for contour plots.

Results and Discussion

Review of Isolated Monosaccharides. Structural knowledge of glucose (21) and fructofuranose (22,23) is central to work on cellulose, starch and fructans. We have also studied the remainder of the aldohexoses: allose, altrose, galactose, gulose, idose, mannose, and talose (21). For these D-sugars, the conventional chair form, $^4C_1$, was dominant, with several exceptions. α-Idose had the lowest energy in a skew form, as previously proposed (24), and both chairs also have low energy. Non-$^4C_1$ forms are also plausible for α-altrose, α-allose and α-gulose. The β-anomers of these sugars have relative energies for the alternate chair more than two kcal/mol above the minimum, so non-$^4C_1$ forms would be found only under special circumstances. Both anomers of glucose are likely to appear only in the $^4C_1$ form, because the other forms have relative energies higher than five kcal/mol. Still, there is substantial flexibility within the $^4C_1$ form that must be considered.

MM3(92) at $\epsilon=3.0$ predicts the experimental anomeric ratios of the above sugars in aqueous solution with an average accuracy of about 0.5 kcal/mol, with the β-form inordinately favored (20). The simplest interpretation of that result is that the MM3(92) modeling system is fairly accurate and that specific solvent effects are small, both in terms of structural change and energy differences. Another indication of the validity of MM3(92) for use on carbohydrates is that the observed crystal structures containing the very flexible fructofuranose rings are all predicted to have ring shape energies of less than three kcal/mol, with the majority less than one kcal/mol (22,23).

Review of Isolated Disaccharides. MM3(92) energy maps for β-cellobiose and α-maltose are shown in Figures 1 and 2. Except for sucrose (25-27), our modeling studies of disaccharide linkage conformations have succeeded in placing observed crystal structures within contours 2 kcal/mol above the minimum. Another exception is one of the linkages in crystalline cycloamylose hexahydrate (28), which, in the crystal, has an energy nearly three kcal/mol above the minimum found for an isolated disaccharide. Our work included all the possible linkages of glucopyranose, including the three-bond 1,6 linkages (29). Three maps for α-mannose disaccharides have also been reported (30). Yui et al. studied chitobiose in a similar manner (31) and applied the results to predict the solution properties of the polymer chitosan, another rather insoluble polysaccharide. That work, and comparison with results of other workers (for example, see ref. 32), has shown that the treatment of electrostatic interactions is very important, but difficult. Such difficulties gave rise to the recent, despairing comment that the models didn't work "when hydrogen bonding was properly accounted for (32)." However, we surmise that those workers failed to account for the higher dielectric constant of the crystals, compared to vacuum. This caused overemphasis of the electrostatic interactions.

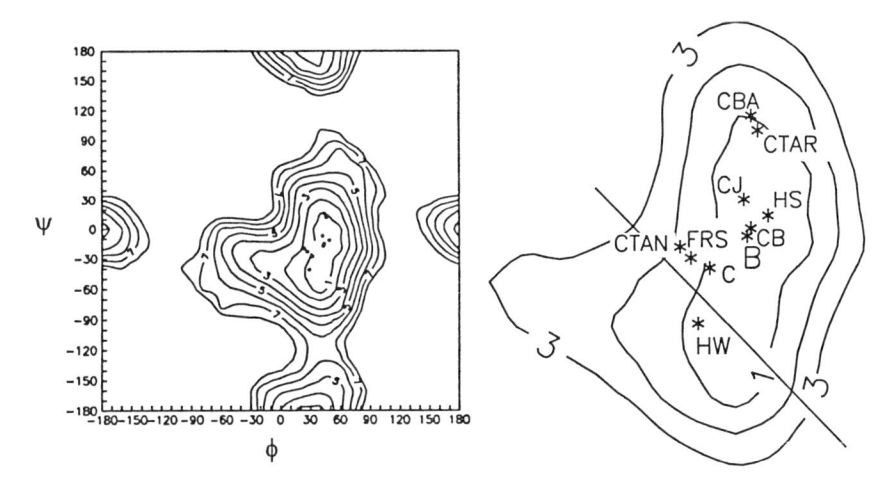

Figure 1 MM3(92) calculated energies for β-cellobiose, ε=3.0. Sixteen combinations of side group orientations were used as initial structures at each point on a 20° grid of φ and ψ, including all combinations of *gt, gg*, r, and c orientations. Points on the map are: C, proposed cellulose I-IV structures from this work. CJ and HW are cellobiose and methyl cellobioside. CBA, CTAN, and CTAR are cellobiose and cellotriose acetates, and FRS, CB, B and HS are various lactose structures. See reference 30 for the references.

Crystal structures of molecules related to disaccharides show that their inter-residue linkages have substantial flexibility. For example, the two linkage torsion angles take values almost continuously over ranges of 90° in crystal structures of maltose-like molecules (Figure 2). We have adopted the hypothesis (33) that, because of the different collections of nonbonded interactions found in the many different crystal lattices, the observed conformations will deviate in a random way from the minimum-energy structure for the isolated molecule. Moreover, if the number of conformations plotted on the energy surface is sufficiently large, their variation in position should be consistent with the Boltzmann probability function.

One exception to this hypothesis would be the comparison of an energy surface for maltose with linkage geometries from maltosyl residues within cycloamyloses which have covalent constraints on linkage geometry. In cycloamyloses the linkage torsions must lead to closure of the macroring, and other shapes that are likely for the isolated maltose molecule might not occur in the cyclic molecule. Another possible cause of deviation from a random distribution would be a preferred set of torsion angles for the various derivatives and complexes of a given molecule that would always lead to packing with the highest density. To avoid problems from covalent constraints, we

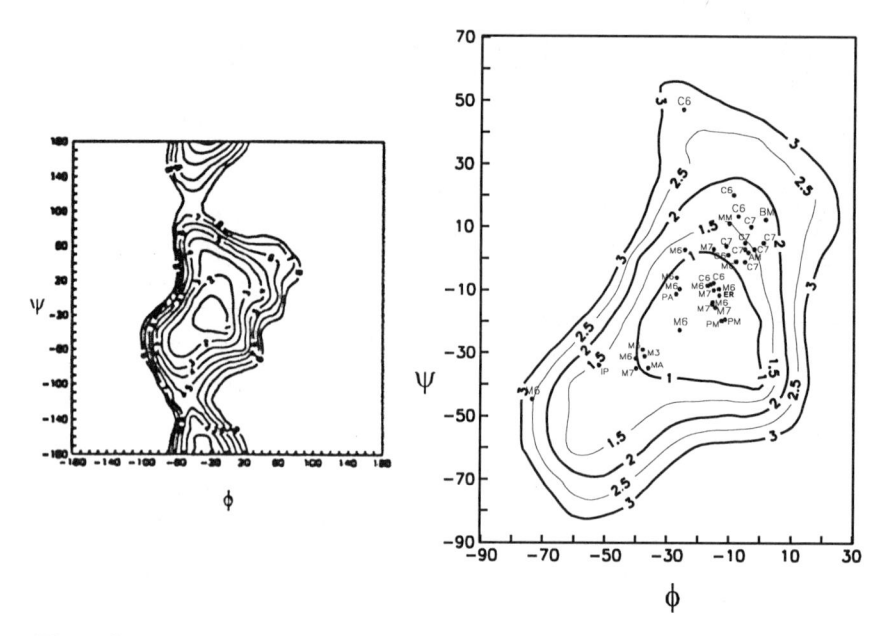

Figure 2 MM3(92) calculated energies for α-maltose, ε=3.0. Twenty-four combinations of side group orientations were used. Compared to the cellobiose map (Figure 1) a second position of the O2H hydrogen atom was used that permits an intramolecular hydrogen bond to the glycosidic oxygen atom. Points discussed in this work include AM (α-maltose), BM (β-maltose monohydrate), MM (methyl β-maltoside), PA (panose), M3 (methyl α-maltotrioside), and M7 (maltoheptaose). See reference 30 for keys to the other codes and their references.

limited the number of maltose geometries from cycloamyloses on the maltose map in Figure 2. A similar safeguard against drawing inappropriate conclusions was not obvious for the happenstance of a limited range of torsion angles being preferred for packing. Ultimately, specific effects from the chemical differences in the range of "related" compounds will defeat an exact application of this hypothesis. Still, even chemically identical linkage torsion angles such as in cyclohexaamylose take a wide range of values as a result of non-bonded interactions. Also, the specific effects from selected, chemically different molecules can be quite small (< 1.0 kcal/mol) so the hypothesis is a useful tool.

A measure of the ability of a system to model crystalline disaccharides is the RMS energy for the crystal structures plotted on the energy surface. On the cellobiose map (Figure 1), all of the structures that were determined by single crystal methods have an RMS energy of 0.87 kcal/mol. Including the lactose crystals (which also have β-1→4 linkages) but omitting the acetates reduces the RMS energy to 0.68 kcal/mol. The map for maltose (Figure 2) has a value of 1.19 kcal/mol for 38 linkage conformations, whereas the maltose maps by other workers, and our previous work with MM2, would have higher RMS energies, perhaps three kcal/mol. Another consideration is that the crystal structures should be distributed in a manner consistent with the energy surface. Within the limitations discussed above, there should be little area of very low energy that is not occupied, just as there should be few structures in the high-energy areas.

Given the map for maltose with a low RMS energy, it is of interest to check the linkage conformations found in the complex of maltoheptaose with rabbit phosphorylase (34). Three of the five resolved linkage conformations (M7 on Figure 2) fall at the center of the 1 kcal/mol contour, and the other two are just outside it, despite 30-40° ranges in the linkage torsion angles. In general, data from protein-carbohydrate complexes must be accepted carefully because the resolution is often inadequate for these purposes.

Review of Model Miniature Carbohydrate Crystals. Model cellulose and other carbohydrate crystals were described in a 1993 paper (20), and further work on the native Iα form of cellulose (35) was published in 1994. In particular, various conformational aspects and the lattice energy of α-maltose were accounted for by study of a minicrystal (30). Miniature crystals of raffinose (27), nystose (22) and gentiobiose (29) have also been described, with the first two being unsatisfactory because of unknown problems with their sucrose linkages (27). Special attention was given to hydrogen bonding in miniature crystals of threitol, erythritol, β-glucose and α-maltose (17). That work showed that the experimentally-12
determined hydrogen bonding arrangements were reproduced fairly well, even in the instances of bifurcated hydrogen bonding. In that work lattice energies, calculated with a dielectric constant of 3.5, were about 80% of the experimental value, with the deficit attributed to the absence of long-range forces in the models (all had fewer than 750 atoms). An exact match of the lattice energy for α-maltose with the experimental result was obtained by using ϵ of 2.5 (26), but we felt that this overemphasized the electrostatic interactions to compensate for the lack of long-range forces.

New Work

Cellobiose Conformation. Crystalline cellobiose (36) has a shape that corresponds to a local minimum on the MM3(92) surface. The main reason for taking another look at its minicrystal model is that it was not completely optimized in the earlier work (20). Although MM3 had reported that the structure was minimized, some surface hydroxyl groups were fluctuating. We are now used to dealing with this peculiar behavior of MM3, and realize that the surface hydroxyl groups must be readjusted and optimization resumed. We took advantage of the need for a new study to increase the size of the minicrystal to 27 cellobiose molecules (Figure 3). In the past, we used some glucose instead of cellobiose molecules to surround the central molecule completely without exceeding the old atom-size limits. Because of the completion of minimization, the increased model size, the change in the calculation of hydrogen bonding energy in MM3(92), and the lower dielectric constant, the calculated lattice energy of cellobiose increased from 61.0 to 65.7 kcal/mol. As an example of the lattice energy calculation, and the partitioning of the total energy terms, the MM3 outputs are shown in Table I. Taking the values from Table I, the calculation is:

Lattice Energy = 29.94 + 293.17 - 257.40 = 65.71 kcal/mol

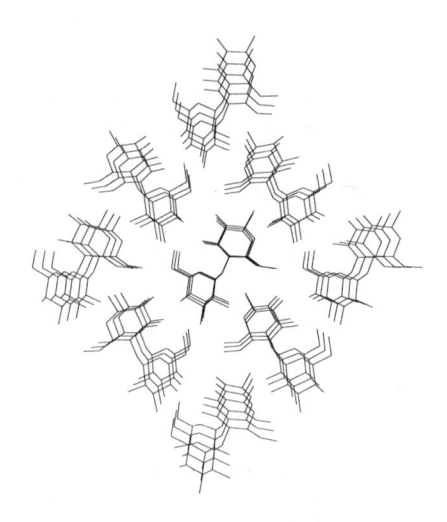

Figure 3 Miniature crystal of cellobiose used in this study. 1215 atoms were used, in 27 molecules. The perspective drawing of the unoptimized structure was done with CHEM-X. Hydrogen atoms are not shown.

Table I. MM3(92) Energies for Cellobiose Minicrystal and Components (kcal/mol)

	Minicrystal	Central Molecule	Outer Shell
Final Steric Energy (total)	257.40	29.94	293.17
Compression	36.92	1.25	35.67
Bending	106.80	4.14	102.65
Bend-Bend	-4.54	-0.17	-4.38
Stretch-Bend	10.85	0.37	10.48
Van der Waals 1,4 Energy	691.55	25.77	665.78
Van der Waals Other	-435.43	-2.30	-387.75
Torsional	-1.35	0.27	-1.62
Torsion-Stretch	-4.40	-0.16	-4.24
Dipole-Dipole	-143.00	0.76	-123.42

Because of the way that hydrogen bonding energy is calculated in MM3, there is no separate summary of hydrogen bonding energy. Part of the hydrogen bonding arises from the dipole-dipole term, and the other part comes from an altered van der Waals expression for atoms that could participate in hydrogen bonding (17). The intermolecular hydrogen bond in the methanol dimer has an MM3(92) energy of 2.23 kcal with $\epsilon = 3.5$. Of this energy, 1.23 kcal is from the dipole-dipole interaction and 0.97 kcal is from the supplemental term reported with the "van der Waals other" values (17).

To study the extent of distortion of the inter-residue linkage, we optimized the central molecule from the minicrystal in isolation. Because of fluctuation, the first attempt at optimization of the central molecule in isolation did not finish successfully but it dropped the energy from 29.9 to 27.0 kcal. Manual rotation of the fluctuating O3 hydroxyl group dropped the energy to 25.4, and subsequent minimization dropped the energy to 25.05 kcal/mol. In the calculations to produce the cellobiose map in Figure 1, the minimum-energy grid point closest to the cellobiose crystal structure was at $\phi, \psi = 40°, 0°$. Unlike the crystal conformation, the hydroxyl groups of the structure from our searching of variation space were reverse-clockwise (nonreducing residue) and clockwise (reducing residue) (19). The energy of that structure, optimized without constraint, was 20.8 kcal, about nine kcal/mol lower in energy than the central molecule of the optimized minicrystal. Table II compares selected structural details of the original crystal structure, the central molecule from the optimized minicrystal, that central molecule optimized further in isolation, and the structure identified by our conformational mapping as the local minimum near the observed crystal structure.

There is apparently very little distortion of the cellobiose molecule by its crystal lattice, and the O3'H···O5 hydrogen bond is modeled exceptionally well. The greatest discrepancies are around the reducing residue's anomeric center. Especially noteworthy are the C1'–O1' bond length, the HO1'–O1'–C1'–O5' torsion angle (on which C1'–O1' depends), and the O5'–C1'–O1' bond angle. Minor problems with parameterization are indicated in situations where the parameters from the structures optimized without neighbors are actually closer to the experimental structure than the minicrystal values. Despite these small problems, the mean atomic displacements from the crystal structure positions resulting from minimization are much smaller for the cellobiose molecule optimized in the center of the minicrystal than for the models optimized in isolation.

Conformations of Cellulose Allomorphs. Historically, the crystalline forms of cellulose have been designated Cellulose I - IV and Cellulose X, which has not been studied recently (37). Within the past two decades, distinctions were discovered between Cellulose III made from Cellulose I, and Cellulose III made from Cellulose II. Thus, both forms of Cellulose III have approximately the same lattice dimensions but have different distributions of diffraction intensities (38), depending on their histories. The structure of Cellulose IV also depends on whether it was made from I or II (39). More recently, native Cellulose I was found to have two distinct crystal structures, Iα and Iβ (40). Iα is found mixed with Iβ in bacterial and algal cellulose, whereas Iβ is the dominant form in higher plants and after steam-annealing algal cellulose.

Table II. Conformational descriptors of cellobiose

Structural Descriptor	Diffraction	Minicrystal	Isolated	Local Min
ϕ_H (°)	42.3	40.9	42.0	45.8
ψ_H (°)	-17.9	-11.1	-12.7	-4.5
ϕ_O (°)	-76.3	-81.7	-79.6	-76.1
ψ_{C5} (°)	-132.3	-130.8	-132.4	-125.2
τ (°)	116.1	114.0	115.0	113.8
O3'···O5 (Å)	2.77	2.73	2.71	2.76
O3H'···O5 (Å)	1.89	1.85	1.86	1.99
C1–O1 (Å)	1.40	1.41	1.41	1.41
C1'–O1' (Å)	1.38	1.42	1.42	1.43
HO1'–O1'–C1'–O5' (°)	-116.3	-93.4	-64.2	180.0
C5–O5 (Å)	1.44	1.44	1.44	1.44
C5'–O5' (Å)	1.44	1.43	1.44	1.44
O5–C1	1.43	1.43	1.43	1.43
O5'–C1'	1.44	1.43	1.43	1.42
C5–O5–C1 (°)	112.4	112.3	112.5	112.0
C5'–O5'–C1' (°)	113.5	112.5	111.6	112.2
O5–C1–O1 (°)	107.5	105.7	104.0	104.9
O5'–C1'–O1' (°)	107.0	102.5	104.4	101.9
Atomic movement (mean) (Å)	-	0.07	0.13	0.19
Atomic movement (RMS) (Å)	-	0.09	0.19	0.25

The previous comparisons of the energies of cellulose polymorphs (20) were in qualitative agreement with the general understanding of the relative energies, once an error in our model of the cellulose Iβ structure was corrected (35). However, our manual reorientation of the surface hydroxyl units during optimization was affecting the total energy for the system, and another approach would be more rigorous. Thus, the tabulated energies that compare the different allomorphs were computed in a different manner. The new energies for comparisons of stabilities result from the summation of the lattice energy (Table IX, below) and the energy of the isolated central molecule without further optimization. This is equivalent to the difference between the minicrystal energy and the energy of the outer shell. Excepting the Iα structure, based on work by Aabloo and French (35), the cellulose structures were all proposals of Sarko and coworkers (38,39,41,42).

Another improvement was the averaging of the structures that have two chains per unit cell. Two miniature crystals for such structures are now used. For example, one

Table III. Comparative energies and linkage torsion angles of cellulose allomorphs

Cellulose Allomorph	Minicrystal Energy	Energy for comparison (per tetramer) [per residue]	ϕ,ψ/ϕ,ψ corner / center
Iα	208.9	-20.69 [-5.17]	31.8, -27.6 / n.a.
Iβ	207.3	-21.36 [-5.34]	31.0, -26.6 / 28.6, -29.0
II	192.9	-25.51 [-6.38]	35.0, -29.5 / 29.6, -26.4
III$_I$	215.9	-18.39 [-4.60]	30.8, -26.6 / 28.7, -29.4
IV$_I$	213.0	-18.27 [-4.57]	33.1, -26.9 / 30.6, -25.9
IV$_{II}$	216.8	-16.20 [-4.05]	27.9, -27.7 / 28.3, -27.6

model structure of Cellulose II was based on four "up" chains at the corners of the unit cell, with three "down" chains. The other Cellulose II model had four "down" chains at the corners and the other chains were "up." Although the ranking of the structures is the same, based on either the total minicrystal energy or the sum of the lattice energy and the central-chain energy, the "comparison energies" from Table III should be the best representations of the energies of the several structures. The only known experimental value for the difference between cellulose Iβ and II is 0.3 kcal/mol of glucose residues (43), compared with our predicted value of about 1.0 kcal/mol of glucose residues. The difference between cellulose Iα and II was 1 kcal/mol of glucose residues, according to the PLMR program (35). Experimental values would be smaller because of amorphous material in the experimental samples.

Previously, it was thought that the 2-fold screw-axis proposed for cellulose would strain the structure unreasonably (44), owing partly to short contacts across the glycosidic linkage between the hydrogen atoms on C1 and C4'. The ϕ and ψ values from our various models are shown in Table III. Their average, ϕ = 30.5°, ψ = -27.6°, is plotted in Figure 1. It is essentially a 2-fold conformation. When the values fall on the line between the ϕ,ψ pairs 180°/-180° and -180°/180°, the structures possess 2-fold (or pseudo 2-fold) screw symmetry. This line passes through the center of the 1 kcal/mol contour of the cellobiose map in Figure 1, showing that any strain from the lowest-energy 2-fold conformations is minor. In the cellulose minicrystals, the lengths of the model central cellobiose units ranged from 10.26 to 10.36 Å, values within experimental error of the reported values of the fiber repeat distances of cellulose (38-42).

Maltodisaccharide Conformations.

Along with the α-maltose structure, the linkage conformations of the crystal structures of β-maltose monohydrate and β-methyl maltoside have distortions. These structures all possess intramolecular hydrogen bonds between O2 and O3'. Part of the reason for the non-minimal energy of the β-structures might be that the energy surface for β-maltose is slightly different from the α-maltose surface. However, the energies of the two β-anomers were even higher on our MM3(90) map of β-maltose (45), relative to its global minimum.

β-Maltose Monohydrate. This disaccharide was studied first with x-rays (46) and subsequently by neutron diffraction in 1977 (47). Its water of hydration bridges O6 to O6' of a neighbor and also bridges O3 atoms of two adjacent molecules. Because the structure was determined by neutron diffraction (R = 0.044), the positions of the hydrogen atoms should correspond closely to the MM3 values. Table IV shows the energies of several models.

In the central molecule of the optimized miniature crystal model, the experimental torsion angle values and hydrogen bond geometries were well preserved (Table V). The first two data columns in Table IV show the breakdown of the overall MM3 energy for both the experimental structure and the central molecule in the model minicrystal. Similar comparisons, based on x-ray instead of neutron diffraction, would typically have much larger energy differences. For example, the unoptimized energy of the structure of methyl β-maltoside from the experimental x-ray diffraction study (R=0.056) (48) was 138 kcal/mol, with 71 kcal from bond stretching and 39 kcal from angle bending. This much higher energy arises because x-ray diffraction locates the centers of electron density of hydrogen atoms, whereas the neutron diffraction and modeling methods determine the locations of the nuclei. To a lesser extent, the dispositions of lone pairs of electrons cause x-ray diffraction studies to give somewhat different positions for oxygen atoms as well. Also, the crystallographic R factor was higher for the methyl maltoside, so it is likely that random errors in the atomic positions are higher.

In Table IV, the largest differences between the components of the initial energy and the intramolecular energy of the molecule optimized in the minicrystal are for the bond length and bond angle terms, with each contributing about five kcal/mol. In particular, the parameterization for O-H lengths in MM3 is for the vapor phase, electron diffraction value of 0.947 Å, whereas the usual bond length by neutron diffraction of single crystals is about 0.97 Å. In the neutron diffraction experiment, neutrons are scattered by the nuclei, while electrons are scattered by charged particles. In x-ray diffraction, the scattering is based on the positions of maximum density of the electron cloud. Because the electron cloud of a hydrogen atom consists of a single electron, the cloud's center is often closer to the oxygen or carbon to which it is bound than is the hydrogen nucleus. Thus the C-H and O-H bond lengths measured by x-ray diffraction would be shortest, bond lengths determined by electron diffraction would have intermediate lengths, and neutron diffraction bond lengths would be longest. Because the O-H stretching constant is fairly large, variation from the MM3 standard values accounts for a big part of the energy discrepancy. C-H bonds pose a similar but smaller problem. The largest single bond-length discrepancy for bonds between carbon and oxygen atoms is for the reducing C1'-O1' bond which has an experimental value of 1.390 Å (Table V). The minicrystal result is much longer, 1.432 Å. In MM3, this bond length is modified to compensate for an anomeric effect, based on the value of the O1'H–O1'–C1'–O5' torsion angle. However, the parameterization for this effect was based on ab initio calculations with small basis sets that may not be accurate enough. The anomeric regions are generally troublesome, as shown by Table V. The

Table IV. Intramolecular MM3(92) Energies for Various β-Maltose Structures, in Sequence Described in Text

Energy	Original	Minicrys	Fully Opt., Isolated	Isolated, Set to -15/-20	Isolated Global Minimum	Isolated, Set to -50/-50	Isolated Secondary Minimum
φ,ψ	4.8, 13.3	4.9, 11.0	1.1, 15.4	-15, -20	-20.3, -26.7	-50, -50	-53.2, -49.2
Total	38.54	27.80	25.00	24.93	23.23	29.09	23.80
Compression	6.21	1.20	1.30	1.34	1.31	1.31	1.35
Bending	9.65	4.95	4.97	4.94	4.26	4.95	5.48
Bend-Bend	0.11	-0.11	-0.13	-0.13	-0.13	-0.13	-0.11
Stretch-Bend	0.09	0.44	0.47	0.47	0.45	0.47	0.46
Van der Waals 1,4	25.98	25.92	25.89	25.90	26.06	26.06	25.93
Van der Waals Other	-2.81	-2.66	-3.37	-3.03	-3.17	3.22	-2.83
Torsional	1.64	0.75	0.21	-0.98	-1.80	-3.68	-3.33
Torsion-Stretch	0.16	-0.17	-0.14	-0.12	-0.11	-0.04	-0.06
Dipole-Dipole	-2.49	-2.51	-4.21	-3.46	-3.66	-3.06	-3.08

model crystal field was of no value in preserving the band lengths and band angles in the original structure.

Table V. Conformational descriptors at the anomeric center of β-maltose monohydrate for the experimental structure, the central molecule in the minicrystal and the central molecule, optimized in isolation

	Original Crystal	Minicrystal	Isolated Model
C1–O1	1.41	1.43	1.43
C5–O5	1.43	1.44	1.44
O5–C1	1.40	1.42	1.42
C5–O5–C1	110.7	114.8	115.0
O5–C1–O1	113.8	107.2	107.3
C1–O1–C4'	117.9	115.3	115.6
C1'–O1'	1.39	1.42	1.42
C5'–O5'	1.42	1.44	1.44
O5'–C1'	1.41	1.43	1.43
C5'–O5'–C1'	112.7	111.7	112.5
O5'–C1'–O1'	108.2	103.8	104.5
H'–O1'–C1'–O5'	-100.9	-95.1	-62.1
Mean atomic movement		0.04	0.09
RMS atomic movement		0.10	0.11

The major contributor to the bending energy is the glycosidic bond, C1–O1–C4', initially at 1.75 kcal/mol. In the optimized minicrystal, this distortion was diminished somewhat, to 1.04 kcal/mol, as the angle changed from 117.8° in the crystal to 115.3° in the model. The C1–O5–C5 and C1'–O5'–C5' bond angles also make large contributions to the energy. The sum of their energies remains roughly constant (1.2 kcal/mol) during minimization because of the requirement of ring formation. Unlike the bond stretching energies, which are diminished to only 1.2 kcal/mol (total) by optimization, five kcal/mol of bending energy remains after minimization in the minicrystal, owing mostly to ring formation and steric crowding at the glycosidic linkage. Exocyclic oxygen atoms often contributed to high initial bending energies that were mostly relieved by optimization. For example, the initial energy of O2–C2–C3 was 0.71 kcal/mol, and the O2–C2–H2 energy was 0.23 kcal/mol. After optimization, the energies were 0.19 and 0.04 kcal/mol, respectively. Because O2 is involved in a strong inter-residue hydrogen bond, the 0.19 kcal can be considered a deformation energy, whereas the drop from 0.71 to 0.19 kcal is probably due to parameterization problems. The C4'–C5'–C6' and C5'–C6'–O6' bond angles were also still distorted after optimization in the minicrystal, apparently by intermolecular hydrogen bonding forces. The O5–C1–O1 and O5'–C1'–O1' angles were both different from the MM3 minimum energy form. We suspect a parameterization problem here.

To learn more about the distortion energies of β-maltose monohydrate, we attempted to optimize the central molecule of the minicrystal without its neighbors. As happened with the MM3 optimization of cellobiose, however, the optimization ended prematurely because of rapid fluctuation of a hydroxyl group in an unstable position. Rotation of the O3 hydroxyl group from the crystal structure position to give an H–O3–C3–C4 torsion of 180° immediately reduced the energy to 26.3 kcal/mol, and optimization went smoothly. After minimization (143 iterations), the isolated molecule was not very different in structure, with an energy smaller by 1.3 kcal/mol. The structure had merely gravitated to a local minimum that is near one found on the MM3(90) map (45).

We then changed φ and ψ to values corresponding roughly to the global minimum on the α-maltose map (-15°, -20°) and optimized further (Tables IV and VI). The manual adjustment lowered the energy by less than 0.1 kcal, but the structure then optimized (298 iterations), decreasing its potential energy by 1.8 kcal/mol. Finally, the φ and ψ values were again manually adjusted with CHEM-X, to -50°, -50° and optimized. The initial (rigid) rotations increased the energy by almost six kcal/mol, but the final structure (after 83 iterations) at this secondary minimum was only 0.6 kcal/mol higher in energy than at the central minimum.

In traversing this region of φ,ψ space, the main variations in optimized energy were shown by the torsion, bending and dipole-dipole terms (Table IV). The crystal structure of maltose occurs in a region favored by hydrogen bonding (compare the various dipole-dipole and van der Waals other terms) but with relatively high torsional energies. Torsional energies at the global minimum are reduced 2.0 kcal, giving that minimum. The strain in the glycosidic bond angle is also relieved here, offsetting the loss of hydrogen bonding stabilization. The secondary minimum (the global minimum with MM3(90) and an extensive treatment of the exocyclic orientations) has much lower torsional energy, but no inter-residue hydrogen-bonding. The energy of bending the glycosidic angle is also larger. Thus, the torsional energy contributions over the 60° ranges vary by 4 kcal/mol and are a dominant force in determining the energy surface. The reduction in energy from 29.1 to 23.8 kcal during the last minimization arose mostly from the contributions to the "van der Waals other" component. This reduction in energy did not correspond to formation of hydrogen bonds (the dipole-dipole contributions did not change much) but only the relief of short contacts. The complexity of intramolecular variation during these changes in molecular geometry shows the importance of relaxed-residue conformational analysis, without which the van der Waals short contacts would have been unrelieved. Also, the balance of the terms of the potential energy function is critical to the correct representation of changes in the torsional, bending, hydrogen bonding and van der Waals energies.

Table VI. Conformational Descriptions of Several Maltose Models

Model	ϕ	ψ	O2···O3'	O2H···O3'	O4--O1
Crystal Structure of β-maltose monohydrate	4.8	13.3	2.79	1.84	4.41
Optimized Minicrystal of β-maltose monohydrate	4.9	11	2.73	1.83	4.35
Optimized Original Mol, isolated β-maltose monohydrate	1.1	15.4	2.85	1.94	4.39
Global Minimum (starting at ϕ=-15°, ψ=-20°)	-20.3	-26.7	3.14	2.37	4.47
Secondary Minimum (starting at ϕ= -50°, ψ=-50°)	-53.2	-49.2	4.49	4.24	4.31
MM3(92) α-Maltose Global (30)	-15.8	-23.7	n.a.	n.a.	n.a.
MM3(90)Global β-Maltose (42)	-54.3	-47.4	n.a.	n.a.	n.a.
MM3(90) Secondary (42)	-28.3	-27.5	n.a.	n.a.	n.a.
MM3(90) Tertiary (42)	0.6	23.7	n.a.	n.a.	n.a.

The reduction in energy when the O3H was rotated from its position as optimized in the minicrystal to form an intramolecular hydrogen bond was 1.5 kcal/mol, similar to the finding for cellobiose, above. This shows the importance of clockwise and reverse-clockwise hydroxyl orientations as found in some of the experimental oligomaltose structures (especially the panose structure) (49), below. Table VI also shows a substantial variation of the O4--O1 distance, larger than we expected.

Methyl β-Maltoside Monohydrate. This crystal structure was solved in 1967 by x-ray diffraction (48). It has an intramolecular O2H···O3' hydrogen bond on one side of the molecule. On the other side, the water of hydration is the receptor for hydrogens donated by both O6 and O6', bridging the two residues. Still, the glucose-glucose linkage conformation (Table VII) is similar to the two other observed maltosyl structures.

Comparison of the minicrystal conformation and the x-ray values is not favorable if based on torsion angles defined with the hydrogen atoms, H1 and H4'. The values plotted on the energy surface (Figure 2) and in Table VII have been calculated by

Table VII. Comparison of methyl maltoside conformations

	Crystal Structure	Minicrystal center	Local Minimum
ϕ	5.6[a]	-10.7	-7.0
ψ	15.9[a]	6.5	7.35
ϕ_{O5}	109.9 (-10.1)[b]	109.5	112.8 (-7.2)[b]
ψ_{C5}	-109.1 (10.9)[b]	-116.3	-115.4 (4.6)[b]
C1–O1–C4'	117.6	115	115.4
C1–O1	1.415	1.427	1.429
C5–O5	1.441	1.441	1.442
O5–C1	1.408	1.424	1.425
C1'–O1'	1.377	1.409	1.411
C5'–O5'	1.439	1.434	1.436
O5'–C1'	1.424	1.432	1.432
C5–O5–C1	114.7	114.9	115.1
C5'–O5'–C1'	111.4	108.2	108.0
O5–C1–O1	111.2	111.1	112.5
O5'–C1'–O1'	107.1	104.6	105.0
O2···O3'	2.823	2.872	2.847
O2H···O3'	1.862 [2.000][c]	2.027	2.593
O4--O1	4.57	4.446	4.402
Mean atom movement		0.089	0.08
RMS movement		0.10	0.099

[a] These experimental values are thought to be in error.
[b] Values obtained by subtracting or adding 120° from the O5–C1–O1–C4' and C1–O1–C4'–C5' torsion angles, respectively.
[c] [Value] resulted from correction of the O2-O2H covalent bond length to 0.97 Å.

subtracting and adding 120°, respectively from the experimental O5–C1–O1–C4' and C1–O1–C4'–C5' torsion angles. The torsion angles based on the original H1 and H4' positions are 5.6 and 15.9° in the crystal, discrepancies of 15.7 and 12.1°, respectively from the extrapolated values. Although this is an extreme example, it points out the pitfall of relying on hydrogen atom positions that are (poorly) determined by room temperature x-ray diffraction. Based on the corrected values, the ψ angle changed 4.6° during optimization of the minicrystal while the φ angle was constant. The interresidue hydrogen bond (O2H···O3') geometry was retained fairly well, except that the position of the hydroxyl hydrogen O2H is also affected by experimental error, leading to a short O2H···O3 distance. Moving the experimentally determined hydrogen atom to give a 0.97 Å O3–H covalent bond length gives a O3H···O2' hydrogen bond length, shown in brackets, close to the MM3 minicrystal value. The O1--O4 distance in the optimized minicrystal was not quite as long as the observed value, but is closer to the bottom of the energy well. Full optimization of the structure produced very little change in linkage conformation because this structure also migrated to a nearby local minimum. The φ and ψ are slightly different from the maltose monohydrate coordinates, because the molecular structures are different and the O6' hydroxyl hydrogen atoms had different orientations. The O1--O4 distance shortened further, to the average experimental value.

Trisaccharide Conformations

Methyl α-Maltotrioside Tetrahydrate. This crystal structure was reported in 1985 (50). X-ray diffraction gave an R of 0.054. Hydroxyl and water hydrogen atoms were not resolved, although a hydrogen bonding scheme was proposed based on interoxygen distances. The conformation of the disaccharide linkages in methyl maltotrioside is nearly identical to those in proposals for double helices of native A and B starch. Unlike the maltose disaccharides above and the panose structure below, there are no inter-residue intramolecular hydrogen bonds. They are impossible at the observed φ,ψ torsion angles.

The preliminary model of the maltotrioside had 546 atoms. After optimization, some hydrogen bonding was inconsistent with the proposal in the crystal structure report. However, subsequent work with a 744-atom model was consistent with the proposed scheme. Still, the RMS atomic movement for the central molecule in the mini crystal was 0.258 Å, quite a bit higher than for the central molecules of the other structures. Thus, we suspect that the hydrogen bonding scheme, especially involving the water, may be disordered. In the reported crystal structure, the water is in clusters. Three of the four waters are not close enough to the carbohydrate to form hydrogen bonds, and so they must form hydrogen bonds with each other. This would be similar to liquid water, and may be the reason that these hydrogens were not found during the crystal study.

Panose. The crystal structure of (anhydrous) panose was solved at room temperature (49). In 1991, it was redone at liquid nitrogen temperature to an R of 0.036 (51). It was only in the latter work that the hydrogen bonding was worked out correctly, and is thus an example of why carbohydrate crystal structures should be done at low temperature. Disorder, giving 18% β-anomer, was not modeled. The central (primed) residue of panose (see Table VIII) is very distorted in a way that affects models of amylose (52). Its O1--O4 distance is the longest known at 4.78 Å, and can be compared with 4.05 Å for the nonreducing residue in α-maltose. The average O1--O4 distance is 4.41 Å for all α-glucose residues, and the glucose chairs in crystalline α-glucose, glucose monohydrate and glucose-urea have O1--O4 distances of 4.494 ± 0.018 Å (52). As was the case for the α-maltose minicrystal (26), the puckering is mostly preserved during optimization of the panose minicrystal. The puckering energy for a glucose molecule that is deformed like the experimental central panose residue is about 2 kcal/mol (21), and the puckering energies of the glucose residues in panose drop to very small values when panose is optimized as an isolated molecule. Another feature of panose is that there is substantial intramolecular hydrogen bonding, including the intra residue rings often found in modeling studies. Recent work shows that such clockwise and counter-clockwise rings may not correspond to hydrogen bonds, but instead result from the absence of repulsive dipole-dipole interactions that are found for other orientations (23). Table VIII shows a number of conformational indicators for the crystal structure, optimized minicrystal, and freely optimized structure.

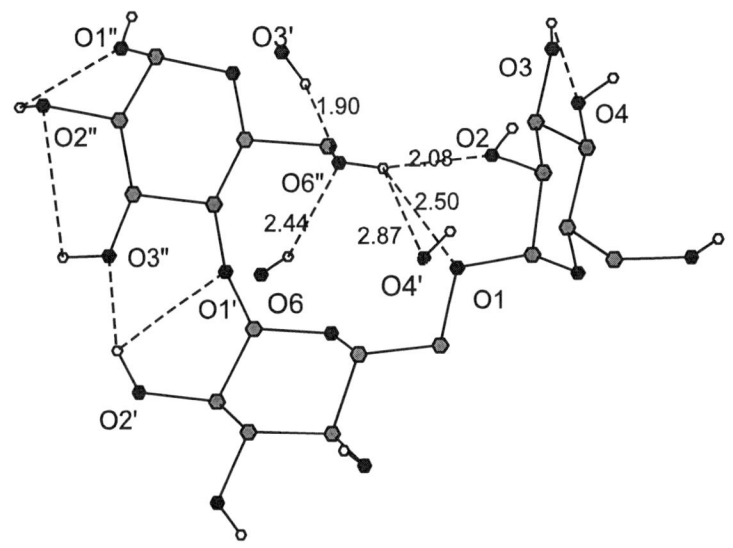

Figure 4 Hydrogen bonding in crystalline panose, including intra-residue bonds, based on uncorrected H···O covalent bond lengths.

An important feature of the panose structure is the orientation of the O6" hydroxyl group. In the crystal, the O5"–C5"–C6"–O6" torsion angle is 132.7°, a nearly eclipsed conformation just inside the "forbidden" *tg* (53) energy well. To an extent this supports the proposed intramolecular hydrogen bonding in native celluloses, based on the *tg* position. In panose, the *tg* conformation is stabilized by a three-centered (bifurcated) intramolecular hydrogen bond where O6H" is donated to both O1 and O2, and is nearly within range (2.87 Å) of O4' on another molecule (Figure 4). O6" also is the receptor for protons from two neighboring hydroxyl groups, so it is extensively

Table VIII. Conformational features of panose

Conformational Feature	Crystal	Minicrystal	Freely Optimized from Minicrystal
$\phi_{1,4}$ (°)	-17.7	-20.8	-19.3
$\psi_{1,4}$ (°)	-20.7	-19.5	-26.9
$\tau_{1,4}$ (°)	113.9	112.0	113.2
$\phi_{1,6}$ (°)	-44.8	-45.9	-43.1
$\psi_{1,6}$ (°)	167.3	168.4	170.2
$\omega_{1,6}$ (°)	-43.0	-46.9	-47.9
$\tau_{1,6}$ (°)	112.0	111.9	113.0
$\tau_{C\,1,6}$ (°)	108.3	109.1	108.4
O2'···O3" (Å)	2.94	3.00	3.03
O2H···O3" (Å)	2.07 [2.15][a]	2.05	2.19
O5"–C5"–C6"–O6" (°)	132.7	143.6	149.7
O6"···O1 (Å)	3.08	2.85	3.09
O6"···O2 (Å)	2.84	3.01	3.14
O6H"···O1 (Å)	2.41 [2.51][a]	2.07	2.46
O6H"···O2 (Å)	1.94 [2.08][a]	2.2	2.21
O1–C1	1.41	1.42	1.42
O1'–C1'	1.41	1.43	1.43
O1"–C1" (Å)	1.39	1.44	1.43
H–O1"–C1"–O5"	68.0	52.0	59.0
O5–C1	1.44	1.42	1.43
O5'–C1'	1.42	1.43	1.43
O5"–C1"	1.45	1.42	1.42
C5–O5–C1	114.3	115.7	114.8
C5'–O5'–C1'	114.5	114.8	114.3
C5"–O5"–C1"	113.1	114.5	115.1
O5–C1–O1	112.4	108.2	108.7
O5'–C5'–O1'	110.8	106.2	107.0
O5"–C5"–O1"	111.7	107.9	108.5
O1--O4 (Å)	4.63	4.61	4.49
O1--O1'	4.78	4.72	4.55
O1'--O1"	4.51	4.48	4.49
Mean atomic movement	-	0.10	0.20
RMS Movement	-	0.12	0.23

[a] [Values] are corrected to O-H distances of 0.97Å

hydrogen bonded. We examined the energetics of O6 rotation with MM3. An energy surface was constructed based on rotations about the C5-C6 and C6-O6 bonds of a glucose molecule with the other hydroxyl groups started in the positions found in the panose crystal. This map (not shown) indicates that the nearly eclipsed position is about 3.2 kcal/mol higher in energy than the global minimum (some 70° in O5"–C5"–C6"–O6" torsion angle away) on that MM3(92) surface. The observed conformation is on a saddle point on the ridge corresponding to the 120° eclipsed conformation on the energy surface. The distortion energy of 3.2 kcal can be compared with the hydrogen bonding energy (2.2 kcal/mol) computed with MM3(92) for the methanol dimer. Thus, the several hydrogen bonds in panose should be strong enough to stabilize the strained conformation that is observed. Still, an 11° change in the O5"–C5"–C6"–O6" torsion angle occurred during optimization of the model, along with changes in the hydrogen bonding, suggesting that the model is not perfect.

Lattice Energies. The lattice energies of the various molecules are shown in Table IX, with and without the water molecules (if present). Comparisons of results from ~700-atom models with experimental values suggest that the lattice energies are underestimated by roughly 20% when long-range forces are not included (26). The relative uniformity (well within the anticipated error in the modeling calculations) of the per-residue lattice energies for the di- and oligosaccharides is not surprising in view of the similarity of the available heat of combustion values. Also shown are the comparably calculated β-glucose and α-maltose lattice energies from our previous MM3(92) work on hydrogen bonding.

The lattice energies of the cellulose models are lower than for the other molecules because of the two-dimensional character of their intermolecular interactions. The third dimension, the fiber axis, is occupied by covalently bound glucose residues along the molecular axis. Therefore, the calculated lattice energy should, as a first approximation, be roughly two thirds that of glucose. Also, there are only three hydroxyl groups per residue, instead of the five on the glucose molecule, so fewer hydrogen bonds could be formed. Additionally, the molecular weight is 10% lower for a glucose residue than for glucose itself, so the dispersive forces would be lower. Lattice energies for carbohydrate alone were lower in the hydrated structures. We speculate that the lower energies may occur when the water bridges parts of the same carbohydrate molecule. In such models, if the central carbohydrate alone is removed from the lattice, neither the central molecule nor the outer shell is stabilized by the water, causing a drop in the lattice energy.

Comparison with Other Work. The RMS atomic movements in the present study can be compared with values presented by Lii and Allinger for cyclic hexapeptides (14). Those molecules have 51 atoms, comparable to 45 in the disaccharides, 48 in the methyl maltoside, and 66 in the trisaccharides, not counting any water molecules. One might expect that our movement values would be higher because, unlike their work (14), our minicrystals were not constrained by a fixed outer cage of atoms during optimization. However, that was not what we found. The RMS movements of the heavy atoms in six different hexapeptides, by MM3(90), ranged from 0.089 Å for comparison with a low-temperature x-ray study to 0.15 Å for a room-temperature study. Normal AMBER values ranged from 0.07 Å (for a room-temperature x-ray study) to 0.22 Å (for a low-temperature study) and 0.31 for another room-temperature

Table IX. Lattice Energies Computed with MM3(92) at $\epsilon=3.5$

Molecule	No. of atoms	Lattice Energy - Water (if any) with central molec. [per residue]	with outer shell (hydrates only)
β-D-Glucose		38.2	n.a.
α-Maltose	729	61.3 [31.65]	n.a.
Cellobiose	1215	65.7 [32.85]	n.a.
Cellulose Iα	609	72.6 [18.15]	n.a.
Cellulose Iβ (averaged)	609	73.6 [18.40]	n.a.
Cellulose II (averaged)	609	75.2 [18.80]	n.a.
Cellulose III$_I$ (averaged)	609	71.2 [17.80]	n.a.
Cellulose IV$_I$ (averaged)	609	70.5 [17.63]	n.a.
Cellulose IV$_{II}$ (averaged)	609	68.7 [17.18]	n.a.
β-Maltose Monohydrate	729	62.5 [31.25]	57.0 [28.5]
Methyl β-Maltoside Monohydrate	705	53.6 [26.80]	52.3 [26.15]
Methyl α-Maltotrioside Tetrahydrate	744	82.5 [27.50]	67.7 [22.6]
Panose	1866	85.2 [28.40]	n.a.

study. AMBER/OPLS values ranged from 0.09 to 0.23 Å. For a set of three cyclic peptides, including the two low-temperature studies, MM3 caused an RMS movement of 0.10 Å while movements of 17 other force fields ranged from 0.09 to 0.30 Å. Thus, our values of 0.05 to 0.12 Å (except for the methyl maltotrioside) suggest that small carbohydrates can be modeled well.

In three of the four carbohydrates studied in detail, there was an increase in RMS movements, going from the minicrystal to the freely optimized molecule started from the crystal coordinates. This is what we expected, leaving us to speculate regarding the methyl maltopyranoside structure, which showed a slightly smaller RMS value as an isolated molecule than as the central molecule in the minicrystal. We can only surmise that the errors in this, the oldest of experimental determinations, were somehow compensated by the errors in MM3, and that the further errors in MM3s interatomic interactions somehow resulted in slight additional deformations.

Conclusions

MM3(92) was used to model cellobiose, six cellulose allomorphs, and four starch-related molecules. Minicrystals, consisting of from 609 to 1866 atoms were studied, as were isolated molecules. Lattice energies, including comparably determined values from earlier work, were in the range of 17 to 38 kcal/mol of glucose residues when calculated at a dielectric constant of 3.5. Use of that value has been previously shown to cause an underestimation of the experimental value by 20% (26), so the present values are probably low by roughly that amount.

The strength of the hydrogen bond in the methanol dimer (2.2 kcal/mol) by MM3(92) at $\epsilon=3.5$ should be comparable to the hydrogen bonding energy, per donor, in the minicrystal models. Compared with the overall lattice energies per glucose residue, then, hydrogen bonding makes only a small, albeit specific, contribution to cellulose. For example, removal of each glucose residue in the cellulose I models breaks only two intermolecular hydrogen bonds and the 2.2 kcal/mol of hydrogen bonds would be small compared with the 18 kcal/mol of residues for the total lattice energy. Removal of a β-glucose molecule from its crystal structure, with five hydrogen bonds per molecule, breaks 10 hydrogen bonds, just more than half of the 38 kcal of lattice energy.

Unlike the α-maltose model, the modeled small molecules in the present work all had crystalline linkage conformations close to local minima. This was true even of the maltotrioside that had a relatively high RMS movement. Changes in the conformation of the maltose linkage cause changes in a delicate relationship between the values of the torsional energy terms, the hydrogen bonding functions, and the glycosidic angle bending term. The exo-anomeric effect favors the values of about -50° for the HC1–C1–O1–C4' torsion angle of axial glycosidic bonds from D-sugars. However, when hydrogen bonding occurs, as observed in α-maltose, β-maltose and methyl β-maltoside crystals, the exo-anomeric effect is counteracted and the O1–C4' and C1–HC1 bonds are nearly eclipsed at the linkage. The energy differences among all three stable β-maltose conformers, and the barriers between them, are all within two kcal/mol, and crystal structures are observed near all three points.

The primary alcohol groups were found in all three of the staggered positions in these models. Proposals for the *tg* conformation have long been on the table for cellulose, and much of the available evidence is consistent with that proposal. In panose, the nearly eclipsed, distorted *tg* conformer is stabilized by extensive hydrogen bonding. The model hydrogen bonds did not hold the structure in exactly the observed conformation. Instead, the O6–C6–C5–O5 torsion angle increased somewhat to be closer to the *tg* minimum at 180°, changing the pattern of hydrogen bonding in the model somewhat. This structure will be a good test case for future potential functions.

The examples of the range of maltose linkage conformations and the intramolecular hydrogen bonding in panose show the importance of the model torsional energy functions. It is necessary to know the energies of intermediate values of the torsion angles, not just their minima and maxima.

Several apparent inadequacies arose in the parameterization of MM3, such as the discrepancies of the reducing end C1–O1 bond lengths and O5–C1–O1 bond angles. The error in nonreducing C1–O1 bond lengths is less than at the reducing end, and therefore, the RMS atomic movement is not affected very much. On the other hand, the O5-C1-O1 bond angles seem defective at both reducing and nonreducing centers and are a primary source of the RMS discrepancies. Such errors would also be cumulative in polymers and should be targeted for elimination. To validate improvements to this and other aspects of modeling software, improved experimental results would be useful.

Acknowledgments

The authors thank Paul Vercellotti who adapted MM3(92) to execute on PC compatible computers. Norman L. Allinger, J. Phillip Bowen, John Brady and Lothar Schäfer gave useful advice. Iain Taylor and Mary An Godshall kindly read the manuscript and gave helpful comments.

Literature Cited

1. Tvaroška, I.; Carver, J. *J. Chem. Res.*, **1991**, 123-144.
2. Zheng, Y.-J.; Le Grand, S. J.; Merz, Jr., K. M. *J. Comput. Chem.*, **1992**, *13*, 772-791.
3. Gundertofte, K.; Palm, J.; Petterson, I.; Stamvik, A. *J. Comput. Chem.*, **1991**, *12*, 200-208.
4. Burkert, U.; Allinger, N. L. Molecular Mechanics, ACS Monograph 177, American Chemical Society, Washington, D.C. 1982.
5. Allinger, N. L.; Yuh, Y. H.; Lii, J.-H. *J. Am. Chem. Soc.*, **1989**, *111*, 8551-8566.
6. Allinger, N. L.; Rahman, M.; Lii, J.-H. *J. Am. Chem. Soc.*, **1990**, *112*, 8293-8307.
7. MM3 is available from Tripos Associates, Technical Utilization Corporation and the Quantum Chemistry Program Exchange.
8. Dowd, M. K.; Reilly, P. J.; French, A. D. *J. Comput. Chem.*, **1992**, *13*, 102-114.
9. Dowd, M. K.; French, A. D.; Reilly, P. J. *Carbohydr. Res.*, **1992**, *233*, 15-34.
10. Dowd, M. K.; Reilly, P. J.; French, A. D. *J. Carbohydr. Chem.*, **1993**, *12*, 449-457.
11. Jimenez-Barbero, J.; Noble, O.; Pfeffer, C.; Pérez, S. *New J. Chem.*, **1988**, *12*, 941-946.
12. Tran, V. ; Buleon, A.; Imberty, A.; Pérez, S. *Biopolymers*, **1989**, *28*, 679-690.
13. Hooft, R. W. W.; Kanters, J. A.; Kroon, J. *J. Comput. Chem.*, **1991**, *12* 943-947.
14. Lii, J.-H.; Allinger, N. L. *J. Comput. Chem.*, **1991**, *12*, 186-199.
15. Rao, V.S.R.; Sundararajan, P.R.; Ramakrishnan, C.; Ramachandran, G. N. *Conformation in Biopolymers*, Vol. 2, Academic Press, London, 1963.
16. Williams, D. E.; Stouch, T. R. *J. Comput. Chem.*, **1993**, *14*, 1066-1076.
17. French, A. D.; Miller, D. P. *ACS Symp. Ser.*, **569**, 1994, 235-251.
18. French, A. D.; Dowd, M. K. *J. Comput. Chem.*, **1994**, *15*, 561-570.

19. French, A. D.; Tran, V.; Pérez, S. *ACS Symp. Ser.*, **1990**, *430*, 191-212.
20. French, A. D.; Aabloo, A.; Miller, D. P. *Int. J. Biol. Macromol.*, **1993**, *15*, 30-36.
21. Dowd, M. K.; French, A. D.; Reilly, P. J., *Carbohydr. Res.*, **1994**, *264*, 1-19.
22. French, A. D.; Mouhous-Riou, N.; Pérez, S. *Carbohydr. Res.*, **1993**, *247*, 51-62
23. French, A. D.; Dowd, M. K.; Reilly, P. J. *J. Mol. Struct.* (Theochem), in press.
24. Ferro, D. R.; Provasoli, A.; Ragazzi, M. *Carbohydr. Res.*, **1992**, *228*, 439-443.
25. French, A. D.; Newton, S. Q.; Cao, M.; Schäfer, L. *Carbohydr. Res.*, **1993**, *239*, 51-60.
26. French, A. D.; Dowd, M. K. *J. Mol. Struct. (Theochem)*, **1993**, *286*, 183-201.
27. van Alsenoy, C., French, A. D.; Newton, S. Q.; Cao, M.; Schäfer, L. *J. Am. Chem. Soc.*, **1994**, *116*, 9590-9595.
28. Lindner, K.; Saenger, W. *Acta Crystallogr. Sect. B*, **1982**, *38*, 203-210.
29. Dowd, M. K.; Reilly, P. J.; French, A. D. *Biopolymers* **1994**, *34*, 625-638.
30. Dowd, M. K.; French, A. D.; Reilly, P. J. *J. Carbohydr. Chem.*, **1995**, 14, 589-600.
31. Yui, T.; Kobayashi, H.; Kitamura, S.; Imada, K. *Biopolymers*, **1994**, *34*, 203-208.
32. Kroon-Batenburg, L. M. J.; Kroon, J.; Leeflang, B. R.; Vliegenthart, J. F. G. *Carbohydr. Res.* **1993**, *245*, 21-42.
33. Bartenev, V. N.; Kameneva, N. G.; Lipanov, A. A. *Acta Crystallogr.* **1987**, *B43*, 275-280.
34. Goldsmith, E.; Sprang, S.; Fletterick, R. *J. Mol. Biol.*, **1982**, *156*, 411-427.
35. Aabloo, A.; French, A. D. *Macromol. Theory Simul.*, **1994**, *3*, 185-191.
36. Chu, S. S. C.; Jeffrey, G. A. *Acta Crystallogr. Sect. B*, **1968**, *24*, 830-838.
37. French, A. D.; Bertoniere, N. R.; Battista, O. A.; Cuculo, J. A.; Gray, D. G. *Kirk-Othmer Encyclopedia of Chem. Tech. 4th Edition*, **1993**, *5*, 476-496.
38. Sarko, A.; Southwick, J.; Hayashi, J. *Macromolecules*, **1976**, *9*, 857-863.
39. Gardiner, E. S.; Sarko, A. *Can. J. Chem.*, **1985**, *63*, 173-180.
40. Sugiyama, J.; Vuong, R.; Chanzy, H. *Macromolecules*, **1991**, *24*, 4108.
41. Woodcock, C.; Sarko, A. *Macromolecules*, **1980**, *13*, 1183-1187.
42. Stipanovic, A. J.; Sarko, A. *Macromolecules*, **1976**, *9*, 851-857.
43. Rånby, B. G. *Acta Chem. Scand.*, **1952**, *6*, 101-115.
44. Atalla, R. H. *ACS Symp. Ser.*, **1979**, *181*, 55-69.
45. Dowd, M. K.; Zeng, J.; French, A. D.; Reilly, P. J. *Carbohydr. Res.*, **1992**, *230*, 223-244.
46. Quigley, G. J.; Sarko, A.; Marchessault, R. H. *J. Am. Chem. Soc.*, **1970**, *92*, 5834-5839.
47. Gress, M. E.; Jeffrey, G. A. *Acta Crsytallogr. Sect. B*, **1977**, *33*, 2490-2495.
48. Chu, S. S. C.; Jeffrey, G. A. *Acta Crystallogr. Sect. B*, **1967**, *23*, 1038-1049.
49. Imberty, A.; Pérez, S. *Carbohydr. Res.*, **1988**, *181*, 41-55.
50. Pangborn, W.; Langs, D.; Pérez, S. *Intl. J. Biol. Macromol.*, **1985**, *7*, 363-369.
51. Jeffrey, G. A.; Huang, D.-b. *Carbohydr. Res.*, **1991**, *222*, 47-55.
52. French, A. D.; Rowland, R. S.; Allinger, N. L. *ACS Symp. Ser.*, **1990**, *430*, 120-140.
53. Marchessault, R. H.; Pérez, S. *Biopolymers*, **1979**, *18*, 2369-2374.

RECEIVED September 11, 1995

Chapter 3

Glycosidases, Large and Small: T4 Lysozyme and *Escherichia coli* β-Galactosidase

R. H. Jacobson[1], R. Kuroki[2], L. H. Weaver, X-J. Zhang[3], and B. W. Matthews[4]

Institute of Molecular Biology, Howard Hughes Medical Institute and Department of Physics, University of Oregon, Eugene, OR 97403

T4 lysozyme and *E. coli* β-galactosidase provide contrasting examples of glycosidases. Their three-dimensional structures are shown to be substantially different, not only with regard to size, but in other respects as well.

β-galactosidase is a tetramer of four identical subunits. Within each subunit the 1023-amino acid polypeptide chain folds into five sequential domains plus an extended segment at the amino terminus. Each of the four active sites in the tetramer is formed by elements from two different subunits. In contrast, T4 lysozyme (T4L) is a monomeric protein of 164 amino acids with the active site at the junction of the amino-terminal and carboxy-terminal domains.

The mutation Thr 26 → Glu in the active site cleft of phage T4 lysozyme produces an enzyme that cleaves the cell wall of *Escherichia coli* but leaves the product covalently bound to the enzyme. The crystalline complex is non-isomorphous with wild-type T4L and analysis of its structure shows a covalent linkage between the product and the newly-introduced Glu 26. The covalently-linked sugar ring is substantially distorted, suggesting that distortion of the substrate toward the transition state is important for catalysis, as originally proposed by Phillips. It is also postulated that the adduct formed by the mutant is an intermediate consistent with a double displacement mechanism of action (in the mutant) in which the glycosidic linkage is cleaved with retention of configuration as originally proposed by Koshland.

---

[1]Current address: Department of Biochemistry, University of California, 401 Barker Hall, Berkeley, CA 94720
[2]Current address: Central Laboratories for Key Technology, Kirin Brewery Company, Ltd., 1–13–5 Fukuura, Kanazawa-ku, Yokohama 236, Japan
[3]Current address: Oklahoma Medical Research Foundation, 825 NE 13th Street, Oklahoma City, OK 73104
[4]Corresponding author

0097–6156/95/0618–0038$12.00/0
© 1995 American Chemical Society

The lysozymes and β-galactosidases provide contrasting examples of glycosidases. Hen egg-white lysozyme (HEWL), the prototypical lysozyme, is a small monomeric enzyme of 129 amino acids. In contrast, the β-galactosidase from *Escherichia coli* is a larger tetrameric enzyme (MW = 465,412 Da). In this report we briefly describe the recently-determined structure of β-galactosidase (*1*) and contrast it with the much smaller enzyme, bacteriophage T4 lysozyme (T4L).

β-Galactosidase

β-galactosidase is comprised of four identical monomers, each of 1,023 amino acids (*2,3*). There are four independent active sites but the full tetramer is required for activity (*4*). Similar to lysozyme, β-galactosidase catalyzes the hydrolysis of β1-4 glycosidic linkages, but its substrates can be much simpler and more varied and include compounds that provide the basis for a variety of colorimetric assays (*5,6*).

The structure of *E. coli* β-galactosidase has been determined from crystals with four tetramers per asymmetric unit (space group $P2_1$; $a = 107.9$Å, $b = 207.5$Å, $c = 509.9$Å, $\beta = 94.7°$). The presence of 16 monomers per asymmetric unit complicated the crystallographic analysis. On the other hand, once the relative arrangement of the tetramer had been determined, averaging over the 16 crystallographic copies of the monomer greatly enhanced the quality of the electron density map (*1*).

Each monomer (Figure 1) is composed of five essentially compact domains plus approximately 50 residues at the amino-terminus that are relatively extended. The first domain, residues 51-217, consists of twelve β-strands and five short segments of helix. The core of the domain corresponds to a "jelly-roll" β-barrel. The second domain exhibits a fibronectin-type III fold. This topology is similar to that of immunoglobulin constant domains, except that here the fourth strand of the barrel hydrogen bonds to the opposite sheet. The connection between the fourth and third strands of the barrel (residues 272-288) forms a protruding loop that extends across a subunit interface to the neighboring monomer and contributes to the active site (see below). The third domain forms a distorted "TIM" barrel and contains the catalytic site. Unlike typical TIM barrels that consist of eight β/α repeats, the core of the α/β barrel in β-galactosidase is irregular in that the fifth helix is absent and the sixth parallel strand of the barrel is distorted. This irregularity results in a space or opening that is occupied by the loop that extends from the second domain of the two-fold related monomer and forms part of the active site. The fourth domain, residues 628-736, is topologically similar to the second domain and also forms a seven-stranded fibronectin type III β-barrel. If the second and fourth domains are superimposed, 76 α-carbon atoms superimpose within 3.4Å. The fifth domain, residues 737-1023, exhibits a novel topology. The core of the domain consists of an 18-stranded, anti-parallel sandwich. The faces of the sandwich each contain nine strands and exhibit a large twist. From the first strand to the last strand in each sheet, the direction rotates by over 90°. A search of the PDB structural data base (*7*), kindly carried out by Dr. Liisa Holm using the program "DALI" (*8*), showed that this sandwich somewhat resembles the seed storage

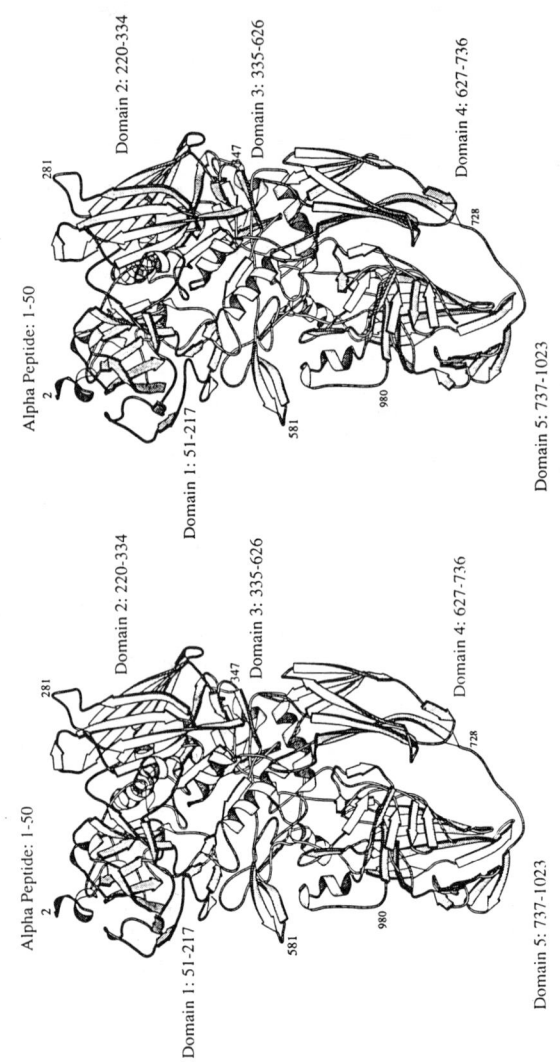

Figure 1a. Stereo ribbon diagram of the β-galactosidase monomer showing the domain organization of the chain. The TIM barrel (Domain 3) forms the core of the monomer and is surrounded by the four other largely β structures. Visible at the upper right is the extended loop (residues 272-288) that reaches across the activating interface.

protein phaseolin (9), however the two strands at the edge of the sheets are topologically permuted. In addition to the twisted sheets, the fifth domain also contains an irregular assembly made up of segments connecting the β-strands and including residues 776-819, 960-982, and 990-1012. This assembly packs against the back of the first β-sheet and positions Trp 999 at the active site.

The four subunits making up the β-galactosidase tetramer (Figure 1b) form two different monomer:monomer contacts that we refer to as the "activating" interface and the "long" interface. The "activating" interface includes contacts between residues near the amino-termini, and also includes two helices from each monomer that are packed together to form a four-helix bundle. In addition, an extended loop reaches across this interface to complete the active site of the neighboring monomer. The "long" interface consists of two regions of contact, one of which includes residues near the middle of the sequence and the other involves contacts within the C-terminal fifth of the sequence.

The Active Site of β-Galactosidase

E. coli β-galactosidase cleaves β1-4 linkages of β-galactosides with retention of configuration at the anomeric center and is believed to function through a double-displacement mechanism (10,11). Unlike many other glycosidases, β-galactosidase requires the presence of Mg^{2+} for maximal activity (4,12). Previous biochemical studies have characterized the residues Glu 461, Met 502, Tyr 503, Glu 537, to be important for catalytic function, or to be near the active site (13-15). These residues are found in close proximity to one another, and are located around a deep pit formed at the end of the TIM barrel (Figure 2). This pocket is therefore identified as the substrate binding site. Its location, at the C-termini of the β-sheet strands, is consistent with other known structures that contain the TIM barrel motif (16,17).

At present no structure of an enzyme/substrate-analogue complex is available so the specific interactions between enzyme and substrate remain to be determined. Nevertheless, the location of the disaccharide binding site and residues likely involved in substrate recognition are apparent. The active site (Figure 2) is formed primarily from residues within the third domain, but residues from distant regions of the sequence also contribute. Among homologous β-galactosidase sequences the residues that form the active site pocket are very highly conserved (1). These include two tryptophan residues (Trp 568, Trp 999) that help form the hydrophobic walls of the pocket as well as several potential hydrogen bond donors and acceptors that are present in the vicinity (Asn 102, Asp 201, His 357, His 391, His 540, Asn 604). Three polar residues (Glu 416, His 418, and Glu 461) that are conserved in all of the sequences coordinate what appears to be a magnesium ion. This is consistent with site-directed substitutions showing Glu 461 to be important both for catalysis and magnesium binding (15). Also, Gebler et al. (14) have used a derivatized substrate to trap a covalent intermediate, suggesting that the carboxylate of Glu 537 is the nucleophile. Glu 537 lies at the bottom of the pocket with the carboxylate aligned by hydrogen bonds with the hydroxyl group of Tyr 503 and the guanido group of Arg 388. The loop that extends across the activating interface also contributes to the integrity of the active site. It does

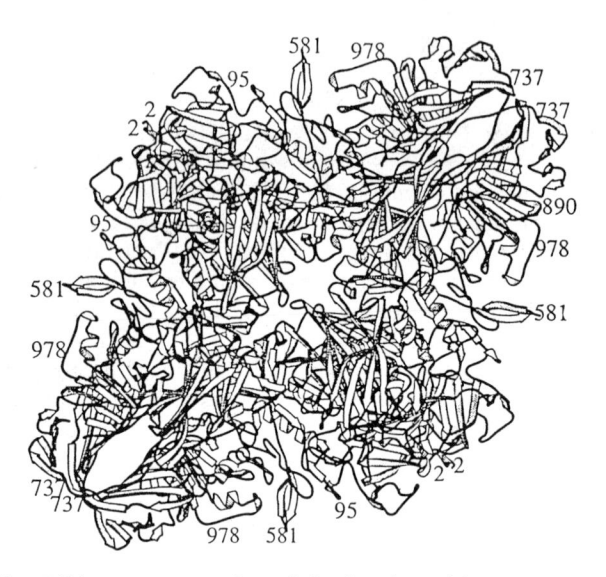

Figure 1b. Ribbon representation of the β-galactosidase tetramer showing the largest face of the molecule. Formation of the tetrameric particle results in two deep clefts that run across opposite faces of the molecule and each contain two active sites.

Figure 2. The active site of β-galactosidase. Glu 416, His 418 and Glu 461 appear to coordinate a bound magnesium ion. Glu 537, identified as the putative nucleophile (14), is situated on the opposite side of the cavity and is oriented through hydrogen bonds with Tyr 503 and Arg 388. Residues from the TIM barrel in Domain 3 of one subunit are drawn with solid bonds; residues from Domain 2 of the other subunit are drawn with open bonds.

not appear to provide catalytic residues, *per se*, but to stabilize the main-chain in the vicinity of the magnesium binding ligands.

T4 Lysozyme

The amino acid sequence of T4L is not obviously related to that of HEWL but the three-dimensional structures have parts in common, especially in the vicinity of the active site cleft (*18*). If the active site regions of the two enzymes are aligned, Glu 11 of T4L superimposes on the catalytically essential Glu 35 of HEWL, and Asp 20 of T4L approximately corresponds with the other key residue of HEWL, Asp 52. Therefore, it has been assumed that the two enzymes share a common catalytic mechanism (*18*).

In the process of making mutants of T4 lysozyme it was observed that the substitution, Thr 26 → Glu (T26E), produced an enzyme inactive at neutral pH (*19*). This was unexpected since Thr 26 is located within the active site cleft of T4L, but was thought not to be critical for catalysis because some alternative amino acids are tolerated at this position (*20*).

Mutant T26E was constructed and purified by standard procedures followed by cation exchange and reversed phase chromatography (Figure 3a). More than 30 cycles of Edman degradation showed the N-terminal region of the major component to have the same amino acid sequence as the T26E mutant. Electron ion spray mass spectrometry showed the mass to be 19,548 Da (T26E + 918). The major component of *E. coli* cell wall (NAM-NAG)-LAla-DGlu-DAP-DAla, has a molecular weight of 940 (NAM, N-acetyl muramic acid; NAG, N-acetyl glucosamine; DAP, diaminopimelic acid) (*21,22*). Allowing 18 daltons for the loss of a water of condensation, and 4 daltons for the difference between the formula molecular weight of the protein and its actual state of ionization, the purified material corresponds to a covalent adduct between the mutant T26E and the saccharide. The mass of the second component identified by chromatography (18622 Da) (Figure 3a) corresponds to the (uncomplexed) mutant, T26E. The third component (Figure 3a) has a mass of 39084 Da, corresponding to a pair of lysozyme-adduct molecules linked to each other by the same (NAG-NAM)-peptide-(NAM-NAG) linkage as present in cell walls of *E. coli*.

The monomeric form of the adduct was crystallized in a form non-isomorphous with wild-type and the structure determined and refined to a residual of 16.3% at 1.9Å resolution (*19*). The refined structure showed a disaccharide of NAG-NAM bound in subsites C and D, and a peptide of LAla-DGlu-DAP-DAla extending across an open groove on the surface of the molecule between α-helices 108-113 and 126-134 (Figure 3b). This mode of binding is as anticipated from early studies of low activity mutants of T4 lysozyme (*23*). Interactions between the peptide moiety and the enzyme are critical for catalysis since T4L, unlike HEWL, will not hydrolyze oligosaccharides such as chitin that lack a peptide substituent (*24*).

Previously, no saccharide has been observed in the D subsite of T4L. The sugar in this subsite is distorted. This is clearly seen in the superposition of the NAM bound in subsite D on the NAG bound in subsite C (Figure 3c). The NAM ring adopts a sofa form with atoms C1, C2, C4, C5 and O5 nearly

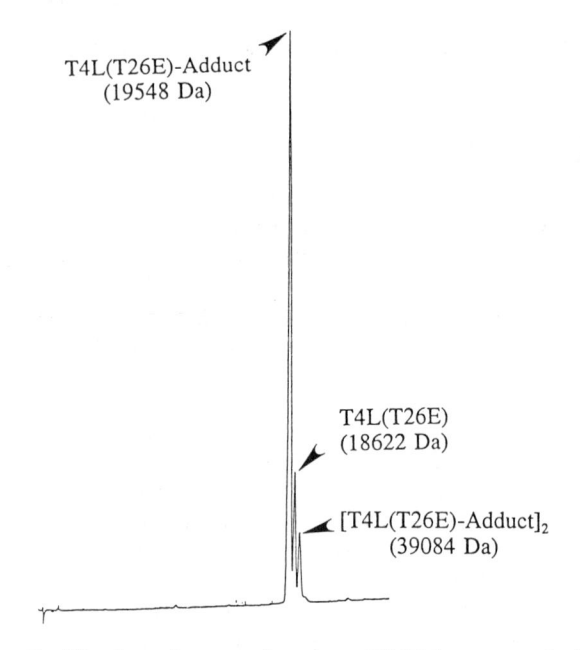

Figure 3a. Purification of mutant lysozyme T26E by reversed phase chromatography. (Applied Biosystems 130A HPLC, Aquapore RP300 column, gradient elution from 10% to 80% of B phase in 40', mobile phase A 0.1% trifluoroacetic acid, mobile phase B 70% CH$_3$CN, 0.08% trifluoroacetic acid, flow rate 100 µl/min, detection at 220 nm.) The mass of the species present in each of the three peaks, determined by mass spectroscopy (see text) are indicated.

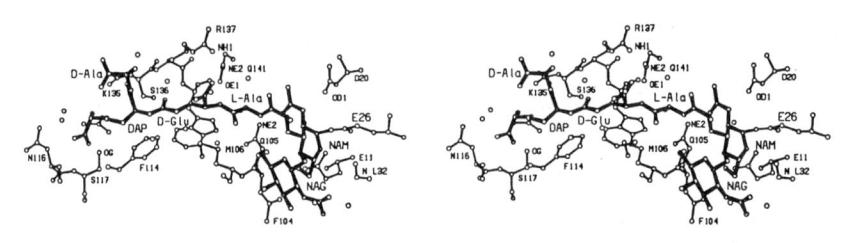

Figure 3b. Stereo drawing showing the peptidoglycan adduct (thicker bonds) bound to T4 lysozyme (thinner bonds). (Reproduced with permission from ref. *19*. Copyright 1993 American Association for the Advancement of Science.)

coplanar (0.07Å rms discrepancy). (In the full chair configuration the rms discrepancy from coplanar for the same 5 atoms is 0.25Å.) The N-acetyl group at C2 is shifted from axial to equatorial to make hydrogen bonds with Asp 20 and Gln 105. Also the normally equatorial hydroxymethyl group at C5 is shifted toward the axial position where it makes a favorable hydrogen bond to Gly 11. Thus, these multiple interactions with the enzyme seem to favor

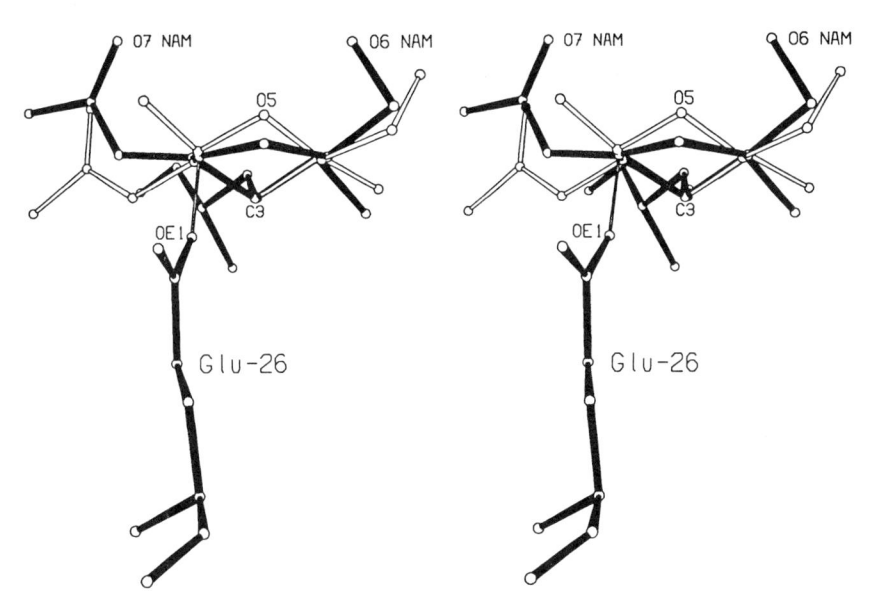

Figure 3c. Superposition of N-acetyl muramic acid (solid bonds) as seen in the D subsite and covalently linked to Glu 26 (also solid bonds) on N-acetyl glucosamine (open bonds) as seen in the C subsite.

distortion of the saccharide bound in subsite D. A similar distortion and interactions favorable to that distortion were seen in the complex of HEWL and NAM-NAG-NAM (25).

The presumed mechanism leading to the covalent adduct is illustrated in Figure 4a. Glu 11 is presumed to donate a proton to the glycosidic oxygen and Glu 26 is optimally located for nucleophilic attack on the C1 carbon, leading to the observed adduct. The resultant adduct is slowly cleaved from the enzyme (data not shown), possibly as illustrated in Figure 4a, although the role of Glu 11 in promoting the attack of the water molecule is uncertain. Nucleophilic attack by a carboxylate in glycosyl bond cleavage, as shown in Figure 4a, is well established for β-glucosidase (26) and β-galactosidase (14). Because the observed adduct is distorted toward an oxocarbonium ion-like conformation it suggests that the same enzyme-substrate interactions stabilize the transition state and, in addition, prevent relaxation of the covalent glycosyl-enzyme intermediate to a stable ground-state conformation. Therefore, the deglycosylation step should be rapid assuming free access by an acceptor group to the strained ring. In the case of the present adduct there is a well ordered water molecule hydrogen bonded to the N-acetyl group and to Glu 11 (Figure 2b). These hydrogen-bonding interactions, plus $O^{\epsilon 2}$ of Glu 26, apparently restrict access of the solvent to C1. The overall mechanism shown in Figure 4a is consistent with the double-displacement reaction envisaged for "configuration retaining" glycosidases in which a glycosyl-enzyme is formed and subsequently hydrolyzed (26,27).

If the enzyme-adduct complex is superimposed on wild-type T4L $O^{\epsilon 1}$ of Glu 26 in the adduct superimposes almost exactly on a solvent molecule that is bound between Thr 26 and Asp 20 in the native structure. Consideration of hydrogen bonding suggests that the lone pair of the bound solvent is directed toward the site occupied by the C1 carbon. This arrangement suggests that in the wild-type enzyme this solvent molecule could attack in what would be a single displacement reaction, inverting the anomeric configuration from equatorial in the substrate to axial in the product of hydrolysis. This hypothesis, that solvent attack occurs from the α-side of the saccharide in native T4 lysozyme, predicts that the anomeric configuration of the substrate is inverted in the product. In contrast, HEWL is known to retain anomeric configuration (*28*). In other words a mechanism for T4L as shown in Figure 4b would necessarily be different from that commonly accepted for HEWL (*25,29*).

Thus the fortuitous mutant lysozyme T26E (*19*) suggests different scenarios for lysozyme activity. The mutant (Figure 4a) could illustrate glycosidases that cleave with overall retention of configuration by a double displacement mechanism. Native T4 lysozyme itself might possibly represent cleavage by a single displacement mechanism (Figure 4b) with inversion of configuration although alternative mechanisms are not excluded. At the same time, the presence of protein-substrate interactions which stabilize a sugar ring conformation similar to an oxocarbonium ion-like transition state can be taken as evidence that the mechanism of action of T4L includes elements similar to those originally postulated by Phillips for HEWL (*29*). Further questions regarding the mechanism of lysozyme are, however, raised by the observation (*30-32*) of lysozymes that are structurally related to the T4 and/or hen egg-white enzymes, yet lack any counterpart to the "catalytic aspartate", which has generally been considered to be a key element in lysozyme catalysis.

β-Galactosidase *versus* Lysozyme

An intriguing question is why *E. coli* β-galactosidase is so big and, in particular, why it is so much larger than most lysozymes. In the tetrameric structure of β-galactosidase different regions contribute either to the active site, to the long interface, or to the activating interface (Figure 1). There is no large segment that could obviously be removed without having a deleterious effect. One has the impression that the protein arose from a fortuitous combination of structural domains that happened to have an activity that was useful to the cell, and became trapped in sequence space at roughly its current size.

In contrast, lysozymes in general, and the T4 and hen egg-white enzymes in particular, are small and monomeric. Recently an intriguing observation was made that a dimeric form of T4 lysozyme was inactive, even though the active site(s) seemed to be completely unobstructed (*33*) (Figure 5).

The natural substrates for lysozyme, bacterial cell walls, are highly crosslinked peptidoglycans. Therefore it may be that lysozyme needs to be small in order to access the susceptible bonds (*33*). β-galactosidase has no such restriction. Therefore the question may not be so much as to why β-galactosidase is large, but whether T4 lysozyme needs to be small.

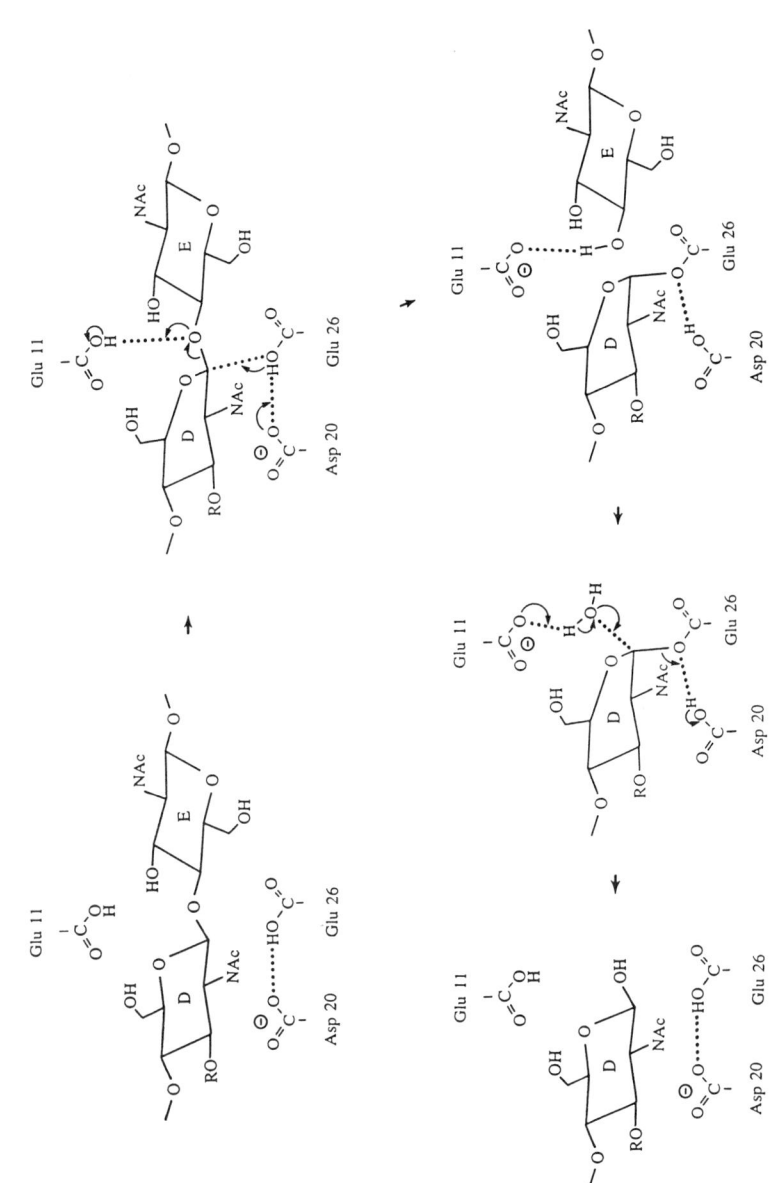

3. JACOBSON ET AL. *Glycosidases, Large and Small* 47

Figure 4a. Proposed steps leading to the adduct formed by T26E lysozyme, and its subsequent breakdown. (Reproduced with permission from ref. *19*. Copyright 1993 American Association for the Advancement of Science.)

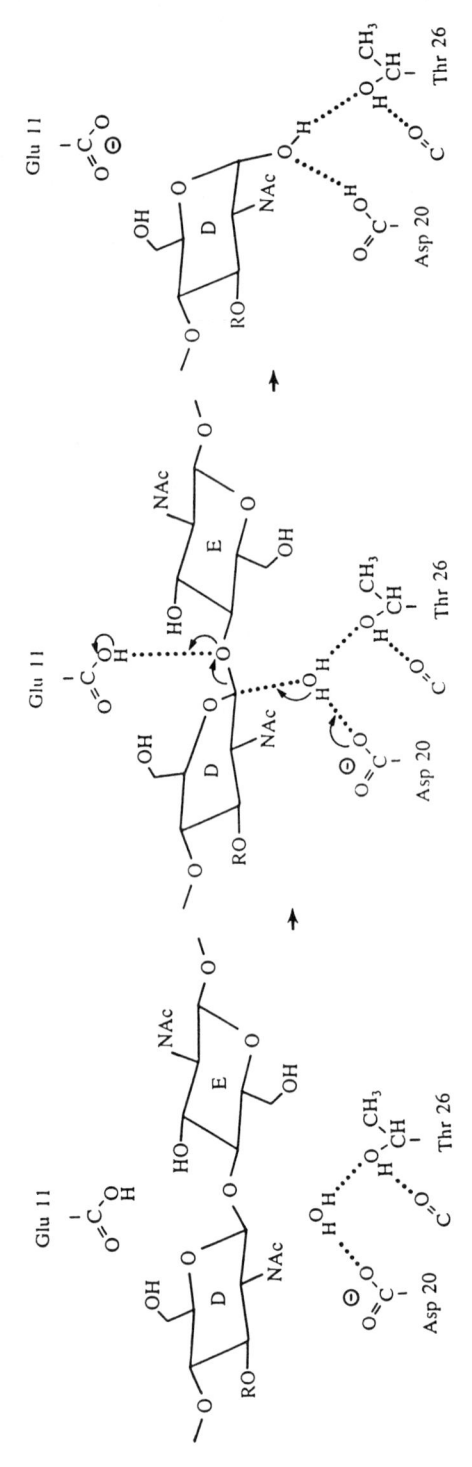

48   ENZYMATIC DEGRADATION OF INSOLUBLE CARBOHYDRATES

Figure 4b. One of several possible mechanisms of action of wild-type T4 lysozyme. Other mechanisms are also consistent with the available data. (Reproduced with permission from ref. *19*. Copyright 1993 American Association for the Advancement of Science.)

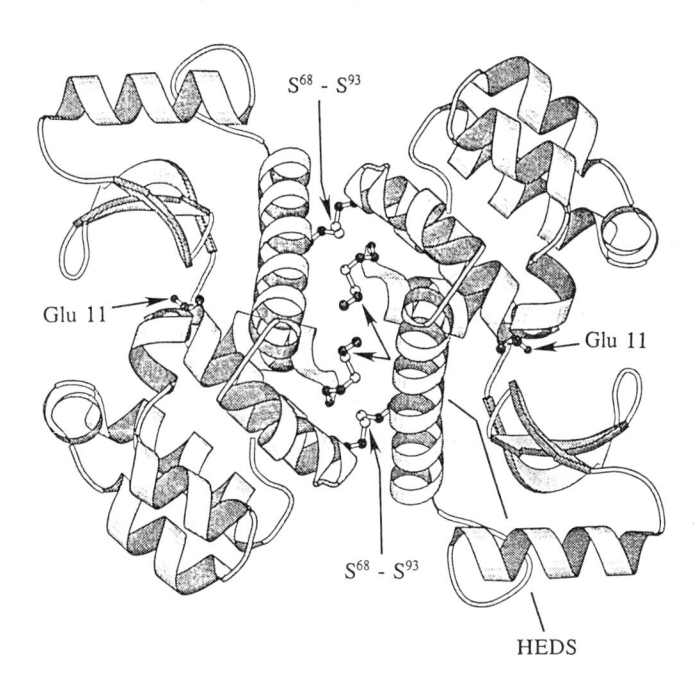

Figure 5. Crosslinked "back-to-back" dimer of T4 lysozyme. (Reproduced with permission from ref. 33. Copyright 1985 Oxford University Press.)

Acknowledgments

The assistance of W.A. Baase, R.F. DuBose, D. McMillen, S. Snow, D.E. Tronrud and J.A. Wozniak is greatly appreciated. This work was supported in part by grants from the NIH (GM21967 and GM20066).

Literature Cited

1. Jacobson, R. H.; Zhang, X-J.; DuBose, R. F.; Matthews, B.W. *Nature* **1994**, *369*, 761-765.
2. Fowler, A.; Zabin, I. *J. Biol. Chem.* **1978**, *253*, 5521-5525.
3. Kalnins, A.; Otto, K.; Ruther, U.; Müller-Hill, B. *EMBO J.* **1983**, *2*, 593-597.
4. Cohn, M. *Bacteriol. Rev.* **1957**, *21*, 140-168.
5. Lederberg, J. *J. Bact.* **1950**, *60*, 381-399.
6. Wallenfels, K.; Weil, R. *The Enzymes*; 3rd Ed., Vol. 7; Academic Press: London, UK, **1972**; Vol. 7, pp. 617-663.
7. Bernstein, F. C.; Koetzle, T. F.; Williams, G. J. B.; Meyer, E. F. Jr.; Brice, M. D.; Rodgers, J. R.; Kennard, O.; Shimanouchi, T.; Tasumi, M. *J. Mol. Biol.* **1977**, *112*, 535-542.
8. Holm, L.; Sander, C. *J. Mol. Biol.* **1993**, *233*, 123-138.

9. Lawrence, M. C.; Suzuki, E.; Varghese, J. N.; Davis, P. C.; Van Donkelaar, A.; Tulloch, P. A.; Colman, P. M. *EMBO J.* **1990**, *9*, 9-15.
10. Koshland, D. E. Jr. *Biol. Rev.* **1953**, *28*, 416-436.
11. Sinnott, M. L. *Chem. Rev.* **1990**, *90*, 1171-1202.
12. Ullmann, A.; Monod, J. *Biochem. Biophys. Res. Commun.* **1969**, *35*, 35-42.
13. Ring, M.; Huber, R. E. *Arch. Biochem. Biophys.* **1990**, *283*, 342-350.
14. Gebler, J. C., Aebersold, R.; Withers, S. G. *J. Biol. Chem.* **1992**, *267*, 11126-11130.
15. Cupples, C. G.; Miller, J. H.; Huber, R.E. *J. Biol. Chem.* **1990**, *265*, 5512-5518.
16. Brändén, C. *Quart. Rev. Biophys.* **1980**, *13*, 317-330.
17. Farber, G. K. *Curr. Opin. Struct. Biol.* **1993**, *3*, 409-412.
18. Matthews, B. W.; Grütter, M.G.; Anderson, W.F.; Remington, S. J. *Nature* **1981**, *290*, 334-335.
19. Kuroki, R.; Weaver, L. H.; Matthews, B. W. *Science* **1993**, *262*, 2030-2033.
20. Rennell, D.; Bouvier, S. E.; Hardy, L. W.; Poteete, A. R. *J. Mol. Biol.* **1991**, *222*, 67-87.
21. Ghuysen, J-M. *Bacteriology Rev.* **1968**, *32*, 425-464.
22. Schleifer, K. H.; Kandler, O. *Bacteriology Rev.* **1972**, *36*, 407-477.
23. Grütter, M. G.; Matthews, B. W. *J. Mol. Biol.* **1982**, *154*, 525-535.
24. Jensen, H. B.; Kleppe, G.; Schindler, M.; Mirelman, D. *Eur. J. Biochem.* **1976**, *66*, 319-325.
25. Strynadka, N. C. J.; James, M. N. G. *J. Mol. Biol.* **1991**, *220*, 401-424.
26. Koshland, D. E. Jr. *Biol. Rev.* **1953**, *28*, 416-436.
27. Kempton, J. B.; Withers, S. G. *Biochemistry* **1992**, *31*, 9961-9969.
28. Dahlquist, F. W.; Borders, C. L.; Jacobson, G.; Raftery, M. A. *Biochemistry* **1969**, *8*, 694-700.
29. Phillips, D. C. *Proc. Natl. Acad. Sci. USA* **1967**, *57*, 484-495.
30. Weaver, L. H.; Grütter, M. G.; Remington, S. J.; Gray, T.M.; Isaacs, N. W.; Matthews, B. W. *J. Mol. Evol.* **1985**, *21*, 97-111.
31. Thunnisen, A-M. W. H.; Dijkstra, A. J.; Kalk, K. H.; Rozeboom, H. J.; Engel, H.; Keck, W.; Dijkstra, B. W. *Nature* **1994**, *367*, 750-753.
32. Weaver, L. H.; Grütter, M. G.; Matthews, B. W. *J. Mol. Biol.* **1995**, *245*, 54-68.
33. Heinz, D.; Matthews, B. W. *Prot. Engin.* **1994**, *7*, 301-307.

RECEIVED August 17, 1995

Chapter 4

Barley Seed α-Glucosidases: Their Characteristics and Roles in Starch Degradation

C. A. Henson[1,2] and Z. Sun[2]

[1]Cereal Crops Research Unit, Agricultural Research Service, U.S. Department of Agriculture, and [2]Department of Agronomy, University of Wisconsin, 1575 Linden Drive, Madison, WI 53706

> The identities of enzymes involved in degradation of seed starch are well established, as are the characteristics of several of the individual enzymes. However, the interactions between the different enzymes as they function together to hydrolyze starch are not well understood. The development of a reconstituted starch hydrolyzing system using components isolated from seeds allowed the identification of an interaction between α-amylases and α-glucosidases that resulted in synergistic hydrolysis of native starch granules. The relative importance of this interaction, as well as interactions with β-amylase and starch debranching enzyme, was influenced by whether or not the crystalline structure of starch was disrupted. Examination of these four enzymes as they function in a multistep biochemical process revealed that the importance of their independent actions, in addition to their interactions, varied depending upon the structure of the starch being hydrolyzed.

Cereal seeds are the primary source of calories for most humans and are a major food of livestock. Starch from cereal seeds is the raw material for industries that produce ethanol and mono- and oligosaccharide syrups and myriad derived products. Due to the importance of starch, its breakdown sugars and subsequently derived products, cereal seeds are some of the most important agriculture products in the world and starch degrading enzymes are among the most important commercial enzymes. Besides being important to industry, starch degradation is critical to the viability of cereal seeds as starch, typically 50-70% of the seed's dry weight, is the primary source of respirable substrates until the seedling is fully autotrophic. In the following sections the roles of starch degrading enzymes in germinating cereal seeds will be examined and characteristics of barley seed α-glucosidase will be presented. Emphasis is placed on α-glucosidase, which has been less thoroughly studied than amylases, due to recent studies indicating it is important in starch degradation.

Starch Degrading Enzymes in Germinating Cereal Seeds - The complete degradation of starch in germinating cereal seeds results from the concerted action of α-amylase, β-amylase, debranching enzyme and α-glucosidase. Complete hydrolysis depends on these enzymes' abilities to independently function in hydrolysis of starch granules or granule breakdown products and on their interactions with each other as part of a multienzyme system. The ability of each enzyme to initiate hydrolysis of raw starch granules is generally understood. Early studies showed that α-amylase initiated raw (native) starch granule hydrolysis whereas β-amylase and debranching enzyme did not initiate hydrolysis (1). As late as 1989 it was still accepted that only α-amylase could initiate raw starch granule hydrolysis (2). We recently demonstrated that barley seed α-glucosidase can initiate attack of raw starch granules and that this attack is independent of the presence of α-amylase (3). This has been verified by Sissons and MacGregor (4). The generally accepted roles of β-amylase and debranching enzymes are that they function to hydrolyze glucan fragments released by the action of α-amylase and that they may attack starch granules after sites are made available by previous action of α-amylase. Hydrolysis of released glucan fragments, which is necessary for complete degradation of starch to glucose, is also performed by α-amylase and α-glucosidase.

Several different approaches have been used to determine the relative importance of α-amylase, β-amylase, debranching enzyme and α-glucosidase as they function together to degrade native starches. The impact of changing the activity of just one enzyme upon an intact seed's ability to degrade starch has been studied with mutants and with chemicals that specifically inhibit the activity of only one of these four carbohydrases. Simultaneous changes in activities of two to four of the carbohydrases have been explored with two approaches. One is the use of reconstituted starch hydrolyzing systems containing different combinations of enzymes. Another approach used was statistical and mathematical analyses of starch hydrolysis by genotypes varying in activities of these four enzymes. A brief review of some of these studies follows.

The mathematical and statistical approach we used to investigate degradation of starch in germinated barley seeds varying widely in their activities of the four carbohydrases relied upon linear regression and path coefficient analysis (5). Linear regression was used to quantitate the relationships between the four carbohydrases and the seeds' abilities to degrade starch. Path coefficient analysis, a method which quantitatively evaluates the individual, or independent, contributions and the interactive, or dependent, contributions of each component in a biological system, was used to analyze the relative importance of the four carbohydrases in starch degradation (5). When raw starch was the substrate provided to extracts of germinated seeds, α-amylase and α-glucosidase activities were positively correlated, significant at the 1% level, (r= 0.962 and 0.670, respectively) with seed hydrolysis of the starch. In contrast, β-amylase and debranching enzyme activities were not significantly correlated (r= -0.283 and -0.141, respectively) with the seeds' abilities to hydrolyze raw starch. Path analysis revealed that the primary contribution of α-amylase to raw starch hydrolysis was direct, or independent of interactions with the other carbohydrases. In contrast, only 7% of the total contribution by α-glucosidase was due to direct or noninteractive action upon the starch granules. The bulk (93%) of the contribution made by α-glucosidase to raw starch hydrolysis was dependent upon its interaction with α-amylase. The activities of α-amylase and α-glucosidase were highly correlated (r= 0.671) with each other.

Studies of *in vitro* starch hydrolyzing systems constructed of raw starch and hydrolases isolated from barley seeds showed that starch digestion by α-amylase alone was slightly less than digestion by α-amylase and β-amylase combined (6). The two amylases combined with debranching enzyme gave hydrolysis rates 25% above the two amylases (7). These three enzymes resulted in hydrolysis rates approximately 70% of *in situ* rates. Maeda and coworkers concluded that a factor, perhaps another enzyme, was still missing from their *in vitro* system.

Early studies with an *in vitro* system consisting of corn starch, pancreatic α-amylase and Apsergillus extracts showed that Aspergillus enzymes enhanced starch degradation by α-amylase (8). It was later shown that the enhancement was due to α-glucosidase (maltase) present in the Aspergillus preparation (9). Work in the 60's proved that barley α-glucosidase was expressed during seed germination and that it could degrade boiled soluble starch in addition to other maltodextrins (10). Taken together these early studies indicated a potential role for barley seed α-glucosidase in starch degradation. Hence, we used α-amylase, α-glucosidase and raw starch isolated from barley to study the potential roles of α-glucosidase (3). We showed that α-glucosidases independently initiate hydrolysis of raw starch granules, as discussed previously. Additionally, we demonstrated that hydrolysis by the combined action of α-amylase and α-glucosidase was synergistic. We attributed the synergism to three factors. First, α-amylase activities may be enhanced by α-glucosidase because the latter degrades the maltose produced by α-amylase. This would prevent maltose accumulating to concentrations that are inhibitory to α-amylase. A second factor contributing to the synergism is that some of the free glucose and total reducing sugar produced during starch granule hydrolysis would result from hydrolysis of glucan fragments released from starch granules rather than from direct attacks upon the starch granules. A third possibility is that α-glucosidase may generate attack sites that were previously unaccessible to α-amylase. Data supporting this third possibility are scanning electron micrographs showing that hydrolysis by the two enzymes together results in an increased number of visible attacks upon the granule surfaces (3). The micrographs also showed that attacks occurred more often upon the flat surfaces of the granules when the two enzymes were combined than when the enzymes function independently.

The *in vitro* system used by Sun and Henson (3) contained a ratio of α-amylase/α-glucosidase of 33:1 and the synergism varied from 1.7 to 10.7 fold depending upon hydrolysis time, the isozymes of α-amylase and α-glucosidase combined and whether the end product detected was free glucose or total reducing sugars. This ratio of α-amylase to α-glucosidase was selected based on the maximal amounts of the two enzyme activities we extracted from the germinated seeds used in these studies. We examined the effect of varying the ratio of α-amylase to α-glucosidase from 50:1 to 1:5 on the synergy between the two enzymes. We used high pI α-amylase and high pI α-glucosidase as they constitute the bulk of the α-amylase and α-glucosidase activity in barley seeds. We found that the optimal ratio of high pI α-amylase to high pI α-glucosidase was 15:1 and that synergy was marginal when ratios were from 5:1 to 1:5. Sissons and MacGregor (4) recently showed that 1:1 ratios of amylase and glucosidase resulted in a 1.3-fold synergy for combined high pI isoforms at 36 h of hydrolysis when glucose production was the end product detected. They observed a 1.5-fold synergy for combined high pI isoforms. Since the ratio of α-amylase to α-glucosidase activities found in germinated seeds of Morex, a common malting barley, is 33:1 there is potential for increasing the extent of synergism by increasing the level of α-glucosidase above what is present in Morex seeds. A 3-fold range in α-glucosidase activity was found in germinated seeds of 41 barley genotypes indicating there may be sufficient genetic potential for developing germplasm with enhanced expression of α-glucosidase (5).

The *in vitro* synergism exhibited by a constant ratio of enzymes is influenced by several factors. As α-glucosidase activity always produces glucose while α-amylase hydrolysis of starch granules, at least in the early stages of granule degradation, results in less than 2% of the end product being free glucose (11), the extent of synergism is considerably influenced by the choice of end product measured. The degree of synergism is also influenced by the choice of substrates used to determine the activities of α-amylase and α-glucosidase added to the *in vitro* system. Im and Henson (12) have shown that high pI α-glucosidase efficiently hydrolyzes maltose, maltotriose, maltotetraose and maltopentaose as determined by the ratio of V_{max} to K_m (6.7, 12.8, 6.3 and 5.2, respectively). In contrast, the artificial substrate *p*-nitrophenyl-α-D-glucoside is a relatively poor substrate (see Table 1). The absolute activities of α-amylase would also be influenced by the choice of substrates used to determine activity. For a thorough discussion of the multiple factors impacting on the synergism between α-amylase and glucoamylase, see Shevel'kova and Sinitsyn (13).

Changing a single enzyme activity in an intact seed has been studied by germinating seeds in the presence of a specific enzyme inhibitor. Wheat seeds germinated in the presence of miglitol (*N*-hydroxyethyl-1-deoxynojirimycin; Bay m 1099), a specific inhibitor of α-glucosidase activity, exhibited normal germination rates and no inhibition of total amylase activity (14). These seeds exhibited about one-third of the α-glucosidase activity found in the controls and the initial rate of starch degradation was decreased 48% compared to control seeds. As germination proceeded, seeds treated with miglitol degraded starch at normal rates even though their α-glucosidase activities remained depressed. Miglitol treated seeds showed normal levels of α-amylase during the early stages of starch degradation but elevated levels were detected later during germination when the treated seeds showed normal rates of starch degradation. These data indicate wheat seed α-glucosidase significantly contributes to the early stages of starch hydrolysis and that germinating seeds develop the ability to enzymatically compensate for the decreased contribution of α-glucosidase to early stages of starch degradation.

The significance of changing a single enzyme activity upon an intact seed's ability to germinate and degrade starch has also been studied by generation of mutants deficient in synthesis of seed β-amylase. Such mutants have been generated in barley (15), rye (16) and soybean (17). Mutant seeds of all three species germinated or malted (a process dependent upon germination) normally. These studies indicate that β-amylase is either not involved in the normal degradation of starch that occurs during seed germination or that its absence can be compensated for by the actions of other enzymes. Levels of α-amylase were normal, not elevated, in the barley, rye and soybean seeds deficient in β-amylase. Levels of debranching enzyme and α-glucosidase activities were not reported.

The emerging picture for the early stages of raw starch hydrolysis in germinating cereal seeds is that α-amylase is the most important enzyme, α-glucosidase is the second most important enzyme, debranching enzyme is the third most important, and β-amylase is not important. That α-amylase is the most important of these enzymes has been accepted for many years partially because it was, until recently, thought to be the only enzyme initiating hydrolysis of raw starch granules. The noncritical nature of β-amylase in seed germination has been a developing concept for several years and, now, multiple experimental protocols support this interpretation. The importance of α-glucosidase in cereal seed germination has only been rigorously investigated in the past five years and much is still to be learned. It is important to realize that an enzyme's importance, or lack thereof, during early stages of germination and starch degradation may change as development progresses. Changes in an enzyme's relative importance could result from changes in the concentrations of preferred substrates or of glucans which are inhibitory. For example, as starch degradation progresses there are more small dextrins containing α-1-6 linkages, which are the preferred substrates for the debranching enzyme present in germinating barley seeds (18). Hence, debranching enzyme's activity and relative importance may be enhanced as starch degradation progresses. Additionally, an enzyme's physiological role can also change as a function of seedling development, as Konishi et al. (14) showed for α-glucosidase in germinating wheat seeds. They found that during early stages of germination α-glucosidase functioned to enhance rates of starch hydrolysis by α-amylase while in later stages of seedling development α-glucosidase's role was to ensure that sufficient levels of glucose were available for synthesis of the sucrose needed by the growing embryonic tissues. Repression of α-glucosidase function during later stages of seed germination resulted in significant inhibition of seedling growth and development.

Characteristics of barley seed α-glucosidase - Barley seed α-glucosidases are acidic α-glucosidases with pH activity optimum of 4.3 for the low pI and 4.1 for the high pI isoform (Figure 1). As the pH of a germinating barley seed endosperm is approx. 4.5-5.5 the activity optima reported for the barley α-glucosidases is appropriate for physiological relevance. Most plant α-glucosidases studied have acidic pH activity optimum. An exception is an α-glucosidase from pea chloroplasts that exhibits maximal activity at pH 7.0, which is near the stromal pH expected of chloroplasts exposed to darkness (19). We demonstrated that this chloroplastic α-glucosidase can initiate degradation of the starch stored in pea chloroplasts (20).

Some kinetic parameters of the high and low pI α-glucosidases, purified as described by Sun and Henson (3), are reported in Table I. Based on the specificity constants, ratio of V_{max} to K_m, for maltose and *p*-nitrophenyl-α-glucoside (*p*NPG), maltose is a much better substrate for both isoforms than is *p*NPG, although the latter

Fig. 1. Effect of pH on the hydrolytic activity of barley seed α-glucosidases. The pH of the reaction mixtures was varied with 0.2M sodium citrate buffer. The substrate was maltose and released glucose was measured with glucose-6-phosphate dehydrogenase and hexokinase.

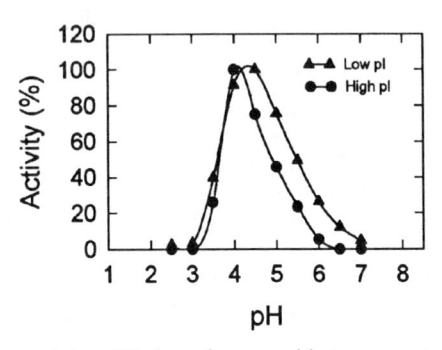

is often used to determine α-glucosidase activity. We have been unable to separate the maltose hydrolyzing function from the pNPG hydrolyzing function for either isoform. Hence, we conclude that the two activities are a function of the same protein. Germinated barley seeds do, however, have α-glucosidases with pNPG hydrolyzing activities that do not hydrolyze maltose (21). α-Glucosidases from buckwheat and sugarbeet in addition to those from mammals and bacteria hydrolyze a wide variety of substrates including both maltose and artificial substrates (22-26).

Table I. Kinetic characteristics of barley seed α-glucosidases

Substrate or Inhibitor	K_m high pI/low pI	V_{max} high pI/low pI	K_i high pI/low pI	Inhibition high pI/low pI
Maltose	2.4 mM 1.91 mM	1.23 0.54		
pNPG	1.0 mM 3.5 mM	0.11 0.13		
erythritol			10.7 mM 18.3 mM	competitive
castano-spermine			0.11 mM 0.09 mM	competitive

All kinetic parameters were determined from Lineweaver-Burk plots of maltose or pNPG hydrolysis. Glucose produced was measured with glucose-6-phosphate dehydrogenase and hexokinase. Inhibiton studies used maltose as the substrate. V_{max} = μmol glucose or μg p-nitrophenolate released from substrate/min/ml enzyme preparation.

Inhibition of mammalian, microbial and plant α-glucosidases by erythritol and castanospermine have been frequently reported and likely result from their structural similarities to glucose. Erythritol has a configuration identical to glucose from carbons 3-6 and castanospermine is a tetrahydroxyindolizidine alkaloid that resembles maltose in its stereochemistry (27). The K_i value for inhibition of a mixture of high and low pI α-glucosidases from barley by erythritol were previously reported to be 12.2 mM (28). This value is similar to the inhibition constants for purified high pI α-glucosidase even though those previously reported resulted from the combined action of both isoforms. Inhibitors can be used to elucidate the importance of α-glucosidases *in situ* (14) and in industrial processes, to establish the number of isoforms in crude tissue extracts, to distinguish between α-glucosidases and exoglucanases, to compare α-glucosidases between species, to "label" the enzyme for use in identifying and sequencing portions of the active site, and to elucidate details of the kinetic mechanism. We are presently using ^{3}H-conduritol-B-epoxide, an active site label of α-glucosidases, to determine the number of α-glucosidase isoforms present in germinated barley seeds and to locate peptide fragments containing portions of the active site.

The thermal stability of barley seed α-glucosidase is important because the conversion of barley seed starch to fermentable sugars during industrial processing typically takes place at temperatures of 65-73°C. Thermal characteristics of barley seed high and low pI α-glucosidases are shown in Figure 2.

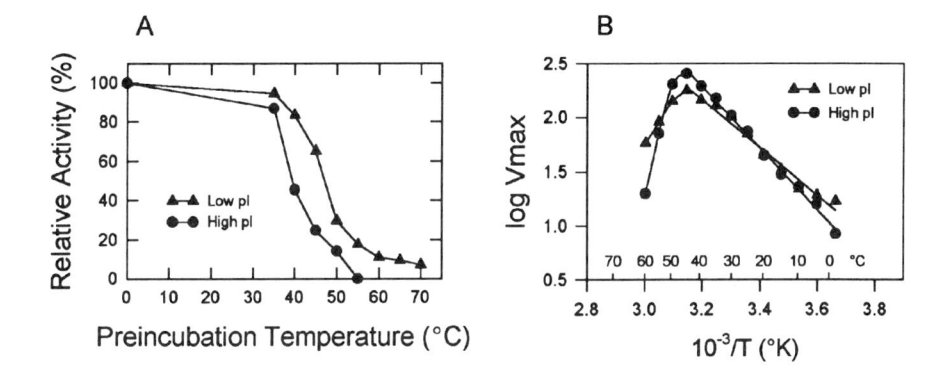

Figure 2A. Effects of a 10 min exposure of barley seed α-glucosidases to different temperatures on the residual activity measured at 37°C. Activities were determined with maltose as the substrate. Figure 2B. Arrhenius plots of barley seed α-glucosidase activities. The enzymes and substrates were separately incubated for 10 min at the indicated temperatures prior to assays at the same temperatures. Activities were determined with maltose as the substrate, which was saturating at all assay temperatures.

The low pI α-glucosidase appears slightly more thermal stable than the high pI isoform although neither isoform appears sufficiently thermostable to function throughout industrial mashing procedures conducted at elevated temperatures. The residual activity of the low pI α-glucosidase decreased to 50% at about 47°C, whereas that of the high pI α-glucosidase decreased to 50% at about 39°C (Figure 2A). Arrhenius plots of the barley seed α-glucosidases (Figure 2B) showed that the low pI isoform had an activation energy of 6.2 kcal/mol. The high pI isoform had an activation energy of 11.7 kcal/mol. The deviation from linearity at higher temperatures is attributed to enzyme denaturation.

α-Glucosidases are now recognized as important in starch degradation during cereal seed germination. Their roles in industrial starch conversion processes still need to be thoroughly examined, but our data indicate they may be limited by a lack of thermostability. Barley seed α-glucosidases have fairly recently begun to be critically examined and much is still to be learned.

Literature Cited

1. Dunn, G. *Phytochemistry* **1974**, *13*, 1341.
2. Beck, E.; Ziegler, P. In *Ann. Rev. Plant Physiol. Plant Mol. Biol.*; Briggs, W.R.; Jones, R.L.; Walbot, V., Eds.; Annual Reviews, Inc.; Palo Alto, CA, **1989**; Vol. 40: pp.95-117.
3. Sun, Z.; Henson, C.A. *Plant Physiol.* **1990**, *94*, 320.
4. Sissons, M.J.; MacGregor, A.W. *J. Cereal Sci.* **1994**, *19*, 161.
5. Sun, Z.; Henson, C.A. *Arch. Biochem. Biophys.* **1991**, *28*, 298.
6. Maeda, I.; Kiribuchi, S.; Nakamura, M. *Agric. Biol. Chem.* **1978a**, 42, 259.
7. Maeda, I.; Nikuni, Z.; Taniguchi, H.; Nakamura, M. *Carbohydr. Res.* **1978b**, *61*, 309.
8. Blish, M.T.; Sandstedt, R.M.; Mecham, D.K. *Cereal Chem.* **1937**, *14*, 605.
9. Schwimmer, S. *J. Biol. Chem.* **1945**, *161*, 219.
10. Jorgensen, O.B. *Acta Chem. Scand.* **1964**, *18*, 1975.
11. MacGregor, E.A.; MacGregor, A.W. In *New Approaches to Research on Cereal Carbohydrates*; Hill, R.D.; Munck, L., Eds.; Elsevier Science; Amsterdam, **1985**; pp.149-160.
12. Im, H.; Henson, C.A. *Carbohydr. Res.* **1995**, in press.
13. Shevel'kova, A.N.; Sinitsyn, A.P. *Biokhimiya* **1993**, *58*, 1548.
14. Konishi, Y.; Okamoto, A.; Takahashi, J.; Aitani, M.; Nakatani, N. *Biosci Biotech. Biochem.* **1994**, *58*, 135.
15. Allison, J.H. *J. Inst. Brew.* **1978**, *84*, 231.
16. Daussant, J.; Zbaszyniak, B.; Sadowski, J.; Waitroszak, I. *Planta*, **1981**, *151*, 176.
17. Hildebrand, D. F.; Hymowitz, T. *Physiol. Plant.* **1981**, 53, 429.
18. Manners, D.H.; Hardie, D. Grahame. *MBAA Tech. Quart.* **1977**, *14*, 120.
19. Beers, E.P.; Duke, S.H.; Henson, C.A. Plant Physiol. **1990**, *94*,738.
20. Sun Z.; Duke, S.H.; Henson, C.A. *Plant Physiol.* **1995**, *108*, 211.
21. Stark, J.R.; Yin, X.S. *J. Inst. Brew.* **1987**, *93*, 108.
22. Chiba, S.; Kanaya, K.; Hiromi, K.; Shimomura, T. *Agric. Biol. Chem.* **1979**, *43*, 237.
23. Matsui, H.; Chiba, S.; Shimomura, T. *Agric. Biol. Chem.* **1978**, *42*, 1855.
24. Suzuki, Y.; Ueda, Y.; Nakamura, N.; Abe, S. *Biochem. Biophys. Acta* **1979**, *566*, 62.
25. Suzuki, Y.; Shinji, M.; Eto, N. *Biochem. Biophys. Acta* **1984**, *278*, 281.
26. Chiba, S.; Hibi, H.; Shimomura, T. *Agric. Biol. Chem.* **1976**, *40*, 1813.
27. Kanya, K.; Chiba, S.; Shimomura, T.; Nishi, K. *Agric. Biol. Chem.* **1976**, *40*, 1929.
28. Jorgensen, B.B.; Jorgensen, O.B. *Biochim. Biophys. Acta.* **1969**, *146*, 167.

RECEIVED November 1, 1995

Chapter 5

Three-Dimensional Structure of an Endochitinase from Barley

Jon D. Robertus and John Hart

Department of Chemistry and Biochemistry, University of Texas, Austin, TX 78712

Endochitinases hydrolyze chitin, cleaving glycosidic bonds within the insoluble polymer. Chitin is extremely abundant, being found in the exoskeletons of many insects, crustacea and fungi. Chitinases are of commercial interest since they can be used to create chitin oligomers useful in the chemical, pharmaceutical, and food industries. We have determined the X-ray structure of the chitinase from barley. The model has been refined to high accuracy against 1.8 data. This chitinase has 10 α-helices comprising 47% of the structure. It has a very pronounced cleft, assumed to be the chitin binding and catalytic site. A number of residues, which are conserved within the chitinase family, cluster in this cleft. Presumably they play key roles in chitin binding and hydrolysis. Barley chitinase reveals an ancient resemblance to lysozyme; they share a common central core despite lacking any obvious amino acid sequence similarity. In barley chitinase, Glu 67 appears to act as the proton donor and Glu 89 as the general base in the catalytic mechanism.

Chitin has been called the second most common organic molecule in the world, after cellulose. The annual production of this renewable biosource has been estimated at between 10 and 100 billion tons (*1*). The polymer is found in the exoskeletons of insects and crustaceans like crabs and lobsters and is also a major component of many fungal cell walls. Chitin has a number of clinical and industrial uses. It is used in paper processing and the food and cosmetic industries. Chitosan, derived from chitin by acid hydrolysis, is being used in water treatment and in burn dressings. Oligomers of chitin and chitosan can act as bioactivators. They induce defense responses in higher plants and exhibit some antitumor effects in animals. A wide range of uses for the two polymers is reviewed in reference (2), the proceedings of the 5th International Conference on Chitin and Chitosan.

Chemically, chitin is a linear, non-branched polymer of β-1→4 linked units of N-acetylglucosamine (NAG). Figure 1 shows a segment of the polymer. Hydrogen bonds donated from the C3 hydroxyl to O5 in the pyranose ring add greatly to the tensile strength of the molecule. Chitin is a close relative of cellulose, which has a hydroxyl at C2 in place of the acetylated nitrogen. Also related to chitin is chitosan, which results from hydrolysis of the acetyl group. The presence of the this amine group causes chitosan to be protonated and behave as a poly cation.

Chitinases hydrolyze chitin. Exochitinases remove NAG units from the non-reducing end of the polymer. This class of enzyme is fairly common in bacteria but occurs only rarely in higher plants. Far more common are endochitinase which cleave chitin internally. These are found in higher plants, fungi, and bacteria. A few endochitinases also exhibit weak lysozyme activity. That is, they are able to cleave the β-1→4 bond between the N-acetyl muramic acid and NAG found in peptidoglycan. (Conversely, some lysozymes exhibit chitinase activity.)

Endochitinases are generally monomeric proteins between 25 and 40 KDa. Four classes have been proposed based on amino acid sequences (3,4). Figure 2 shows a sequence alignment of several representative chitinases from classes I, II, and IV. These classes have homologous catalytic domains, of about 26 KDa. This domain contains the chitin substrate binding and catalytic functions. Class II enzymes, like that from barley, contain only this catalytic domain. Class IV enzymes have three deletions within this main unit, but clearly belong to the same family (4). Class I and IV endochitinases also have an N-terminal cysteine-rich domain of about 50 residues. It is homologous to wheat germ agglutinin (WGA), and presumably binds the enzyme to chitin, holding it to the insoluble polymer. The N-terminal domain is linked to the chitinase domain by a glycine/proline-rich hinge segment. Class III chitinases show no apparent sequence similarity to enzymes in class I, II, or IV. They are relatively rare in higher plants, but common in fungi.

Plants have no immune system and have evolved other methods to defend themselves against pathogens such as viruses, bacteria, and fungi. Plants under attack can induce genes which express a battery of defense proteins, known as pathogenesis-related (PR) proteins (5,6). These include ribosome inactivating proteins (RIPs) (7), pore-forming polypeptides that insert themselves into fungal cells (8), and lytic enzymes such as β-1,3 glucanases and chitinases (9,10). In fact these proteins often work synergistically within a single host. This is true in barley (Hordeum vulgare) where three antifungal activities from a RIP, endoglucanase and chitinase have been found (11).

Our laboratory has been examining the structure and action of representatives of many of these plant defense proteins. We have solved the structure of plant RIPs including ricin (12) and PAP (13), and crystallized the anti-

5. ROBERTUS & HART   Three-Dimensional Structure of an Endochitinase

Figure 1: The molecular structure of the chitin polymer.

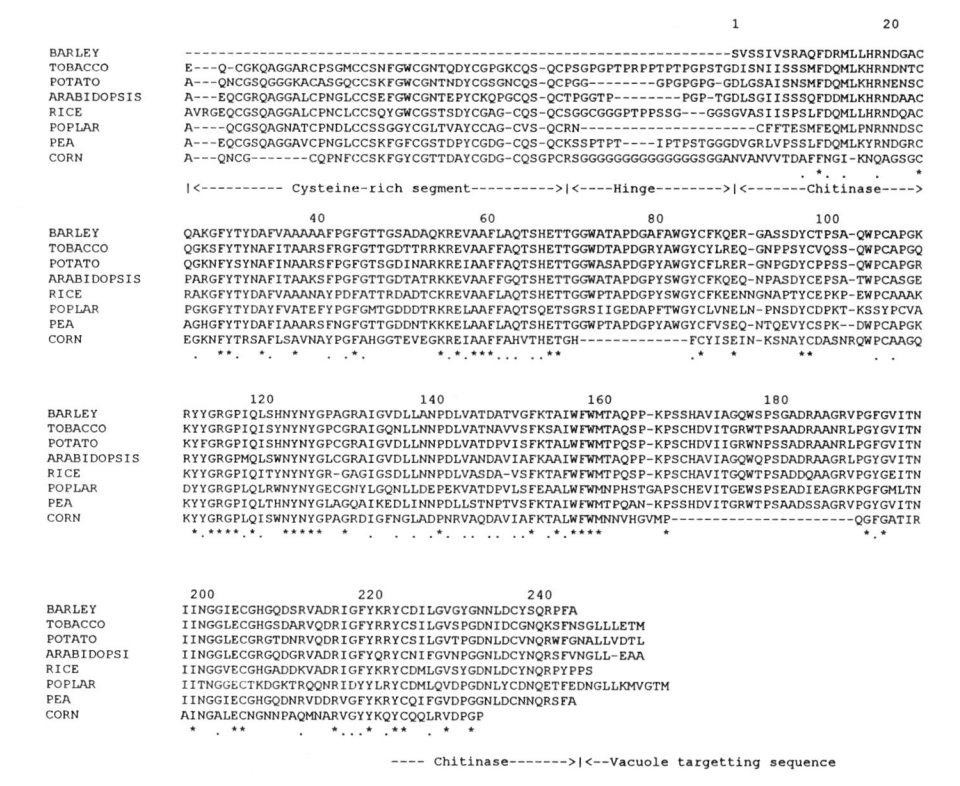

```
                                                                               1                   20
BARLEY      ----------------------------------------------------------------------SVSSIVSRAQFDRMLLHRNDGAC
TOBACCO     E---Q-CGKQAGGARCPSGMCCSNFGWCGNTQDYCGPGKCQS-QCPSGPGPTPRPPTPTPGPSTGDISNIISSSMFDQMLKHRNDNTC
POTATO      A---QNCGSQGGGKACASGQCCSKFGWCGNTNDYCGSGNCQS-QCPGG--------GPGPGPG-GDLGSAISNSMFDQMLKHRNENSC
ARABIDOPSIS A---EQCGRQAGGALCPNGLCCSEFGWCGNTEPYCKQPGCQS-QCTPGGTP--------PGP-TGDLSGIISSSQFDDMLKHRNDAAC
RICE        AVRGEQCGSQAGGALCPNCLCCSQYGWCGSTSDYCGAG-CQS-QCSGGCGGGPTPPSSG---GGSGVASIISPSLFDQMLLHRNDQAC
POPLAR      A----QCGSQAGNATCPNDLCCSSGGYCGLTVAYCCAG-CVS-QCRN--------------------CFFTESMFEQMLPNRNNDSC
PEA         A---EQCGSQAGGAVCPNGLCCSKFGFCGSTDPYCGDG-CQS-QCKSSPTPT----IPTPSTGGGDVGRLVPSSLFDQMLKYRNDGRC
CORN        A---QNCG-------CQPNFCCSKFGYCGTTDAYCGDG-CQSGPCRSGGGGGGGGGGGGGGSGGANVANVVTDAFFNGI-KNQAGSGC
                                                                . *.     .    *
            |<--------- Cysteine-rich segment--------->|<----Hinge-------->|<-------Chitinase---->

                       40                    60                    80                    100
BARLEY      QAKGFYTYDAFVAAAAAFPGFGTTGSADAQKREVAAFLAQTSHETTGGWATAPDGAFAWGYCFKQER-GASSDYCTPSA-QWPCAPGK
TOBACCO     QGKSFYTYNAFITAARSFRGFGTTGDTTRRKREVAAFFAQTSHETTGGWDTAPDGRYAWGYCYLREQ-GNPPSYCVQSS-QWPCAPGQ
POTATO      QGKNFYSYNAFINAARSFPGFGTSGDINARKREIAAFFAQTSHETTGGWASAPDGPYAWGYCFLRER-GNPGDYCPPSS-QWPCAPGR
ARABIDOPSIS PARGFYTYNAFITAAKSFPGFGTTGDTATRKKEVAAFFGQTSHETTGGWATAPDGPYSWGYCFKQEQ-NPASDYCEPSA-TWPCASGE
RICE        RAKGFYTYDAFVAAANAYPDFATTRDADTCKREVAAFLAQTSHETTGGWPTAPDGPYSWGYCFKEENNGNAPTYCEPKP-EWPCAAAK
POPLAR      PGKGFYTYDAYFVATEFYPGFGMTGDDDTRKRELAAFFAQTSQETSGRSIIGEDAPFTWGYCLVNELN-PNSDYCDPKT-KSSYPCVA
PEA         AGHGFYTYDAFIAAARSFNGFGTTGDDNTKKKELAAFLAQTSHETTGGWPTAPDGPYAWGYCFVSEQ-NTQEVYCSPK--DWPCAPGK
CORN        EGKNFYTRSAFLSAVNAYPGFAHGGTEVEGKREIAAFFAHVTHETGH-------------FCYISEIN-KSNAYCDASNRQWPCAAGQ
             . **.  *. *    .*.     *.*.***... **             .         .  *       **      . .

                       120                   140                   160                   180
BARLEY      RYYGRGPIQLSHNYNYGPAGRAIGVDLLANPDLVATDATVGFKTAIWFWMTAQPP-KPSSHAVIAGQWSPSGADRAAGRVPGFGVITN
TOBACCO     KYFGRGPIQISYNYNYGPCGRAIGQNLLNNPDLVATNAVVSFKSAIWFWMTAQSP-KPSCHDVITGRWTPSAADRAANRLPGYGVITN
POTATO      KYFGRGPIQISHNYNYGPCGRAIGVDLLNNPDLVATDPVISFKTALWFWMTPQSP-KPSCHDVIIGRWNPSSADRAANRLPGFGVITN
ARABIDOPSIS RYYGRGPMQLSWNYNYGLCGRAIGVDLLNNPDLVANDAVIAFKAAIWFWMTAQPP-KPSCHAVIAGGWQPSDADRAAGRLPGYGVITN
RICE        KYYGRGPIQITYNYNYGR-GAGIGSDLLNNPDLVASDA-VSFKTAFWFWMTPQSP-KPSCHAVITGQWTPSADDQAAGRVPGYGEITN
POPLAR      DYYGRGPLQLRWNYNYGECGNYLGQNLLDEPEKVATDPVLSFEAALWFWMNPHSTGAPSCHEVITGEWSPSEADIEAGRKPGFGMLTN
PEA         KYYGRGPIQLTHNYNYGLAGQAIKEDLINNPDLLSTNPTVSFKTAIWFWMTPQAN-KPSSHDVITGRWTPSAADSSAGRVPGYGVITN
CORN        KYYGRGPLQISWNYNYGPAGRDIGFNGLADPNRVAQDAVIAFKTALWFWMNNVHGVMP--------------------QGFGATIR
            *.****.*. ***** * .   .  .   .*. . ..  * .*.****  *     .    .  *      *                 *.*

                       200                   220                   240
BARLEY      IINGGIECGHGQDSRVADRIGFYKRYCDILGVGYGNNLDCYSQRPFA
TOBACCO     IINGGLECGHGSDARVQDRIGFYRRYCSILGVSPGDNIDCGNQKSFNSGLLLETM
POTATO      IINGGLECGRGTDNRVQDRIGFYRRYCSILGVTPGDNLDCVNQRWFGNALLVDTL
ARABIDOPSIS IINGGLECGRGQDGRVADRIGFYQRYCNIFGVNPGGNLDCYNQRSFVNGLL-EAA
RICE        IINGGVECGHGADDKVADRIGFYKYCDMLGVSYGDNLDCYNQRPYPPS
POPLAR      IITNGGECTKDGKTRQQNRIDYYLRYCDMLQVDPGDNLYCDNQETFEDNGLLKMVGTM
PEA         IINGGIECGHGQDNRVDDRVGFYKRYCQIFGVDPGGNLDCNNQRSFA
CORN        AINGALECNGNNPAQMNARVGYYKQYCQQLRVDPGP
             *  . **        . *...**. *  *
                                            ---- Chitinase------->|<--Vacuole targetting sequence
```

Figure 2: The amino acid sequence alignment of several plant endochitinases. Numbers above the lines refer to barley chitinase, and an * below the line marks invariant residues. All the plant chitinases, except barley, have a cysteine-rich N terminal domain which anchors the enzyme to the chitin matrix. Adapted from ref. 16.

fungal pore forming protein zeamatin (8). In this vein we were interested in examining the structure of chitinases, known to have anti-fungal properties, such as the one from barley.

The gene coding for barley endochitinase has been cloned and the amino acid sequence determined (11). The barley chitinase is a class II enzyme with 243 residues and a MW of 26 KDa. As mentioned earlier, class II enzymes correspond to the major catalytic domain of class I and IV enzymes.

Barely chitinase was isolated using published procedures (7). Initial crystals in a tetragonal space group were adequate for structural determination, but were somewhat unstable (14). Far superior crystals in monoclinic space group $P2_1$ were then obtained which readily led to a 2.8 crystal structure (15). The structure was recently refined to 1.8 Å resolution, accompanied by a hypothetical mode of substrate binding and a mechanism of action (16).

Figure 3 shows a ribbon drawing of the barely chitinase. The molecule is a compact globular structure approximately 40 x 45 x 42 Å. It has three disulfide bonds; the cysteine pairs are 23 to 85, 97 to 105, and 204 to 236. The molecule has little or no β-sheet, but has ten helical segments labeled A - J, which comprise approximately 47% of the structure. A prominent cleft runs the length of the molecule and is thought to be the chitin substrate binding site.

Chitinase action clearly resembles that of lysozyme, in that structurally similar carbohydrates are hydrolyzed. Indeed some lysozymes can hydrolyze chitin readily. The mechanism of lysozyme has been thoroughly studied and is well understood. A review is given in reference (17). The best studied example is the 124 residue hen egg white lysozyme (HEWL). It binds hexa saccharides and will act on NAG_6 with about half the efficiency of $(NAG-NAM)_3$. Hydrolysis occurs between the fourth and fifth sugars, called D and E, from the non-reducing end. The reaction mechanism is facilitated by Glu 35 which protonates the leaving group and by the carboxylate of Asp 52 which electrostatically stabilizes the oxycarbonium transition state.

Superficially, chitinase does not resemble HEWL, which is about half its size. Amino acid sequence comparisons reveal no obvious similarity between chitinase and any other type of protein, including any lysozyme. However, the lysozyme fold appears to be extremely ancient, and it is common to find lysozymes which are structurally related, even though they lack any sequence homology. Two examples of this phenomenon are the comparison between HEWL and the T4 lysozyme (18) and the recently elucidated comparison between HEWL and the C terminal domain of the soluble lytic transglycosylase (SLT) from *E. coli* (19).

We have analyzed the chitinase structure in light of hen egg white lysozyme (HEWL) model (16). Our results show that chitinase has a central core region which is topologically the same as that for HEWL. This is shown as a least squares superposition of the proteins in Figure 4. The superposition of barley chitinase with HEWL shows that Glu 67 is in the same position in the active site cleft as Glu 35 of HEWL. Presumably it plays the same role, protonating the leaving sugar group (E) in the hydrolytic reaction. There is no carboxylate in a position

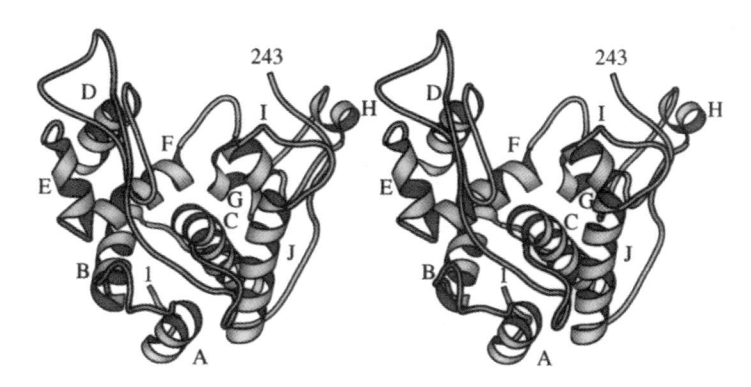

Figure 3: Stereo ribbon drawing of barley chitinase. Residue 1 marks the amino terminus and 243 marks the C terminus. The 10 α-helices are labeled A-J. Adapted from ref. 16.

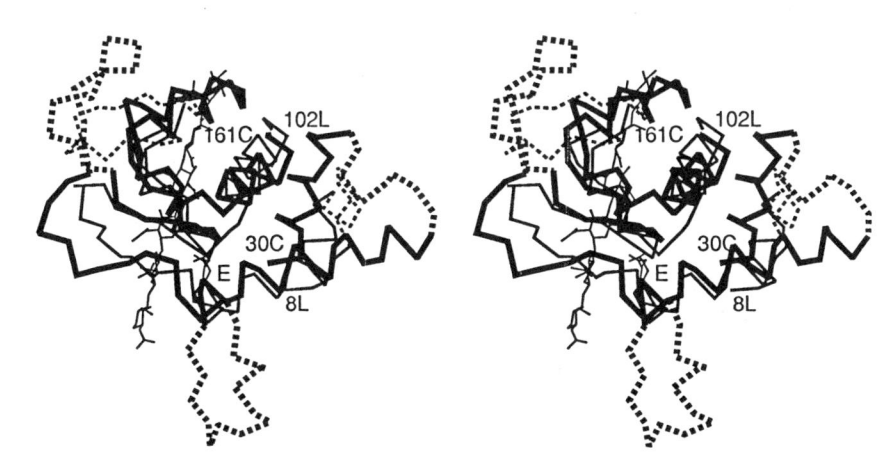

Figure 4. Comparison of structural similarities between barley chitinase and HEWL. Chitinase is much larger than HEWL, however important elements of secondary structure, forming the substrate binding and active site areas, are similiar. The residue 30 to 161 segment of the chitinase Cα backbone is shown in the heavy bonds, superimposed with the 8 to 101 segment of HEWL. The binding of a hexasaccharide is shown in light bonds. The crucial catalytic glutamates, 67 in chitinase and 35 in HEWL are also shown in light bonds. Adapted from ref. 16.

equivalent to the Ap 52 of HEWL, but model building suggests that chitinase Glu 89 is important in the mechanism. There is space between it and the model substrate to suggest Glu 89 acts a base to polarize the water and facilitate an attack on the back side of sugar D. We have suggested that chitinase, unlike HEWL, may catalyze hydrolysis of chitin with inversion of the substrate hand (16). This is consistent with biochemical results from yam chitinase, a homolog of the barely enzyme (20).

In addition to our structural work on barley chitinase, our laboratory has recently obtained useful crystals of a chitosanase isolated from *Streptomyces* N174 (21). The crystals diffract well and we hope to produce a 2.5 Å structure in the near future. This should allow a more detailed analysis of the chitosanase mechanism of action and a structural comparison with chitinases.

Acknowledgments

We wish to thank Drs. Michael Ready and Arthur Monzingo and Edward Marcotte for help with sequence alignment, preparation of the drawing, and helpful discussions. This work was supported by grants GM 30048 and GM35989 from the National Institutes of Health and by a grants from the Foundation for Research and the Welch Foundation.

Literature Cited

1. Gooday, G. W. In *Enzymes and Biomass Conversion*, Leatham, G. F., and Himmel, M. E. Eds; American Chemical Society: Washington, DC, 1991; pp. 478-485
2. *Advances in Chitin and Chitosan*, Brine, C. J., Sanford, P. A., and Zikakis, Eds.; Elsevier Applied Science, London, 1992.
3. Shinshi, H., Neuhaus, J.M., Ryals, J., & Meins, F. *Plant Mol. Biol.* 1990, *14*, 357.
4. Collinge, D.B., Kragh, K.M., Mikkelsen, J.D., Nielsen, K.K., Rasmussen, U., & Vad, K. *Plant Journal*, 1993, *3*, 31.
5. Bowles, D.J., Gurr, S.J., Scollan, C., Atkinson, H.J., & Hammond-Kosack, K.E. In *Biochemistry and Molecular Biology of Plant-Pathogen Interactions* Smith, C.J., Ed.; Clarendon Press: Oxford, 1991 pp. 225-236.
6. Rasmussen, U., Giese, H., & Mikkelsen, J.D. *Planta*, 1992, *187*, 328.
7. Roberts, W.K. & Selitrennikoff, C.P. *Biochim. Biophys. Acta*, 1986, *880*, 161.
8. Roberts, W.K. & Selitrennikoff, C.P. *J. Gen. Microbiol.* 1990, *136*, 1771.
9. Abeles, F.B., Bosshart, R.P., Forrence, L.E., & Habig, W.H. *Plant Physiol.* 1970, *47*, 129.
10. Boller, T., Gehri, A., Mauch, F., & Vogeli, U. *Planta*, 1983, *157*, 22.
11. Leah, R., Tommerup, H., Svendsen, I., & Mundy, J. *J. Biol. Chem.* 1991, *266*, 1564.
12. Monzingo, A. F. and Robertus, J. D. *J. Mol. Biol.* 1992, *227*, pp. 1136-1145.
13. Monzingo, A. F., Collins, E. J., Ernst, S., R., Irvin, J. D. and Robertus, J. D. *J. Mol. Biol.* 1993, *233*, 705.
14. Hart, P. J., Ready, M. P. and Robertus, J. D. *J. Mol. Biol.* 1992, *225*, 565.

15. Hart, P. J., Monzingo, A. F., Ready, M. P., Ernst, S. R., and Robertus, J. D. *J. Mol. Biol.* **1993**, *229*, 189.
16. Hart, P. J., Pfluger, H. D., Monzingo, A. F., Hollis, T., and Robertus, J. D. *J. Mol. Biol.* **1995**, *248*, 402.
17. Imoto, T., Johnson, L. N., North, A.C., Phillips, D.C., and Rupley, J.A. In *The Enzymes* Boyer, P.D., Ed., 3rd ed; Academic Press: N.Y., N.Y., 1972, Vol.7.
18. Matthews, B.W., Remington, S.J., Grutter, M.G., and Anderson, W.F. *J. Mol. Biol.* **1981**, *147*, pp. 545-558.
19. Thunnissen, A.W.H., Dijkstra, A.J., Kalk, K.H., Rozeboom, H.J., Engel, H., Keck, W., and Dijkstra, B.W. *Nature*, **1994**, *367*, 750.
20. Fukamizo, T., Koga, D., and Goto,S. *Biosci. Biotech. Biochem.* **1995**, *59*, 311.
21. Marcotte, E., Hart, P. J., Boucher, I., Brzezinski, and Robertus, J. D. *J. Mol. Biol.* **1993**, *232*, 995.

RECEIVED August 17, 1995

Chapter 6

Existence of Separately Controlled Plastic and Conserved Phases in Glycosylase Catalysis

Edward J. Hehre

Department of Microbiology and Immunology, Albert Einstein College of Medicine, 1300 Morris Park Avenue, Bronx, NY 10461-1602

Stereochemical studies of the reactions of glycosylases with small nonglycosidic substrates have provided deep new insights into the catalytic scope and mechanism of these enzymes. The reactivity of individual glycosylases with both α and β anomers of a glycosyl fluoride and/or with prochiral (enolic) glycosyl substrates has revealed that their catalytic groups act with different stereochemistry on substrates of different configuration; that other protein structures control product configuration without regard to that of substrate. Our newest findings add independent support for the existence of separate plastic and conserved phases in the catalytic process. Crystallographic studies of β-amylase acting on β-maltose and maltal indicate how this enzyme may effect topological control of product configuration. Recent modelling of similar oxocarbonium ion-type transition state structures for α-D-glucosyl fluoride hydrolysis by glucoamylase and α-glucosidase (with inversion or retention of configuration, respectively) indicates the absence of discernable linkage between the transition state structure and the stereochemical outcome of these reactions.

Studies of the reactions of glycosidases and glycosyltransferases with small nonglycosidic substrates have in recent years brought a new and deeper understanding of the catalytic capabilities, anomeric specificity, and mechanisms of these enzymes. The theoretical basis for the studies arose following emergence of the concept of glycosyl (vs glycosyloxy) group transfer (*1*) and the subsequent partition of carbohydrate enzymology between the traditional hydrolase and newer group transferase paradigms. The accumulation of new findings with glycosidases and glycosyltransferases in separate streams with little effect on each other could hardly be construed as the mark of a mature, cohesive scientific discipline. Koshland's (*2*) assignment of direct single (or double) nucleophilic displacement mechanisms for all hydrolytic and group transfer reactions that invert (or retain) substrate configuration was a watershed; but this did not provide a unifying definition of enzymes that mobilize glycosyl groups as distinguished from all other

0097-6156/95/0618-0066$12.00/0
© 1995 American Chemical Society

enzymes. In early papers (*3-5*), the present author proposed that glycosidases and glycosyltransferases form an inclusive class of glycosylases all of whose reactions effect a simple stoichiometric chemical change, glycosyl-X + H-X' ⇌ glycosyl-X' + H-X, whereby group -X, initially glycosylated, somehow becomes protonated while group -X', initially protonated, somehow becomes glycosylated; the possibility of carbonium ion mediation was noted. This equation said something new: namely, that a compound need only to have the ability to bind appropriately to an enzyme, and to provide a glycosyl residue in exchange for a proton, to serve as a substrate. Nonglycosidic compounds of one type, the reducing sugars, had been known to undergo limited condensation (hydrolysis reversal) reactions by glycosidases; but other truncated forms - glycosyl fluorides, glycals, and related (exocyclic) enolic glycosyl compounds - would now be tested and found to be substrates for many glycosylases. The reactions promoted with them have provided much basic evidence for views about the catalytic scope and functioning of glycosylases that depart from conventional beliefs.

Glycosyl Fluorides and the Widened Catalytic Abilities of "Strict Hydrolases".

For example, various glycosidases which had been considered to be strict hydrolases lacking glycosyl transferring ability were shown in our laboratory to catalyze nonhydrolytic (glycosyl transfer) reactions with an appropriate glycosyl fluoride. Crystalline α-amylases from a range of sources were found (*6*) to promote extremely rapid fluoride release from α-maltosyl fluoride and to convert this small substrate to malto-oligosaccharides in great preference to hydrolyzing it. Enzyme preparations incubated with 5.6 mM α-maltosyl fluoride under conditions providing 75-80% substrate utilization, for instance, yielded three to five times more maltosaccharides (tri- to hexasaccharide) than maltose. The use of α-maltosyl fluoride by α-amylase as an acceptor in preference to water at 10^4 times higher concentration indicates that, because of the substrate's small "aglycone", two molecules rather than one bind simultaneously in tandem at the reaction center and function effectively as donor and acceptor. The extremely high reactivity of α-amylases with α-maltosyl fluoride was recently reencountered in work from Withers' laboratory primarily directed to describing the slow reaction catalyzed by the human pancreatic enzyme with α-2-deoxy-2-fluoro-D-glucosyl fluoride (*7*).

Further studies demonstrated that other enzymes traditionally believed to be absolutely limited to catalyzing hydrolysis and reverse condensation reactions -this time with inversion of configuration - also promote glycosyl transfer reactions with a glycosyl fluoride as donor substrate, *i.e.*, with β-maltosyl fluoride in the case of β-amylase, or β-D-glucosyl fluoride for glucoamylase, glucodextranase, and trehalase (Figure 1). Actually, each of these supposedly strict hydrolases for α-D-glucosidic substrates was shown to use both α and β anomers of the appropriate glycosyl fluoride. By following the reactions kinetically and by ^{1}H NMR spectroscopy (also, in most cases, by analyzing the isolated transient intermediate transfer product), we learned that each enzyme directly hydrolyzes the α-glycosyl fluoride with inversion, but uses the β anomer as a glycosyl donor to a carbohydrate acceptor to form a transient α-linked transfer product which then undergoes hydrolysis with inversion (*8-11*).

A new instance of this behavior was recently reported by Withers and his associates (*12*) who showed that glycogen debranching enzyme (amylo-1,6-glucosidase)

Beta Amylase *(8)*
α-Maltosyl F + H$_2$O ⟶ β-maltose + HF
β-Maltosyl F + β-maltosyl F ⟶ β-maltotetraosyl F [1] + HF
β-maltose + β-maltosyl F ⟵ H$_2$O +

Glucoamylase, Glucoamylase *(9)*
α-D-Glucosyl F + H$_2$O ⟶ β-D-glucose + HF
β-D-Glucosyl F + Me α-D-glucoside ⟶ Me α-maltoside + HF
                                    Me α-isomaltoside + HF
β-D-glucose + Me α-D-glucoside ⟵ H$_2$O +

Trehalase *(10, 11)*
α-D-Glucosyl F + H$_2$O ⟶ β-D-glucose + HF
β-D-Glucosyl F + α-D-glucose ⟶ α,α-trehalose [2] + HF
β-D-glucose + α-D-glucose ⟵ H$_2$O +

Glycogen Debranching Enzyme *(12)*
α-D-Glucosyl F + H$_2$O ⟶ β-D-glucose + HF
β-D-Glucosyl F + cycloheptaose ⟶ α-D-glucocycloheptaose + HF
β-D-glucose + cycloheptaose ⟵ H$_2$O +

[1] Postulated transient transfer product was not observed. [2] α-D-Glucopyranosyl α-D-xylopyranoside was formed with α-D-xylose as the acceptor.

Figure 1: Stereocomplementary Hydrolytic and Nonhydrolytic Actions of "Inverting Hydrolases".

hydrolyzes α-D-glucosyl fluoride to form β-D-glucose, and uses β-D-glucosyl fluoride as a glucosyl donor in the presence of a cyclodextrin acceptor to form an α-linked transfer product. We have also reported an example involving the obverse stereochemistry, in that the β-xylosidase of Bacillus pumilus which hydrolyzes β-xylobiose and β-D-xylosyl fluoride with inversion very slowly promotes xylosyl transfer from α-D-xylosyl fluoride to xylose to form xylobiose (*13*). It would appear from the above series of findings that, in contrast to the long early period when all carbohydrases were universally held to have only hydrolytic activity, one may now wonder whether any glycosylase is so limited in scope as to be defined as a Hydrolase in the historical sense.

Catalytic Group Functional Flexibility. It is noteworthy that, in the above stereochemically complementary hydrolytic and nonhydrolytic reactions, the favored glycosyl fluoride anomer is hydrolyzed but the "wrong" anomer is used as a donor in a glycosyl transfer to an acceptor other than water. The reactions thus resemble the stereocomplementary glycoside hydrolysis and reverse sugar-condensation reactions catalyzed by enzymes of this type, except that the reactions with glycosyl fluorides are not demonstrably reversible since they release HF as a product which binds tightly to water with a high heat of hydration. In condensations representing reversals of hydrolysis, such as in maltotetraose synthesis from two molecules of β-maltose by β-amylase or in maltose synthesis from two molecules of β-D-glucose by glucoamylase (*14*), the enzyme's catalytic groups function in an opposite way than for hydrolysis, as required by the principle of microscopic reversibility. However, since the reactions catalyzed by an enzyme with the α and β anomers of a glycosyl fluoride clearly are not reversals of each other, the demonstration of nonhydrolytic reactions from the "wrong" anomer (as in Figure 1) reveals the potential of the catalytic groups of glycosylases to function flexibly even in situations where such flexibility is not required by the principle of microscopic reversibility.

Stereochemistry of Reactions Catalyzed with Prochiral Glycosyl Substrates.

Strong independent evidence of the functional flexibility of the catalytic groups in a number of well known glycosylases has come from studies of reactions catalyzed with appropriate enolic glycosyl substrates, reported from our laboratory and that of Professor J. Lehmann of the University of Freiburg (*15-23*). These studies also provide many examples of glycosylase-catalyzed reactions which show that the process of catalysis comprises two separately controlled parts: a "plastic" phase involving the different catalytic group responses to substrates of different configuration at the reactive carbon atom, and a "conserved" phase in which product configuration is controlled by protein structures independently of substrate configuration. Figure 2 compares the stereochemistry of reactions catalyzed by various glycosylases with glycosidic substrates versus that of reactions with enolic glycosyl donors; the latter substrates are mostly glycals with a double bond between carbon atoms 1 and 2 but in two cases involve D-gluco-octenitol ((Z)-3,7-anhydro-1,2-dideoxy-D-gluco-oct-2-enitol) which has an exo-cyclic double bond at the reactive glucosyl carbon atom. A major point is that, with each listed enzyme, the direction of protonation generally assumed for its glycosidic substrates differs from the direction determined for protonating the enolic substrate. These strong indications of the functional flexibility of the catalytic groups of the particular glycosylases

Enzyme	Protonation of substrates Glycosidic (Assumed)	Enolic (NMR)		Product Config., each case	Ref.
C. tropicalis, buckwheat, rice, A. niger α-glucosidase	↑	↓	Glucal	ALPHA	(15, 19)
Sweet potato, soybean β-amylase	↑	↓	Maltal	BETA	(17, 22)
Rice, A. niger α-glucosidase	↑	↓	Octenitol	ALPHA	(20)
T. reesei trehalase	↑	↓	Octenitol	ALPHA	(20)
Sweet almond β-glucosidase	↓	↑	Glucal	BETA	(15)
E. coli β-galactosidase	↓	↑	Galactal	BETA	(16)
I. lacteus, A. niger, T. reesei CBHI cellulase	↓	↑	Cellobial	BETA	(18, 21)
T. reesei CBHI cellulase	↓	↑	Lactal	BETA	(21)

Protonation: ↑ from below the plane of the ring; ↓ from above the plane of the ring.

Figure 2: Plastic and Conserved Steric Features in the Hydrolysis of Glycosidic versus Enolic Substrates by Glycosylases.

are further enhanced by Lehmann's demonstration that E. coli β-galactosidase protonates the endo-cyclic double bond of D-galactal, and the differently oriented exo-cyclic double bond of D-galacto-octenitol, from opposite sides of the ring (*16, 23*). On the other hand, despite the plasticity of substrate protonation which is evident in the paired reactions of all the enzymes of Figure 2, each enzyme yielded products of matching configuration from both the glycosidic and enolic substrate. Since the enitols have neither α or β configuration, the particular *de novo* configuration of the hydration product formed by a given enzyme would have had to be dictated by some structural factor(s) inherent to the particular protein. The presence of these same elements would ensure that the stereochemical outcome of reactions of the same enzyme with chiral substrates also would be topologically controlled and not derived from substrate configuration by a fixed mode of catalytic group functioning, as implied by the terms "inverting" or "retaining" enzymes, or single or double displacements.

Determination of the Direction of Enzymic Protonation of a Glycal. Figure 3 compares the steric course of the hydration of cellobial catalyzed by exo-cellulases Ex-1 from Irpex lacteus (*18*), and also CBH I from Trichoderma reesei (*21*), with the presumed course of the hydrolysis of cellulose. A key difference between the reactions with the two substrates is that whereas the enzymes are generally assumed to protonate the β-glycosidic oxygen atom of cellulose from "above", they are observed to protonate the double bond of cellobial from "below" (*18,21*). Figure 3B illustrates the method, developed by my colleague Professor C. F. Brewer, whereby the important parameter of the direction of enzymic protonation of a substrate may be unequivocally determined when glycals are the substrates. With glycals, the incoming proton is not lost to solvent as with glycosides, but remains bound to a carbon atom of the reaction product. When hydration is carried out in D_2O, using an enzyme whose exchangeable protons have been replaced by deuterons, the deuteron interacting with substrate can be localized by ^1H NMR spectroscopy. Despite the apparent difference in the protonation of cellobial versus cellulose, the transition state in cellobial hydration differs from that in cellulose hydrolysis only in carrying a proton (shown as a deuteron) at the equatorial position at C-2 rather than a hydroxyl group. The final chemical step, unlike the plastic substrate protonation step, is conserved and leads to a product of the same (β) anomeric configuration from both substrates. The findings (*18,21*) apparently gave the first experimental indication of the presence of a transition state with significant oxocarbonium ion character in a reaction catalyzed by exo-cellulases that hydrolyze cellulose with retention of configuration.

Glycosylation Reactions Catalyzed with Exo-cyclic Enolic Substrates. Reported findings concerning the steric outcome of hydration and transfer reactions catalyzed by various enzymes with 2,6-anhydro-l-deoxy-D-gluco- or D-galacto-hept-l-enitol (heptenitols), or with (Z)-3,7-anhydro-l,2-dideoxy-D-gluco- or D-galacto-oct-2-enitol (octenitols), are summarized in Figure 4. The results show that, with each enzyme, the anomeric configuration of the hydration product formed from the enitol matched that of the product formed on hydrolysis of its glycosidic substrates. This conservation of steric outcome is observed whether the enitol and glycosidic substrates are protonated similarly, as they are by α- and β-galactosidase, or protonated differently as they are by α- and β-glucosidase. Transfer products of α-configuration are formed from D-gluco-heptenitol by glycogen phosphorylase, as they are from the enzyme's natural substrates (*27*). The

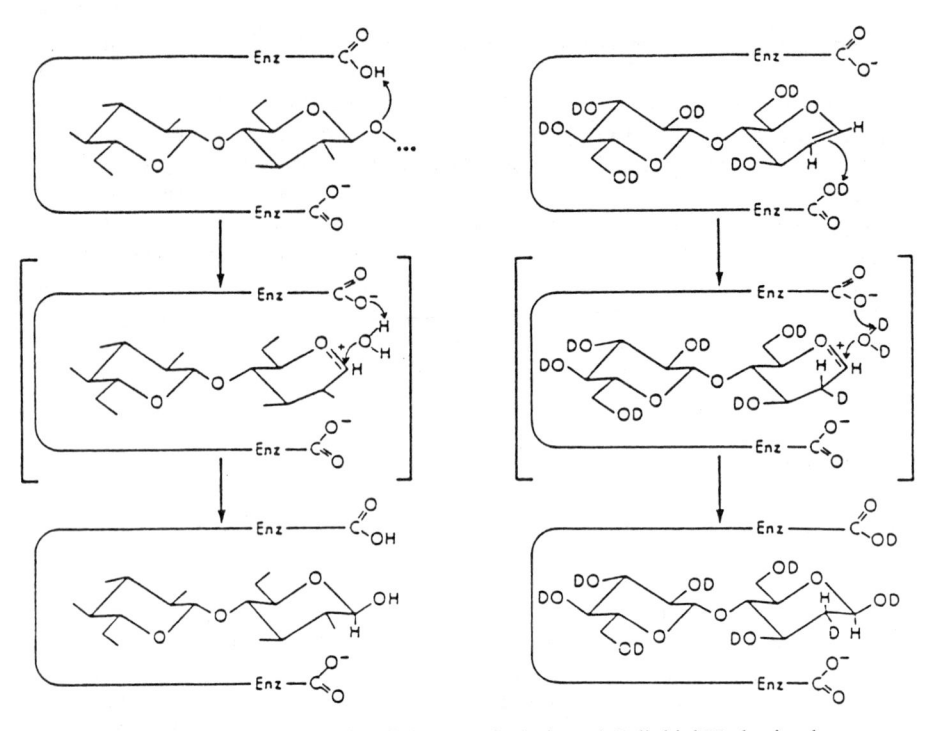

72   ENZYMATIC DEGRADATION OF INSOLUBLE CARBOHYDRATES

Figure 3: Comparison of Cellulose Hydrolysis and Cellobial Hydration by Irpex lacteus Exocellulase Ex-1. Reproduced from ref. 18.

α-Glucosidase Candida, rice			ALPHA	H, T[1]	(25, 26)
Phosphorylase muscle, potato			ALPHA	T	(27)
β-Galactosidase E. coli			BETA	H	(24)
β-Glucosidase almonds			BETA	H	(25)
Glucodextranase Arthrobacter	ALPHA	T (25)	BETA	H	(26)
α-Glucosidase A. niger, rice			ALPHA	H	(20)
α-Galactosidase coffee beans			ALPHA	H	(28)
β-Galactosidase E. coli			BETA	H	(23)
Trehalase T. reesei			BETA	H	(20)

[1] H = Hydration product; T = Transfer product

Figure 4: Conserved Configuration of Products Formed in Reactions Catalyzed with Exo-cyclic Enolic Substrates (Hept- and Oct-enitols).

heptenitol and its α-D-glucoheptulosyl phosphate transfer product have had an important role, as ligands, in elucidating the active site structure of glycogen phosphorylase in crystallolographic studies (*29, 30*). The formation of α-D-heptulosyl transfer products (but of β-D-heptulose) from heptenitol by glucodextranase is comparable to that enzyme's formation of α-D-glucosyl transfer products from β-D-glucosyl fluoride (but of β-D-glucose) from α-D-glucosyl fluoride (*9*).

New Evidence for the Existence of Separately Controlled Plastic and Conserved Phases in Glycosylase Catalysis.

Professor Sinnott (*31-33*) has viewed the complementary stereochemistry of reactions catalyzed with α- and β-glycosyl fluorides by glucodextranase and other "inverting" enzymes (Figure 1) as good evidence for Koshland's single nucleophilic displacement mechanism (*2*) which dictates that product configuration will be opposite to that of substrate. We have proposed a basically different explanation for the findings with the paired glycosyl fluorides that, in addition, accounts for the observed conservation of the stereochemical outcome of reactions catalyzed with enolic substrates. Namely, we envision that glycosylase catalysis comprises a "plastic" phase involving different responses of an enzyme's catalytic groups to substrates of different anomeric configuration, and a "conserved" phase in which product configuration is controlled by protein structures independently of substrate configuration (*19, 20, 34-36*). With regard to the results reported with enolic glycosyl substrates, the criticism has been made (*32*) that enol ethers are reactive compounds which are labile on simple proximity to a carboxyl function; comparable in lability, for example, to the prostaglandin-like compound of Bergman et al (*37*). Unlike the latter compound, however, the hept- and oct-enitols which have been used as substrates have a stabilizing hydroxyl group adjacent to the reactive carbon atom which makes for a substantial enhancement of stability. Above all, the results of concurrent control experiments leave no doubt whatever of the sufficient stability of all of the enolic compounds under conditions of their use as substrates (*e.g., 21, 22*). Independent new evidence that strongly supports our unconventional concept of the process of catalysis by glycosylases has been obtained in several recent studies as indicated below.

The Conversion of β-D-Glucosyl Fluoride to α-D-Glucose by α-Glucosidases.

We have recently determined that enzymes which hydrolyze natural substrates with configurational inversion are not the only glycosylases able to act on both anomers of a glycosyl fluoride (*35,36*). The α-glucosidases of Aspergillus niger, ungerminated rice, and sugar beet seeds, and the β-glucosidase of sweet almonds also were found to use both α- and β-D-glucosyl fluoride. Each enzyme rapidly hydrolyzes the favored anomer and very slowly hydrolyzes the "wrong" anomer, forming α-D-glucose from both substrates in the case of the α-glucosidases, or β-D-glucose from both in the case of the β-glucosidase. These findings extend the previously reported evidence, from reactions catalyzed by the same glucosidases with enolic substrates, showing that the catalytic groups of each of the enzymes react differently to substrates depending on the latters' anomeric configuration. This time, unlike the situation with the inverting enzymes (Figure 1), the reactions catalyzed by the glucosidases with the disfavored D-glucosyl fluoride do not parallel their (a→a) reverse glucose-condensation syntheses, or (a→a) glucosyl transfer reactions catalyzed with glucosidic substrates.

Figure 5 illustrates a model (36) that includes the mechanistic features needed to account for the stereochemical course of the reactions catalyzed by the α-glucosidases with substrates of three different anomeric types: A, methyl α-D-glucopyranoside, representing α-D-glucosyl substrates; B, D-glucal, representing enolic substrates; and C, β-D-glucosyl fluoride, the only known substrate of β configuration for α-glucosidases. We assume a pair of oppositely disposed carboxyl groups at the reaction center, and that substrates A-C bind with a similar orientation to them.

For the initial chemical event, we envision one of the lone pair on the ring oxygen assisting in the cleavage of the C1-O or C1-F bond, or in the stereospecific protonation of C-2. Each of the events leads to a comparable oxocarbonium ion intermediate or transition state whose formation may be electrostatically stabilized by negatively charged carboxylates. A second, common, catalytic step is the attack on the oxocarbonium ion by a water molecule. This step occurs with conserved stereochemistry to yield a product of α-D configuration in each case. We propose again that this occurs by the protein structure that orients the reactant water molecule and an appropriately positioned base, such as the ionized carboxylate group. The findings indicate that use of the term "retaining enzyme" for glucosidases lacks the full generality this term had been assumed to have.

Crystallographic Indications of Means of Structural Control of Steric Outcome.
X-Ray diffraction studies of glycosylases have begun to provide clues as to possible ways whereby protein structural elements may control the stereochemical outcome of catalyzed reactions. One mode has been suggested (38) to operate in the case of exo-cellulase CBH II of Trichoderma reesei whose catalytic center has been located in a tunnel-like cavity. When the nonreducing end of a cellulose chain has entered and filled the tunnel, hydrolysis of the penultimate β-1,4 linkage occurs at the reaction center; the hydrolytic product, α-cellobiose, is then extruded to solvent as the shortened substrate advances to again fill the tunnel. The authors indicate that a narrow tubular passage roughly orthogonal to the tunnel may have the function of directing reactant water from the protein surface to reach the catalytic center from the α direction (38).

A different possible mode is envisioned for β-amylase, based on recently described (39) highly refined structures of competent soybean enzyme crystals separately treated with β-maltose or maltal. The findings hold main interest for glycosylase catalysis *per se* as β-amylase is thought to have little importance in plants as a starch hydrolyzing enzyme; its long known inhibitory effect on starch phosphorylase (40-42) suggests a possible role in regulating the latter's activity (43). β-Amylase, which effects the hydrolytic cleavage of β-maltose from the nonreducing ends of α-1,4-linked chains of starch and maltosaccharides, does not hydrolyze maltose. It does, however, convert β-maltose to maltotetraose to a very small degree by the condensation of two molecules of bound substrate, with the enzyme protonating the β-maltose donor from above the plane of the ring in accord with the principle of microscopic reversibility (14). β-Amylase has been shown (17, 22) to slowly catalyze the hydration of maltal (α-D-glucopyranosyl-1,4-D-glucal) and, as indicated in Figure 2, it protonates starch and maltal differently yet hydrolyzes each type of substrate to form a product (maltose or 2-deoxymaltose, respectively) of the same β configuration. Based on its catalysis of starch hydrolysis with

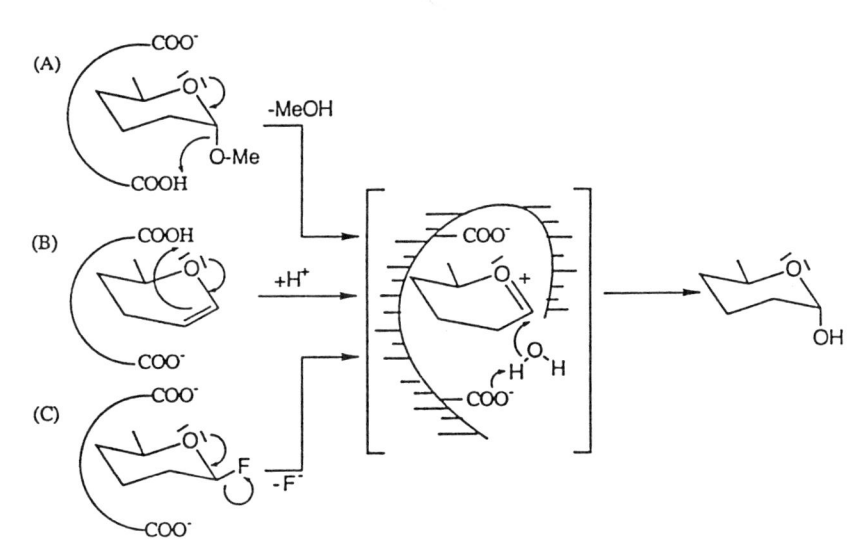

Figure 5: Proposed Mechanisms of Reactions of α-Glucosidases with Substrates Differing in Configuration at C-1. Reproduced with permission from ref. 36. Copyright 1993 Elsevier Science.

inversion, Koshland (2) initially assumed that β-amylase acts by a direct single nucleophilic mechanism; later, Koshland et al. (44) and Thoma (45) proposed a carbonium-ion mediated mechanism with a postulated back-sided approach of water. However, until our recent elucidation of the stereochemistry of maltal hydration (22), no experimental evidence existed to show the contribution of substrate configuration or of protein structure to the determination of stereochemical outcome. We had earlier demonstrated that maltal hydration is characterized by an unusually large solvent deuterium kinetic isotope effect indicative of a single proton undergoing change in forming an oxocarbonium ion-type transition state or intermediate (17); and recent X-ray crystallographic studies (39) suggest how β-amylase structures may possibly control the approach of water to the reaction center.

A deep active site pocket in the protein is found to bind two molecules of β-maltose in tandem (and some maltotetraose condensation product at the same sites), or two molecules of 2-deoxymaltose, the maltal hydration product - in each case via some 26 H-bonds and 36 van der Waals contacts (39). The pocket is normally open for substrate binding and product release, but one of its walls is a mobile hinged loop that closes down on substrates. The closed loop contributes to ligand binding and also forms a hydrophobic surface across the reaction center, shielding it from solvent. This surface extends over the area between the bound pair of β-maltose (or 2-deoxymaltose) residues; over the central α-glycosidic linkage of maltotetraose; and over the enzyme's catalytic groups which comprise the carboxyl group of Glu l86 on the α side of the reaction center and the carboxyl group of Glu 380 on the β side (39). Measurements show zero solvent access to the side chains of these catalytic residues as well as all potentially reactive carbon and oxygen atoms of the sugars at the catalytic center. This means that the water molecule involved in the reactions of starch hydrolysis or maltal hydration must exist beneath this hydrophobic surface. An ordered water molecule is, indeed, found within H-bonding distance of the carboxyl group of Glu 380 in the unliganded β-amylase. In the enzymic hydration of maltal, that water molecule is displaced by the equatorial O-1 atom of the bound β-2-deoxymaltose reaction product. Moreover, when maltotetraose is formed by the enzymically catalyzed condensation of β-maltose, the byproduct is found as ordered water in the same position as in the unliganded enzyme, H-bonded to the carboxyl group of Glu 380. No ordered water is present in the vicinity of Glu 186 which is located on the α side of the reaction center.

An apparently similar situation is discernable in the crystallographic structure of Aspergillus awamori glucoamylase complexed with 1-deoxynogirimycin (46). Here, a bound water molecule is found at the reaction center, across from the putative general acid catalyst, Glu 179, thereby ensuring that hydrolysis or hydration would always result in a product of β-anomeric configuration. Although the mechanism whereby solvent is purged from the active site is not clear, Harris et al. (46) suggest that a transient compression of the active site structure may allow water to slide by the inhibitor as it enters the close-fitting recess of the active site pocket.

Similar Transition States in Enzymic Hydrolysis of a Glucosyl Substrate with Inversion or Retention.

Finally, we have gained new evidence of a still different kind for the independent topological control of product configuration by glycosylases, through studies of transition states for the hydrolysis of α-D-glucosyl fluoride by enzymes providing opposite stereochemical outcome (47). α-Secondary tritium kinetic isotope effects and kinetic constants were measured over a wide pH range for the hydrolysis of this substrate catalyzed by α-glucosidases and glucoamylases with the production of α-D-glucose and β-D-glucose, respectively. At pH values where intrinsic $^{\alpha\text{-T}}V/K$ isotope effects presumably are expressed for the hydrolysis with sugar beet seed α-glucosidase (pH 9) and Rhizopus niveus glucoamylase (pH 4.8), we determined β-secondary tritium and primary ^{14}C kinetic isotope effects. All three kinetic isotope effects are most consistent with an S_N1 mechanism for the hydrolysis by each enzyme. Based upon the values of the combined isotope effects, transition state structures having a narrow range of acceptable geometries were modelled for α-D-glucosyl fluoride hydrolysis by the sugar beet α-glucosidase and R. niveus glucoamylase using the modified BEBOVIB-IV program as reported by Schramm (48-50) for reactions of AMP nucleosidase.

The transition state structures for both the α-glucosidase and glucoamylase catalyzed reactions bear significant oxocarbonium ion character, with the D-glucosyl residue having a flattened $^4C_1$ conformation and a C1-O5 bond order of 1.92, despite the opposite anomeric configuration of the D-glucose produced by the action of the two enzymes. Some differences do exist between the transition states; nevertheless, their general similarity indicates that the transition states of glucosylase catalyzed reactions does not predict the stereochemical outcome of the reactions, nor can the stereochemical outcome be used to predict the transition state. These results support the previously reported evidence for the separate topological control of product configuration by protein structures in these and other glycosylases.

Literature Cited

1. Hehre, E. J. *Adv. Enzymol.* **1951**, *11*, 297-331.
2. Koshland, D. E., Jr. *Biol. Rev.* **1953**, *28*, 416-426.
3. Hehre, E. J.*Bull. Soc. Chim. Biol.* **1960**, *42*, 1713-1714.
4. Hehre, E. J.; Genghof, D. S.; Okada, G. *Arch. Biochem. Biophys.* **1971**, *142*, 382-393.
5. Hehre, E. J.; Okada, G.; Genghof, D. S. *Advances in Chemistry Series* **1973**, No. *117*, American Chemical Society, Washington, D. C., pp. 309-333.
6. Okada, G.; Genghof, D. S.; Hehre, E. J. *Carbohydr. Res.* **1979**, *71*, 287-298.
7. McCarter, J. D.; Adam, M. J.; Braun, C.; Namchuk, M.; Tull, D.; Withers, S. G. *Carbohydr. Res.* **1993**, *249*, 77-90.
8. Hehre, E. J.; Brewer, C. F.; Genghof, D. S. *J. Biol. Chem.* **1979**, *254*, 5942-5940.
9. Kitahata, S.; Brewer, C. F.; Genghof, D. S.; Sawai, T.; Hehre, E. J. *J. Biol. Chem.* **1981**, *256*, 6017-6026.
10. Hehre, E. J.; Sawai, T.; Brewer, C. F.; Nakano, M.; Kanda, T. *Biochemistry* **1982**, *21*, 3090-3097.
11. Kasumi, T.; Brewer, C. F.; Reese, E.T.; Hehre, E. J. *Carbohydr. Res.* **1986**, *146*, 39-49.
12. Liu, W.; Madsen, N. B.; Braun, C.; Withers, S. G. *Biochemistry* **1991**, *30*, 1419-1424.
13. Kasumi, T.; Tsumuraya, Y.; Brewer, C. F.; Kersters-Hilderson, H.; Claeyssens, M.; Hehre, E. J. *Biochemistry* **1987**, *26*, 3010-3016.
14. Hehre, E. J.; Okada, G.; Genghof, D. S. *Arch. Biochem. Biophys.* **1969**, *135*, 75-89.
15. Hehre, E.J.; Genghof, D. S.; Sternlicht, H.; Brewer, C. F. *Biochemistry* **1977**, *16*, 1780-1787.
16. Lehmann, J.; Zieger, B. *Carbohydrate Res.* **1977**, *58*, 73-78.
17. Hehre, E. J.; Kitahata, S.; Brewer, C. F. *J. Biol. Chem.* **1986**, *261*, 2147-2153.
18. Kanda, T.; Brewer, C. F.; Okada, G.; Hehre, E. J. *Biochemistry* **1986**, *25*, 1159-1165.
19. Chiba, S.; Brewer, C. F.; Okada, G.; Matsui, H.; Hehre, E. J. *Biochemistry* **1988**, *27*, 1564-1569.
20. Weiser, W.; Lehmann, J.; Chiba, S.; Matsui., H.; Brewer, C. F.; Hehre, E. J. *Biochemistry* **1988**, *27*, 2294-2300.
21. Claeyssens, M.; Tomme, P.; Brewer, C.F.; Hehre, E.J. *FEBS* **1990**, *263*, 89-92.
22. Kitahata, S.; Chiba, S.; Brewer, C. F.; Hehre, E. J. *Biochemistry* **1991**, *30*, 6769-6775.
23. Lehmann, J.; Schlesselmann, P. *Carbohydr. Res.* **1983**, *113*, 93-99.
24. Brockhaus, M.; Lehmann, J. *Carbohydr. Res.* **1977**, *53*, 21-31.
25. Hehre, E. J.; Brewer, C. F.; Uchiyama, T.; Schlesselmann, P.; Lehmann, J. *Biochemistry* **1980**, *19*, 3557-3564.
26. Schlesselmann, P.; Fritz, H.; Lehmann, J.; Uchiyama, T.; Brewer, C. F.; Hehre, E. J. *Biochemistry* **1982**, *21*, 6606-6614.
27. Klein, H. W.; Helmreich, E. J. M. *Curr. Top. Cell Reg.* **1985**, *16*, 281-294.

28. Weiser, W.; Lehmann, J.; Matsui, H.; Brewer, C. F.; Hehre, E. J. *Arch. Biochem. Biophys.* **1992**, *292*, 493-498.
29. McLaughlin, P. J.; Stuart, D. I.; Klein, H. W.; Oikonomakos, N. G.; Johnson, L. N. *Biochemistry* **1984**, *23*, 5862-5873.
30. Johnson, L. N.; Acharya, K. R.; Jordan, M. D.; McLaughlin, P. J . *J. Mol. Biol.* **1990**, *211*, 645-661.
31. Sinnott, M. L. in *"Enzyme Mechanisms"* (Page, M. I.; Williams, A., Eds.), **1987**, pp. 259-297, Royal Society of Chemistry, London.
32. Sinnott, M. L. *Chem. Rev.* **1990**, *90*, 1171-1202.
33. Konstantidinis, A.; Sinnott, M. L. *Biochem. J.* **1991**, *279*, 587-593.
34. Hehre, E. J. *Denpun Kagaku (Jpn. J. Starch Sci.)* **1989**, *36*, 197-205.
35. Hehre, E. J.; Matsui, H.; Brewer, C. F. *Carbohydr. Res.* **1990**, *198*, 123-132.
36. Matsui, H.; Tanaka, Y.; Brewer, C. F.; Hehre, E. J. *Carbohydr. Res.* **1993**, *250*, 45-56.
37. Bergman, N-Å.; Jansson, M.; Chiang, Y.; Kresge, A. J.; Yin, J. *J. Org. Chem.* **1987**, *52*, 4449-4450.
38. Rouvinen, J.; Bergfors, T.; Teeri, T.; Knowles, J. K. C.; Jones, T. A. *Science* **1990**, *249*, 380-386.
39. Mikami, B.; Degano, M.; Hehre, E. J.; Sacchettini, J. C. *Biochemistry* **1994**, *33* 7779-7787.
40. Porter, H. K. *Biochem. J.* **1950**, *47*, 476-482.
41. Nakamura, M. *J. Agric. Chem. Soc.* Japan **1951**, *24*, 302-309.
42. Fuwa, H. *Arch. Biochem. Biophys.* **1957**, *70*, 157-168.
43. Chang, T-C.; and Su, J-C. *Plant Physiol.* **1986**, *80*, 534-538.
44. Koshland, D. E.; Yankeelov, J. A.; Thoma, J. A. *Fed. Proc.* **1962**, *21*, 1031-1038.
45. Thoma, J. *J. Theor. Biol.* **1968**, *19*, 297-310.
46. Harris, E. M. S.; Aleshin, A. E.; Firsov, L. M.; Honzatko, R. B. *Biochemistry* **1993** *32*, 1618-1626.
47. Tanaka, Y.; Tao, W.; Blanchard, J. S.; Hehre, E. J. *J. Biol. Chem.* **1994.** *269*, 32306-32312.
48. Markham, G. D.; Parkin, D. W.; Mentch, F.; Schramm, V. L. *J. Biol. Chem.* **1987**, *262*, 5609-5615.
49. Parkin, D. W.; Mentch, F.; Banks, G. A.; Horenstein, B. A.; Schramm, V. L. *Biochemistry* **1991**, *30*, 4586-4594.
50. Horenstein, B. A.; Parkin, D. W.; Estupian, B.; Schramm, V. L. Biochemistry **1991**, *30*, 10788-10794.

RECEIVED August 30, 1995

Chapter 7

A Mutant α-Amylase with Enhanced Activity Specific for Short Substrates

Ikuo Matsui[1,3], Kazuhoko Ishikawa[1], Eriko Matsui[1], Sachio Miyairi[1], Sakuzo Fukui[2], and Koichi Honda[1]

[1]National Institute of Bioscience and Human Technology, Tsukuba, Ibaraki 305, Japan
[2]Faculty of Engineering, Fukuyama University, Fukuyama, Hiroshima 729-02, Japan

For some types of enzymes, it is assumed that the active site involves a definite number of subsites, each of which specifically interacts with a certain monomer unit in a polymeric substrate. The substrate and product specificities of enzymes are defined by their original subsite structure. If the subsite structure is changed by mutation, the mutant will show a new enzymatic specificity. We determined the subsite structure of *Saccharomycopsis fibuligera* α-amylase (Sfamy) and estimated the major substrate-binding residues of Sfamy. We altered the 210th lysine (K210), one of the assumed components of the major subsites, into arginine (R) and asparagine (N) by site-directed mutagenesis. Replacement of K210 by R strengthened the 7th and weakened the 8th subsite affinities. K210 was found to contribute to both the 8th and the 7th subsites. The catalytic activity of the K210R enzyme for the hydrolysis of maltose (G_2) was three times higher than that of the native enzyme due to an increase in the affinity of the 7th subsite adjacent to the catalytic site, whereas the activity of the K210N enzyme for G_2 was decreased to 1% of that of the native enzyme by a reduction in the 7th subsite affinity.

α-Amylases are major members of starch-hydrolyzing enzymes. They can hydrolyze α1-4 glucosidic bonds of polymer substrate and produce α-anomer-type products. They are also distributed in all living things and are useful enzymes in the industrial fields.
The steric structure of α-amylase binding maltooligosaccharide in a productive form has not been clarified because of the difficulty of isolating the enzyme-substrate (ES) complex. Matsuura *et al.* proposed a substrate binding model for fitting the

[3]Current address: Carlsberg Laboratory, Department of Chemistry, Gamle Carlsberg Vej 10, DK-2500 Copenhagen, Denmark

maltoheptaose (G7) molecule to the active cleft of *Aspergillus oryzae* α-amylase, Taka-amylase A (TAA) molecule (*1*). Analysis of the kinetic parameters and subsite affinities of various mutant enzymes generated by substitution of the amino acid residues that are the possible components of subsites might clarify the role of these residues in recognizing the substrate.

We reported the expression of the *Saccharomycopsis fibuligera* α-amylase (Sfamy) gene (*2*) in *Saccharomyces cerevisiae*, and the enzymatic characteristics and the molecular structure of the gene product Sfamy (*3*). Suganuma et al. reported a simple method for evaluating the subsite affinities and its application to TAA; it is based on the kinetic parameter (k_{cat}/K_m, where k_{cat} and K_m mean the observed molecular activity and the Michaelis constant, respectively) and the cleaved-bond distributions for oligo-meric substrates determined at sufficiently low substrate concentration where transglycosylation and condensation can be ignored (*4*). We determined the strength and distribution of the subsite affinities of Sfamy according to the above method (*5*). The authors proposed a substrate-binding model for Sfamy (*6*) involving a modification of the TAA model on the basis of the steric structure predicted theoretically (*7*) and on the basis of the subsite structure of Sfamy (*5*).

We prepared mutant enzymes carrying a substitution of the 210th lysine (K210). which is one of the possible components of major subsites (the 7th and 8th subsites) (*6*), and analyzed their enzymatic characteristics. We found short substrate-specific enhancement in the catalytic activity of the mutant enzyme (K210R) (*8*). The present paper describes the properties of the mutant enzymes (K210R and K210N) and the contribution of K210 to both the 7th and 8th subsite affinities.

The method to evaluate the subsite affinity.

On the basis of the subsite theory (*9*), Suganuma et al. devised a simple method to evaluate the subsite affinity by utilizing a series of end-labelled substrates (*4*). We determined the subsite affinities of Sfamy according to their method.

When the catalytic site is located between the r-th and (r+1)-th subsites as shown in Fig 1, the (r+i)-th and (r-i)-th subsite affinities (A_{r+i} and A_{r-i}) are defined by equation 1 and 2, respectively.

$$A_{r+i} = RT \ln \left[(k_{cat}/K_M)_n \left(P_i^* / \sum_{i=1}^{n-1} P_i^* \right) \right] / \left[(k_{cat}/K_M)_{n-1} \left(P_{i-1}^* / \sum_{i=1}^{n-2} P_i^* \right) \right] \quad (1)$$

$$A_{r-1} = RT \ln \left[(k_{cat}/K_M)_n \left(P_{n-1-i}^* / \sum_{i=1}^{n-1} P_i^* \right) \right] / \left[(k_{cat}/K_M)_{n-1} \left(P_{n-1-i}^* / \sum_{i=1}^{n-2} P_i^* \right) \right] \quad (2)$$

where P_i^* and P_{i-1}^* mean the radioactivities of i-mer and (i-1)-mer products from the reducing end-labelled n-mer and (n-1)-mer substrates, respectively, each P_{n-1-i}^* means the radioactivity of the (n-1-i)-mer product from the reducing end-labelled n-mer or (n-1)-mer substrate, $\sum_{i=1}^{n-1} P_i^*$ and $\sum_{i=1}^{n-2} P_i^*$ represent the sum of the radioactivities of all products from the n-mer and (n-1)-mer substrates, respectively, and $(k_{cat}/K_m)_n$ and $(k_{cat}/K_m)_{n-1}$ mean the k_{cat}/K_m value for the hydrolysis of n-mer and (n-1)-mer substrates, respectively. Since the kinetic parameters [$(k_{cat}/K_m)_n$ and $(k_{cat}/K_m)_{n-1}$] and the cleaved-bond distributions

$\left(P_i^* / \sum_{i=1}^{n-1} P_i^*, \ P_{i-1}^* / \sum_{i=1}^{n-2} P_i^*, \ P_{n-1-i}^* / \sum_{i=1}^{n-1} P_i^*, \text{ and } P_{n-1-i}^* / \sum_{i=1}^{n-2} P_i^* \right)$ are obtainable by

analysis of the labelled substrates and products, the subsite affinities can be calculated from Eq.s. 1 and 2.

The subsite structure and the substrate-binding model of Sfamy.

According to this method, the subsite affinities of Sfamy were calculated by using the k_{cat}/K_m values and the cleaved-bond distributions as shown in Table I and II. The sum of the affinities for the two subsites adjacent to the catalytic site $(A_r + A_{r+1})$ is obtained by using the k_{cat}/K_m value for maltose (0.53 min^{-1} M^{-1}) (3) from the following equation (10):

$(k_{cat}/K_m)_2 = 0.018 \times k_{int} \times \exp[(A_r + A_{r+1})/RT]$

where k_{int} is the intrinsic rate constant for the hydrolysis; we took k_{int} to be 8.62 x 10^3 min^{-1} (for amylose A, Mr = approx. 2,900) (3). Hence, we have $(A_r + A_{r+1})$=-3.36 kcal/mol.

By these procedures, 11 subsites have been evaluated for Sfamy. Since the catalytic site is present between subsites 6 and 7, the number r is 6. Figure 2 indicates histogram showing the subsite affinities of Sfamy (5). The total affinity of the subsites 6 and 7 adjacent to the catalytic site has a large negative value (A_6+A_7=-3.36 kcal/mol). On the other hand, subsites 5 and 8 adjacent to subsites 6 and 7, respectively, have large positive affinities (A_5=4.23, A_8=3.32 kcal/mol). Subsites 1, 9, and 11 all have moderate positive affinities (~1 kcal/mol) although subsite 10 has a low negative affinity. The distribution and the strength of the subsite affinities are similar to those of a fungus α-amylase, TAA (4). The common feature of the subsite structure in α-amylases may be summarized as follows; the total affinity of the central subsites is a large negative affinities, and the subsites adjacent to them have large positive affinities. This subsite structure is characteristic of the endo-cleavage type enzyme.

Figure 3 shows the Sfamy's steric structure predicted theoretically (7). On the basis of this structure and the substrate-binding model of TAA, we proposed the substrate-binding model of Sfamy (Fig. 4) (6). According to this model, the major subsite 8 might be composed of the 156th tyrosine, the 210th lysine (K210), and 234th glutamine residues. K210 seemed to be located near the both 8th and 7th subsites. However, this is one possible model and nobody know the precise binding manner. The steric structure of α-amylase binding such a long substrate has not been clarified because of the difficulty of isolating the enzyme-substrate (ES) complex.

A mutant α-amylase with enhanced activity specific for short substrates.

Hydrolytic activity changes in point mutants at these residues should give information on their roles in the formation of the 8th subsite affinity. Thus, the K210 residue was replaced to arginine (R) or asparagine (N) by site-directed mutagenesis (Fig. 5). The cleaved-bond distributions and the kinetic parameters (k_{cat}/K_m) of the mutant and wild-type enzymes for serial maltooligosaccharides were determined. The subsite affinities of the mutant and wild-type enzymes were also evaluated by the method described in the previous section.

Compared to the wild-type enzyme, the k_{cat}/K_m values of K210R, for the short substrates such as maltose (G_2) and maltotriose (G_3), exhibit a two- to three-fold increase as shown in Table III. In contrast, the k_{cat}/K_m values of K210N, for substrates shorter than G_5, decrease to less than a few % of those of the wild-type enzyme. Table IV indicates that the mutations, K210R and K210N, increase cleavage frequency at a specific bond of maltotriose, G_3 (the 1st bond from the reducing end side), and cause a shift of the major cleavage site of maltotetraose, G_4 from the 2nd to the 1st bond from the reducing end side. Figure 6 shows the 4th, 5th, and 8th subsite affinities (A_4, A_5, A_8) and the summed affinities for the 6th and 7th subsites adjacent to the catalytic site (A_6+A_7) of the mutant and wild-type enzymes. The mutations decreased affinities, A_4, A_5, and A_8 at varying degrees. The 8th affinity (A_8) was lowered the most of all three subsites, however, affinities, A_4, A_5 were not decreased

G—G—G—G
↓A_{r-1} ↓A_r ↓A_{r+1} ↓A_{r+2}

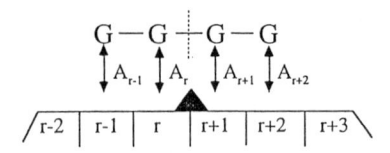

| r-2 | r-1 | r | r+1 | r+2 | r+3 |

Figure 1. Schematic model of a productive ES complex for an α-amylase. The letters, G and A represent a glucose residue of substrate and the affinity of each subsite, respectively. The wedge shows the catalytic site of the enzyme. The enclosures represent the subsites numbered as indicated.

TABLE I. Kinetic parameter k_{cat}/K_m of Sfamy for maltooligosaccharides at pH 5.25 and 25°C

n[a]	k_{cat}/K_m (min$^{-1}\cdot$M^{-1})
3	5.7×10^3
4	1.7×10^6
5	1.5×10^7
6	2.3×10^7
7	4.9×10^7
8	1.1×10^8
9	1.1×10^8

[a] The degree of polymerization of the substrate expressed as glucose units.

TABLE II. **Cleaved-bond distribution in the hydrolysis of maltooligosaccharides by Sfamy.** All reactions were performed in 0.05 M acetate buffer (pH 5.25) at 25°C. The numerals indicate the cleavage frequency of the bond. [S]₀ and [E]₀ represent the concentrations of the substrate and enzyme, respectively. The mark (*) represents a radioactive glucose residue.

		[S]₀ (M)	[E]₀ (M)
n=3	G —— G —— G* 　　0.176　0.824	1.9×10^{-5}	7.3×10^{-7}
n=4	G —— G —— G —— G* 　　0.745　0.255	1.9×10^{-5}	1.8×10^{-8}
n=5	G —— G —— G —— G —— G* 　　0.510　0.490	1.9×10^{-5}	3.7×10^{-9}
n=6	G —— G —— G —— G —— G —— G* 　　0.031　0.685　0.284	1.9×10^{-5}	1.5×10^{-9}
n=7	G —— G —— G —— G —— G —— G —— G* 　　0.370　0.203　0.296　0.125　0.007	1.9×10^{-5}	7.3×10^{-10}
n=8	G —— G —— G —— G —— G —— G —— G —— G* 　　0.125　0.452　0.067　0.091　0.264	2.0×10^{-5}	7.3×10^{-10}
n=9	G —— G —— G —— G —— G —— G —— G —— G —— G* 　　0.081　0.366　0.345　0.075　0.096　0.039	2.0×10^{-5}	7.3×10^{-10}

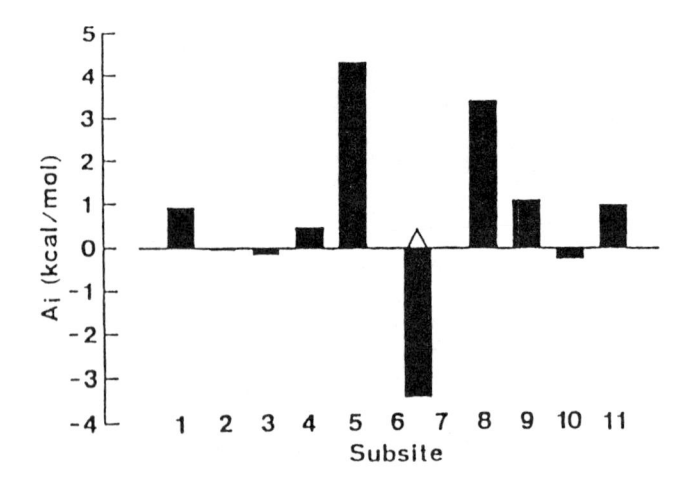

Figure 2. Histograms showing the subsite affinities (A_i) of Sfamy. The wedge represents the catalytic site.

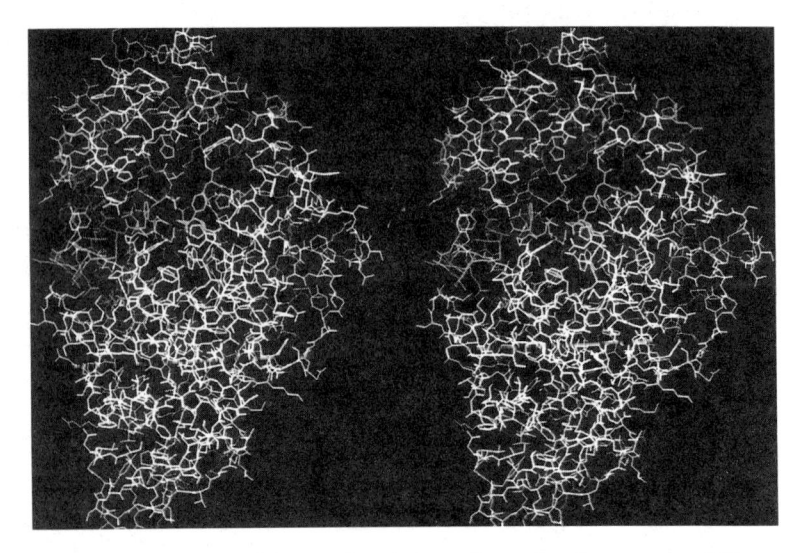

Figure 3. Stereoview of the predicted three-dimensional structure of Sfamy. Dark side chains are conserved completely in all α-amylases and located in the center of the active site.

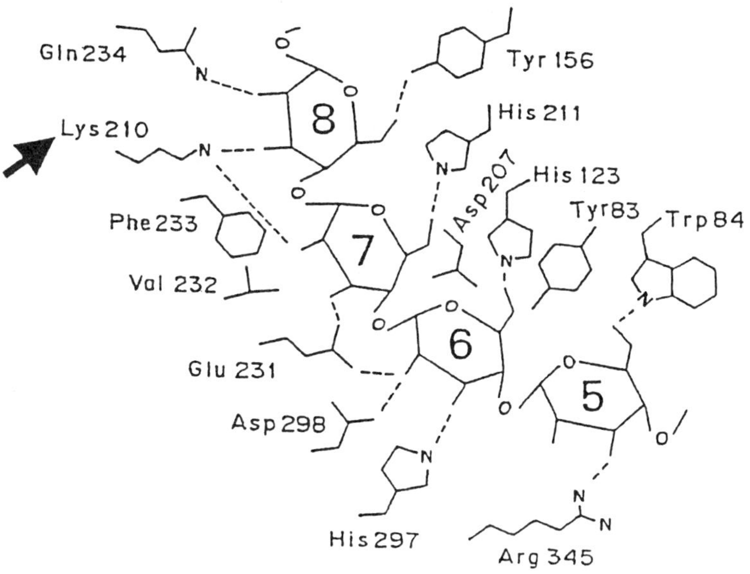

Figure 4. Substrate binding model of Sfamy. The arrow represents the 210th lysine residue replaced. The substrate is hydrolyzed between the 6th and 7th glucose residues. The possible catalytic residues are the 207th aspartic acid, 231st glutamic acid, and 298th aspartic acid.

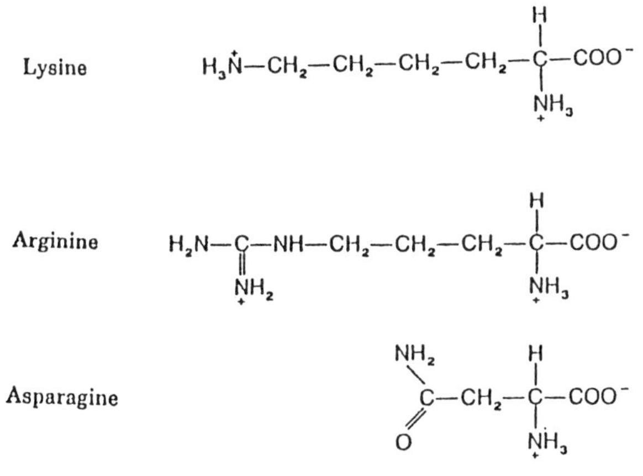

Figure 5. Molecular structures of amino acids.

Table III

Kinetic parameters for the hydrolysis of maltooligosaccharides catalyzed by the K210R, K210N and wild-type enzymes at pH 5.25 and 25°C

n^a	k_{cat} (min^{-1})			K_m (M)			k_{cat}/K_m (min$^{-1}\cdot$M^{-1})		
	K210R	K210N	Wild-type	K210R	K210N	Wild-type	K210R	K210N	Wild-type
2	7.9×10^{-2} (172)^b	2.8×10^{-3} (6)	4.6×10^{-2} (100)	5.1×10^{-2}	4.5×10^{-1}	8.6×10^{-2}	1.5 (283)	6.2×10^{-3} (1)	5.3×10^{-1} (100)
3	—^{c}	—	—	—	—	—	1.2×10^4 (211)	1.9×10^2 (3)	5.7×10^3 (100)
4	—	—	—	—	—	—	8.4×10^5 (49)	1.7×10^4 (1)	1.7×10^6 (100)
5	—	—	—	—	—	—	3.0×10^6 (20)	4.5×10^5 (3)	1.5×10^7 (100)
18	6.4×10^2 (13)	7.6×10^2 (16)	4.9×10^3 (100)	3.3×10^{-5}	1.2×10^{-4}	1.1×10^{-4}	1.9×10^7 (43)	6.6×10^6 (15)	4.4×10^7 (100)

^{a} Degree of polymerization of the substrate expressed as glucose units.
^{b} Percentage based on value of wild-type enzyme.
^{c} Value not determined.

Table IV

Bond-cleavage patterns for the hydrolysis of end-labeled G_3 and G_4 catalyzed by the K210R, K210N and wild-type enzymes

	G_3	G_4
	Frequency[a]	Frequency[a]
K210R	G——G——G*	G——G——G——G*
	0.063 0.937	0.407 0.593
K210N	G——G——G*	G——G——G——G*
	0.084 0.916	0.120 0.266 0.614
Wild-type	G——G——G*	G——G——G——G*
	0.176 0.824	0.745 0.255

$[E]_0$ and $[S]_0$ represent the concentrations of the substrate and enzyme, respectively.
[a] Numbers indicate the normalized cleavage frequencies of the indicated bonds.
*Radioactive glucose residue.

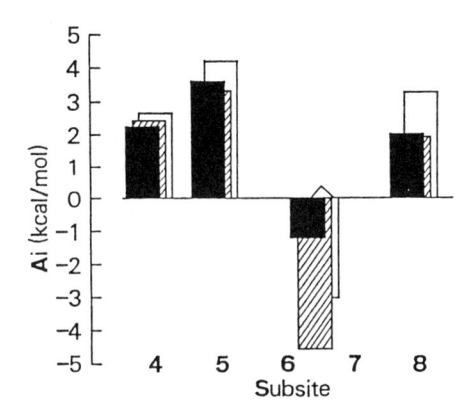

Figure 6. Histograms showing the subsite affinities (A_i) of the mutant and wild-type enzymes. The wedge represents the catalytic site. The columns indicate the affinities of the K210R (filled bars), K210N (hatched bars) and wild-type (open bars) enzymes.

drastically. The effect of the K210R mutation on the total affinity of the central subsites (A_6+A_7) was opposite to that of the K210N: The total repulsion of the central subsites in K210R lowered and those of K210N increased.

As reported previously (3, 7), the Sfamy protein is composed of a main (M) domain and a C-terminal (C) domain, where the M domain is fold into a $(\alpha/\beta)_8$ barrel structure. The K210 is located on a loop called L_4, which is one of the eight loops comprising the M domain. The 211th histidine (H211), which is a possible residue composing the 7th subsite, is also present on the L_4 loop. One of the catalytic residue candidates, the 207th aspartic acid (D207), is present between the C-terminal end of a β-sheet called βA_4 and the N-terminal end of the L_4 loop. As K210 is placed near D207 and H211 on the same loop, the k_{int} value for the hydrolysis and the 7th subsite affinity might be influenced by the mutation at the 210 position. According to the subsite theory (9), the k_{int} value, which is independent of the chain length of the substrate and the binding mode, is identical to the k_{cat} value if the substrate is larger than the substrate binding site. In fact, the mutation decreases the k_{int} values which are identical to the k_{cat} value for amylose A (Degree of polymerization is 18), as shown in Table III. However, the mutation might not affect the 6th subsite affinity, since the 210 residue on the L_4 loop is distant from H123 and H297 on the L_3 and L_7 loops which are presumed to be the major components of the 6th subsite (3, 7). Therefore, the change in the summed affinities of the 6th and 7th subsites caused by the replacement is actually due to the change in the 7th subsite affinity, although evaluation of the separate affinities of the 6th and 7th subsites is not possible because the catalytic residues are located between them. Thus, the influence of the 210th residue can be restricted to the 7th and 8th subsite affinities. Replacement of the K210 residue by an N residue, which has a shorter and neutral side chain, decreases the affinities of the 8th and 7th subsites. Substitution of the residue by R, which possess a longer and positively charged side chain, similarly lowers the 8th subsite affinity, but increases the affinity of the 7th subsite. These results indicate that K210 in the native enzyme contributes to both the 7th and 8th subsite affinities.

In conclusion, it has been shown that for Sfamy the enhanced catalytic activity of K210R specific for short substrates, such as G_2 and G_3, is attributable to a substantial increase in the affinity of the 7th subsite adjacent to the catalytic site. Additionally, it has been shown that the decreased catalytic activity of K210N specific for substrates shorter than G_5 ($G_{\leq 5}$) is achieved by a substantial decrease in the 7th and 8th subsite affinities. This finding provides a guiding principle for the comprehension of the substrate recognition mechanism and a means to improve the substrate and product specificities of the enzymes.

Acknowledgements

We wish to thank Professor H. Umeyama of Kitasato Univ. for the molecular modelling, Dr. I. Yamashita of Hiroshima Univ. for helpful discussion, and Dr. S. Kobayashi of National Food Research Institute, Tsukuba, for his helpful advice and for supplying *Bacillus macerans* transglycosylase. We are also grateful to Dr. Y. Jigami of our institute and Dr. B. Svensson of Carlsberg Laboratory for helpful discussion.

Literature Cited

1. Matsuura, Y., Kusunoki, M., Harada, W., & Kakudo, M. *J. Biochem.* **1984**, *95*, pp. 697-702.
2. Itoh, T., Yamashita, I., &Fukui, S. *FEBS Lett.* **1987**, *219*, pp. 339-342.
3. Matsui, I., Matsui, E., Ishikawa, K., Miyairi, S., Fukui, S., & Honda, K. (1990) *Agric.Biol. Chem.* **1990**, *54*, pp. 20009-2015.

4. Suganuma, T., Matsuno, R., Ohnishi, M., & Hiromi, K. *J. Biochem.* **1978**, *84*, pp. 293-316.
5. Matsui, I., Ishikawa, K., Matsui, E., Miyairi, S., Fukui, S., & Honda, K. *J. Biochem.* **1991**, *109*, pp. 566-569.
6. Matsui, I., Ishikawa, K., Miyairi, S., Fukui, S., & Honda, K. *Biochemistry* **1992**, *31*, pp. 5232-5236.
7. Matsui, I., Yoneda, S., Ishikawa, K., Miyairi, S., Fukui, S., Umeyama, H., & Honda, K. (1994) *Biochemistry* **1994**, *33*, pp. 451-458.
8. Matsui, I., Ishikawa, K., Miyairi, S., Fukui, S., & Honda, K. (1992) *FEBS Lett.*
9. Hiromi, K. *Biochem. Biophys. Res. Commun.* **1970**, *40*, pp. 1-6.
10. Isawa, S., Aoshima, H., Hiromi, K., & Hatno, H. *J. Biochem.* **1974**, *75*, pp. 969-978.

RECEIVED August 17, 1995

Chapter 8

Synergistic Interaction of Cellulases from *Trichoderma reesei* During Cellulose Degradation

Bernd Nidetzky[1], Walter Steiner[2], and Marc Claeyssens[3]

[1]Division of Biochemical Engineering, Institute of Food Technology, University of Agriculture, Peter-Jordan-Strasse 82, A-1190 Vienna, Austria
[2]Enzyme Technology Laboratory, Institute of Biotechnology, Technical University Graz, Petersgasse 12, A-8010 Graz, Austria
[3]Laboratory of Biochemistry, Department of Physiology, Microbiology, and Biochemistry, University of Ghent, K. L. Ledeganckstraat 35, B-9000 Ghent, Belgium

> The synergism between purified cellulolytic components of *Trichoderma reesei* during degradation of different cellulosic substrates was studied and was found to be maximal on crystalline cellulose and to decrease with increasing substrate concentration. Optimal ratios of two cellulase components in synergistic mixtures are calculated from the kinetic constants in a new formal-kinetic model which was derived to describe the synergistic hydrolysis rates. They were found to depend on the total enzyme concentration which reflects the degree of substrate saturation with enzyme. Evidence for a sequential enzymatic attack of cellulose was obtained because pretreatment of cellulose by one enzyme component could increase the activity of a subsequently acting cellulase.

Efficient hydrolysis of crystalline cellulose by fungal cellulase systems is thought to require the synergistic action of endocellulases (EG; E.C. 3.2.1.4) and cellobiohydrolases (CBH; E.C. 3.2.1.91) in combination with at least one ß-glucosidase (ßG; E.C. 3.2.1.21). The latter enzyme is itself not active on the insoluble substrate but degrades cellobiose and higher cello-oligosaccharides. It is concluded that synergism occurs when the rate or the extent of hydrolysis catalyzed by a combination of two or more enzyme components is higher than the sum of the individual enzyme activities. Since relatively little is known about the action of the cellulolytic components on crystalline and semicrystalline cellulose and about their individual roles in hydrolyzing or structurally disintegrating the cellulosic substrate, only simplified mechanistic concepts pertaining to the synergistic interaction of fungal cellulases have been postulated (*1*). The differentiation between 'exo' and 'endo' type cellulases seems to be much less clear-cut than originally thought (2-4) and thus hypothetical models may require re-evaluation.

Two different types of synergistic interaction between fungal cellulases have been identified: Synergism between endoglucanases and cellobiohydrolases, found by several authors (2,5-7), is traditionally interpreted by a sequential mechanism of enzyme action. The endoglucanases initially attacking the amorphous regions of the cellulose, provide new nonreducing polyglucose-chain ends for the subsequent action of the cellobiohydrolases. Another mechanistic model for endo-exo-synergism is based on the results of Ryu et al. (8) and Kyriacou et al. (9) suggesting that competition for adsorption sites on cellulose and a well co-ordinated adsorption and desorption of individual enzyme components lead to an enhanced cellulase turnover on the cellulose surface and consequently to an increased rate of substrate degradation. It was, however, diversely assumed that cellulases occupy distinctly different (8) or also common binding sites on cellulose (9). Whether there is strict correlation between competitive adsorption and synergism during hydrolysis has to be studied more precisely because recently (cross) synergism but not competition for binding sites between a *Thermomonospora fusca* endoglucanase and *Trichoderma reesei* CBH I have been reported (10).

For the synergistic action of two cellobiohydrolases (exo-exo-synergism) (2,9,11-13) two mechanistic concepts have been proposed. One is based on the assumption that a loose enzyme complex of CBH I and CBH II from *T. reesei* is formed in solution and synergistic binding to cellulose is seen, an effect that is maximal in optimal synergistic mixtures of the CBH components (14). Further evidence for other cellulase-cellulase complexes between endo and exo-type components in culture fluids of *T. reesei* and their isolation by isoelectric focusing have been reported by Sprey & Bochem (15).

The second hypothetical concept for exo-exo-synergism, put forward by Wood & McCrae (12), tries to explain synergism between two immunologically distinct CBHs from *Penicillium pinophilum*. It tentatively postulates different orientation of the nonreducing end groups in crystalline cellulose that would require endwise acting CBH components with different stereochemical specificities. The action of CBH I from *T. reesei*, however, was found not to be restricted exclusively to the non-reducing end (4) and endocellulase-activity (2-4,16) of this enzyme was detected. Furthermore, crystallographic studies on the catalytic core protein of CBH II from *T. reesei* showed that a single cellulose chain may enter the active site of the enzyme in two different orientations (17) and the ratio of products obtained during hydrolysis of soluble cellooligosaccharides with CBH II seems to be consistent with this concept (3,16,18).

A third type of synergism, not yet clearly identified, might exist at least in some cases between cellulases and ß-glucosidases (1,10,19). Traditionally, it is assumed that the role of ßG is to hydrolyze cellobiose, which is the principal reaction product of the major *T. reesei* cellulases (20) and which strongly inhibits CBHs, for example CBH I from *T. reesei*. In addition to its function of eliminating product inhibition by cellobiose, a ßG from *Penicillium funiculosum* was suggested to be important for endo-exo-synergism during solubilization of amorphous cellulose (21). Whatever the exact function of ßG in the hydrolysis reaction is, in view of engineering an efficient cellulase system, it has to be considered that optimal mole fractions in synergistic component mixtures will be strongly influenced in the presence of ßG (*10,22*) because inhibition by cellobiose differs at least one order of magnitude in strength for EG and CBH components (*23*).

Less attention has so far been paid to study the synergism between two catalytic core proteins of two individual cellulolytic components, which lack their cellulose-binding-domain (CBD), or else between a core and another intact cellulase (*13,14,19*). In addition to limited proteolysis of an intact cellulase *in vitro*, some core proteins can be found in the culture filtrates of *T. reesei*. It was tentatively assumed that the cores might have an important role in a late stage of cellulose hydrolysis (*24*). One may speculate that (i) diffusion into narrow pores of the partially hydrolyzed cellulose is not possible for an intact cellulase (*25,26*) but could indeed occur for its smaller core protein, and that (ii) the higher specific activities of the cores on soluble substrates may be important for a complete cellulose hydrolysis (*27*).

Synergism during cellulose degradation might be also found at the domain level of one single cellulase. The exact role of the fungal CBD for substrate hydrolysis is unclear, although two hypothetical concepts have been put forward: (i) the CBD serves as an 'anchor' for the attached catalytic domain thereby increasing the local concentration of the core at the cellulose fiber (*28*) and (ii) the CBD, in an active manner, functions as a 'plough' to facilitate single chain liberation from the cellulose crystals (*29*). Therefore, does synergism exist between the catalytic core domain and the CBD and, if so, is it possible to reproduce this synergistic effect using a reconstituted mixture of the isolated domains? The occurrence of CBDs in other hydrolytic (noncellulolytic) enzymes (*30*) is not in favor of an active function of the CBD. In contrast, Din et al. (*31*) and Kilburn et al. (*32*) could show for a bacterial cellulase that (i) the CBD possesses the ability to structurally but not hydrolytically disrupt cellulose fibers and that (ii) there is synergism between the core and the CBD during substrate hydrolysis.

Only few reports appear in the literature dealing with a mathematical analysis of the synergistic hydrolysis of cellulose. All models are restricted to endo-exo-synergism and are based on the mechanistic concept of a sequential enzyme action. The synergistic action of an endo- and exo-acting enzyme system for the depolymerization of a polysaccharide was originally proposed by Suga et al. (*33*) and adopted further for cellulase action by several authors (*34-37*). As yet the most structured model, which however completely neglects exo-exo-synergism, has been described by Converse & Optekar (*38*). It could account for the competitive adsorption onto cellulose (*39*) and also for the lower degrees of synergism at saturation of the substrate with enzyme (*40*).

Results published in the literature with respect to the synergistic action of endo- and exocellulases are not conclusive and sometimes contradictory. This may in part be due to widely varying experimental conditions as well as to the imperative prerequisite of employing individual cellulases purified to homogeneity (*41*). It is, however, widely accepted that synergism between cellulases is dependent on the degree of substrate saturation (*40,42*) and on the physicochemical properties of the substrate itself (*2*). The present report summarizes our results of synergy studies obtained with pure cellulolytic components of *T. reesei* and their core proteins.

Experimental

Chemicals. Whatman filter paper No.1 (Whatman, Maidstone, U.K.) and microcrystalline cellulose, Sigmacell 50 (Sigma, Deisenhofen, Germany), were used as cellulosic substrates. Phosphoric acid- and sodium hydroxide swollen Sigmacell 50 were prepared according to Wood (*43*). Bacterial cellulose was from *Acetobacter xylinum* (*44*). Algal cellulose from *Valonia* was a kind gift of Dr. B. Henrissat (Grenoble, France). 2'-Chloro-4'-nitro-phenyl (CNP)-β-glycosides were prepared as described (*23*). D-Glucono-δ-lactone and galactono-γ-lactone were from Merck (Darmstadt, Germany).

Enzymes. CBH I, its core and its CBD, CBH II, EG I, EG III and its core were prepared from the culture filtrate of *T. reesei* by reported methods (*45-49*). The homogeneity of the individual enzymes was proved by SDS-PAGE, analytical isoelectric focusing and by the very sensitive measurement of the specific activity (50 °C) against small, chromogenic CNP-ß-D-glycosides of lactose, cellobiose and cellotriose (*50*). Under these conditions no contamination of the individual enzymes could be detected, which is an absolute prerequisite for synergy studies. The concentration of the individual cellulases in stock solutions was determined spectrophotometrically (280 nm) using reported molar absorption coefficients (*48,50,51*). ß-Glucosidase from *Aspergillus niger* (Novozym 188; Novo, Bagsvaerd, Denmark) was purified by Biogel P6 gel-filtration (BioRad, Richmond, USA).

Synergy Between Cellulase Combinations During Hydrolysis. If not indicated otherwise all hydrolysis experiments were carried out (i) at 50 °C in 50 mM sodium citrate buffer, pH 4.8 but also in (ii) 50 mM sodium acetate buffer, pH 5.6. The activities of the individual cellulases against filter paper at pH 5.6 are approximately 10-15 % lower than at pH 4.8. To prevent strong product inhibition by cellobiose during cellulose hydrolysis, individual enzymes or combinations dissolved in buffer were supplemented with ß-glucosidase (final activity of 0.2 IU/mL). Two factors reported to be important for the synergistic action of cellulases (*1,2*) were studied and their influence on the initial rate of hydrolysis were evaluated.

    1.) The effects of different types of the cellulose and several substrate concentrations were investigated while employing various binary combinations of cellulases as well as mixture of the four major cellulases CBH I, CBH II, EG I, and EG III (see Table I), each at a constant enzyme concentration.

    2.) The variation of (i) the total enzyme concentration and of (ii) the ratio of the individual cellulases in binary, synergistic enzyme combinations during degradation of 20 g/L filter paper was studied.

    Incubations were at 50 °C and at non mass-transfer-limited conditions (agitation at 200 rpm) as described (*44,50*). Ensuring a complete conversion of soluble hydrolysis products to D-glucose in each sample (*44*) the rates of hydrolysis were calculated from the linear range of D-glucose produced vs. reaction time or, in case of significant deviation from linearity, by taking the amount of D-glucose released after 0.5 hours. The synergistic factors were determined as the ratio of D-glucose produced by a given combination to the sum of the individual activities.

Table I. Synergistic Enzyme Combinations[1]

A	CBH I/CBH II	G	CBH I core/CBH II
B	CBH I/EG I	H	CBH I core/EG I
C	CBH I/EG III	I	CBH I core/EG III
D	CBH II/EG I	J	CBH I core/EG III core
E	CBH II/EG III	K	CBH I/EG III core
F	CBH I/CBH II/EG I/EG III	L	CBH II/EG III core

[1] The concentration of each component was: 1.25 μM CBH I; 1.25 μM CBH I core; 1.25 μM CBH II, 0.4 μM EG I; 0.4 μM EG III; 0.4 μM EG III core.

The addition of ß-glucosidase did not alter the initial hydrolysis rates (1 hour) observed for individual cellulases or combinations of the enzymes, and the ß-glucosidase was shown to exhibit no activity against filter paper.

Enzymic Pretreatment of Filter Paper. The effect of a sequential enzyme action on the cellulosic substrate was studied in a two-step experiment as described by Nidetzky et al. (50). Filter paper (20 g/L) was enzymatically "pretreated" with CBH I or its core protein, CBH II, EG I, or EG III for 1 h at 50 °C (step 1). This treatment was followed by inactivation of the enzyme (95 °C, 90 min) and, after extensive washing of the cellulose with buffer, a subsequent hydrolysis of the "pretreated" filter paper with the same or another cellulase component (step 2) was started.

The activities and the adsorption of the enzyme component added in step 2 on either non-treated or pretreated filter paper were compared. Suitable control experiments proved that (i) the inactivation after step 1 was complete, and that (ii) artefacts due the prolonged heat treatment of the filter paper were absent. Adsorptions of individual cellulases onto filter paper were determined by measuring the individual components' activities on chromophoric CNP-ß-D-glycosides in solution before and after incubation with the cellulosic substrate (1 hour).

Analytical Assays. CBH I and CBH I core, EG I, and EG III activities were determined at 50 °C and pH 5.6 using the CNP-ß-glycosides of lactose (1 mM), cellobiose (1 mM), and cellotriose (0.25 mM) as substrates in the presence of 20 mM glucono-δ-lactone and galactono-γ-lactone to inhibit ßG and ß-galactosidase-activities (50). With no chromogenic substrate for CBH II being available, its activity (and adsorption) was estimated at 50 °C using 20 g/L filter paper as a substrate (1 h reaction time, ß-glucosidase). D-Glucose was determined either by the hexokinase assay (Chemie Linz, Linz, Austria) or the D-glucose-oxidase (GOD)/peroxidase (POD) method (Boehringer, Mannheim, Germany).

Regression Analysis. Parameters in the mathematical models were estimated by nonlinear least square regression using the BMDP statistical software package (BMDP statistical software, Inc., Los Angeles, CA).

Results and Discussion

Effect of the Type and Concentration of the Cellulosic Substrate.

Type of Substrate. Typical initial hydrolysis time-courses of filter paper degradation by individual cellulases are shown in Figures 1 and 2. While the formation of D-glucose by CBH I and CBH I core and CBH II is linear with reaction time (Figure 1), a rapid decrease in reaction rate was observed in case of EG I as well as EG III and EG III core. Apparently, only limited substrate sites are available for the catalytic action of the endocellulases (Figure 2). Limitation in accessible substrate sites may be related to physicochemical properties of the substrate such as specific surface area (52), relative crystallinity and hydratation of the cellulose (2). Furthermore, an intrinsic heterogeneity of the glycosidic bonds in cellulose crystals with respect to their accessibility to enzymatic attack was discussed by Henrissat et al. (2) using a molecular modeling approach.

The activities of the individual cellulases on different types of cellulose (each 20 g/L) are shown in Figure 3. For all enzyme components the rates were highest on easily accessible cellulose. The activities of CBH I and CBH I core were identical on amorphous cellulose whereas the core protein retained only about 50 % of the activity of the intact enzyme on crystalline cellulose (Figure 3). This finding is in agreement with the results of other authors reporting identical or even higher activities of the catalytic cores of fungal and bacterial cellulases on amorphous or derivatized (CM) cellulose as well as on soluble, "artificial" substrates (13,27,54). Comparison of the hydrolytic activities of intact EG III and its core on different cellulosic substrates revealed almost no difference between the two enzyme preparations (Figure 3). Ståhlberg et al. (48), on the other hand, found a significantly lower activity of the core protein against Avicel.

In contrast to the activities of the individual enzymes, a synergistic hydrolytic action between two components was most evident on crystalline cellulose whereas low synergistic factors were calculated using amorphous cellulose as a substrate (Table II). This trend was much less clear in binary enzyme combinations of the core protein of CBH I with other intact cellulases (Table II) and the synergistic factors in these respective combinations were much lower. Because a comparison of intact CBH I and its core revealed the same degree of synergistic action with CBH II, EG I or EG III on amorphous cellulose (Table II), the presence of the CBD does not seem to be an intrinsic prerequisite for synergistic action between two cellulases. This result has also been confirmed using the core protein of EG III (Table II). The differences in the individual hydrolytic activities of CBH I and its core protein on crystalline and semicrystalline cellulose can only partly account for the lower synergistic factors in all mixtures containing the core protein. Hence, it could be speculated that, given non-equivalent binding sites on the cellulosic substrate, there is a different interaction of the intact enzyme or of its core with cellulose which in turn will have a significant influence on the synergistic action with other cellulases. If the core protein was more specifically targeted to substrate sites easily degraded than the intact enzyme, the decrease in the degree of synergism and also the different synergistic factors seen with for CBH I or CBH I core on different types of cellulose could be understood.

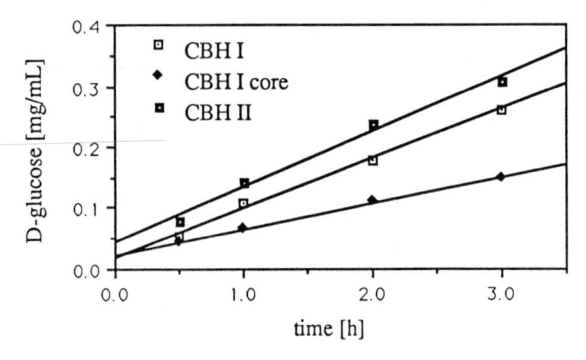

Figure 1. Time-course of hydrolysis of 20 g/L filter paper by *T. reesei* cellobiohydrolases (1.25 µM) in the presence of ß-glucosidase.

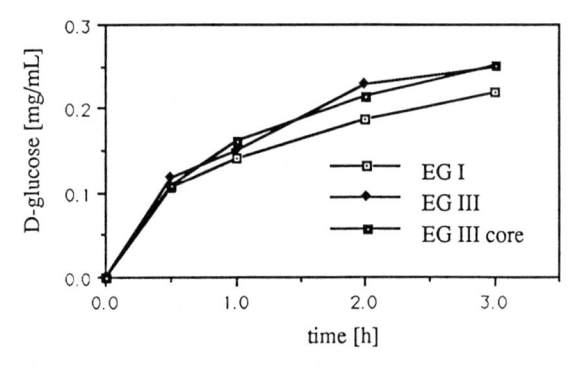

Figure 2. Time-course of hydrolysis of 20 g/L filter paper by *T. reesei* endoglucanases (0.40 µM) in the presence of ß-glucosidase.

Table II. Synergistic Factors of Cellulase Action on Different Celluloses

Enzyme Combination*	Cellulose			
	Bacterial Cellulose	*Valonia* Cellulose	Sigmacell 50	H_3PO_4-swollen Sigmacell 50
CBH I/CBH II	1.94	1.22	2.51	1.40
CBH I/EG I	1.20	1.69	1.59	1.02
CBH I/EG III	2.40	2.48	1.88	1.11
CBH II/EG I	0.99	n.d.	1.44	1.00
CBH II/EG III	1.92	n.d.	1.42	1.28
CBHI/CBH II/ EG I/EG III	2.73	n.d.	2.65	0.88
CBH I core/CBH II	1.11	0.82	1.54	1.33
CBH I core/EG I	1.10	1.45	1.02	1.10
CBH I core/EG III	1.50	1.45	0.78	1.14
CBH I core/ EG III core	1.15	n.d.	0.99	1.18
CBH I/EG III core	1.60	n.d.	1.93	1.12
CBH II/EG III core	1.18	n.d.	1.83	0.95

n.d., not determined; substrate concentration of 20 g/L with the exception of *Valonia* cellulose (5 g/L)
* The concentration of each component was: 1.25 µM CBH I; 1.25 µM CBH I core; 1.25 µM CBH II, 0.4 µM EG I; 0.4 µM EG III; 0.4 µM EG III core.

Synergism in combinations of EG I/EG III or the core of EG III was observed neither on crystalline nor on amorphous cellulose. The hydrolytic activity of two endocellulases was always less than the sum of their individual actions. When the maximal degree of synergism is considered, the most potent two-component combinations were CBH I/CBH II on Sigmacell 50 and CBH I/EG III on bacterial cellulose (Table II). In comparison with all other binary combinations and as judged from the synergistic factors, the synergistic action was even higher in the four-component mixture. This indicates that synergism is not restricted to two-cellulase-systems only, but can be also observed in three or four component mixtures.

Substrate Concentration. As filter paper concentrations are raised, the hydrolysis rates catalyzed by the individual cellulases increase which is most probably due to a more complete enzyme adsorption with decreasing ratios of enzyme-to-substrate (Figure 4). This effect was even more obvious for the case of the core proteins of CBH I and EG III (Figure 4). Two explanations seem valid: (i) The catalytic cores are less tightly adsorbed to crystalline and semicrystalline cellulose than the intact cellulases. While the equilibration-bound fraction of the intact cellulases could be close to 1 already at low substrate concentration, complete adsorption of the

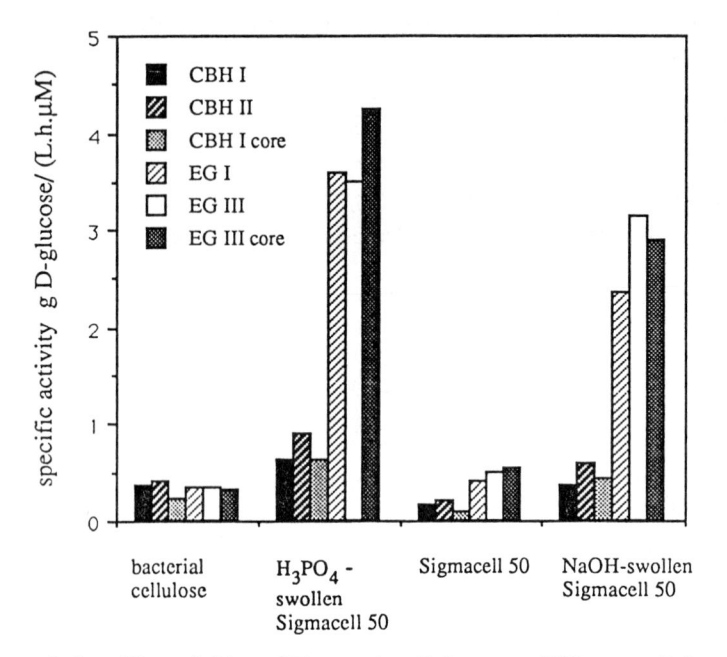

Figure 3. Specific activities of *T. reesei* cellulases on different celluloses at 50 °C.

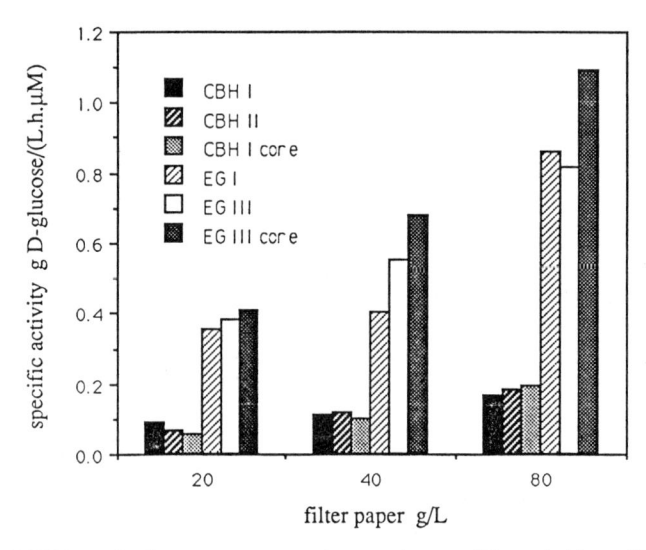

Figure 4. Effect of substrate concentration on the specific activities of *T. reesei* cellulases.

respective core might be achieved only at a large excess of available substrate sites. The lower affinity to cellulose of the core protein of CBH I will change the molar ratio of cellulose-bound CBH I activity to another bound cellulase activity which could be less favorable for synergistic interaction. (ii) Since the probably non-specific interaction of the binding-domain with cellulose is absent, a preferential interaction of the catalytic core with sites easily degraded in semicrystalline cellulose may take place which would be especially so, when, at high substrate concentrations, the number of these sites is not limiting.

The degrees of synergistic action observed for all two-component mixtures decreased when substrate concentration was increased (Table III). This result seems at first glance to conflict with the finding of Woodward et al. (*40,42*) that synergism is maximal at low substrate saturation. Hence, one could expect that the synergistic action should increase at higher substrate concentrations. However, the synergism between two cellulases depends strongly on the type of cellulosic substrate and is low on an easily accessible (amorphous) cellulose (Table II and 3).

Table III. Synergistic Factors of Cellulase Action at Different Substrate Concentrations

*Enzyme Combination**	*Filter Paper g/L*		
	20	40	80
CBH I/CBH II	3.24	2.43	1.93
CBH I/EG I	2.57	2.62	1.85
CBH I/EG III	2.82	3.02	2.02
CBH II/EG I	1.58	1.63	1.16
CBH II/EG III	2.38	2.36	1.76
CBH I/CBH II/ EG I/EG III	3.84	3.71	2.87
CBH I core/CBH II	2.25	2.39	1.82
CBH I core/EG I	1.80	2.09	1.36
CBH I core/EG III	1.72	1.69	1.43
CBH I core/EG III core	1.69	1.67	1.40
CBH I/EG III core	2.91	2.56	2.04
CBH II/EG III core	1.80	1.60	1.28

n.d., not determined; * The concentration of each component was: 1.25 µM CBH I; 1.25 µM CBH I core; 1.25 µM CBH II, 0.4 µM EG I; 0.4 µM EG III; 0.4 µM EG III core.

Filter paper must be considered a heterogeneous substrate in terms of physicochemical properties such as crystallinity, specific surface area or degree of hydratation. At low coverage with enzyme, the interaction of the cellulases with substrate sites easily available for hydrolysis and the special preference of the core proteins for binding onto this type of sites could explain the experimental observations. The similar synergistic factors seen in two component mixtures of CBH I or its core with CBH II at high substrate concentration of 80 g/L are also in favor of this hypothesis. In combination with both endoglucanases and even at non-limiting concentrations of cellulose, however, there was significantly more interaction of intact CBH I than of the catalytic core. A tentative and hypothetical concept in agreement with these results could thus be as follows:

Provided common and distinct substrate sites for the individual cellulolytic components exist on semicrystalline cellulose, the synergistic interaction between the enzymes will depend on the relative concentrations of these sites. A model substrate with a high percentage of common sites appears to be amorphous cellulose and, consequently, synergism is low for all enzyme combinations, although the individual components' activities are maximal on this type of substrate. On semicrystalline cellulose, such as filter paper, the degrees of synergism are high but (i) decrease with an increasing availability of easily degradable sites, i.e. at high substrate concentration, and (ii) obviously depend on the specific interaction of each cellulase component with cellulose which, for an intact cellulase, will be governed by the influence of both the catalytic- and the cellulose-binding-domain.

Effect of Enzyme Concentration and Ratio of Cellulolytic Components.

Individual Cellulolytic Components. Investigation of the synergism between two or more cellulases during cellulose hydrolysis requires precise and statistically significant knowledge on their individual actions (Figure 5). Since the reaction system is perfectly heterogeneous during cellulose hydrolysis and enzyme binding on to the substrate is believed to be a prerequisite prior to hydrolysis, the analysis of initial rate data determined for each individual component was carried out in terms of substrate saturation with enzyme. For a constant concentration of cellulose (20 g/L), this leads to a rate-equation deviating from classical Michaelis-Menten kinetics,

$$V(E) = V_{max}\frac{E}{(K_E+E)} \quad (1)$$

where V(E) is the initial rate (dependent on initial enzyme concentration) in [mg glucose/(mL h)], V_{max} the maximal hydrolyis rate (at substrate saturation) in [mg glucose/(mL h)], E the initial enzyme concentration in [μM], and K_E the half saturation constant with enzyme in μM (*50*).

The kinetic constants V_{max} and K_E for individual, intact cellulases and for the respective core proteins of CBH I and EG III are shown in Table IV and the resulting catalytic efficiencies (V_{max}/K_E) are calculated. The half saturation constants for the cellobiohydrolases were found to be almost two orders of magnitude higher than for the endoglucanases; the maximal hydrolysis rates for the intact enzymes show a trend which is in agreement with the activities of the cellulases on other types of cellulose

(*2,13*). The maximum hydrolysis rates of the endoglucanases on filter paper are 2-3 times lower when compared with those of the cellobiohydrolases (Table IV) but catalytic efficiencies are severalfold higher for the endo-components. A comparison of the kinetic constants of the intact enzymes of CBH I and EG III with those of their core proteins shows that K_E-values are increased 2.3 and 5.1 times, respectively, but within statistical significance maximum hydrolysis rates are not influenced by the proteolytic modification. One may conclude that the *T. reesei* cellulose-binding-domain does not endow the core with any additional "active" property (*29*) at least during filter paper degradation. The catalytic efficiencies of intact CBH I and EG III are, however, 3 and 6 times higher than those of the respective cores and point to a more economic action of the intact components which is in favour of the "anchor" model of the CBD (*27*).

Synergistic Combinations. Because for all cellulases a nonlinear relation of hydrolysis rate and enzyme concentration was found (cf. equation 1 and Figure 5), the theoretical sum of the individual actions is nonlinear too (cf. Figure 6). Hence, analysis of synergism data as described by Henrissat et al. (*2*), who defined the theoretical sum of enzyme action as the tie line between the individual actions, can be performed only when the strict linearity between reaction rate and enzyme concentration has been proved.

The synergistic interaction of the 6 possible binary combinations of intact CBH I, CBH II, EG I, and EG III was investigated at different ratios of the individual enzymes and at different total enzyme concentrations. In 4 cases (Table IV), but not for a CBH II/EG I combination, synergism was significant. To restrict information contained in a large number of hydrolysis rate data to few significant parameters, we derived a formal mathematical model as described in equations (2) and (3). Equation (2) accounts for the individual actions of the enzyme components and equation (3) for the additional, synergistic hydrolysis rate (*50*). The mathematical expression for the synergistic rate (equation (3)) shows formal analogy to sequential, two-substrate reactions where a ternary transitory complex of the enzyme and the two substrates is formed. In equation (3) the two substrate concentrations are substituted by the enzyme concentrations of the binary combination. However, analogy to Michaelis-Menten theory can of course be disputed because theoretical prerequisites such as homogeneity of the reaction system are not fulfilled in the case of cellulose hydrolysis.

$$V(E_1,E_2) = V(E_1) + V(E_2) + V_{syn}(E_1,E_2) \qquad (2)$$

with:

$$V_{syn}(E_1,E_2) = V_{syn,max} \frac{E_1 E_2}{(K_1 K_2 + K_1 E_2 + K_2 E_1 + E_1 E_2)} \qquad (3)$$

where E_1 and E_2 are enzyme 1 and 2 in a binary combination [μM], $V(E_1,E_2)$ the hydrolysis rate of a combination of two cellulases in [mg glucose/(mL h)], $V(E_1)$ and

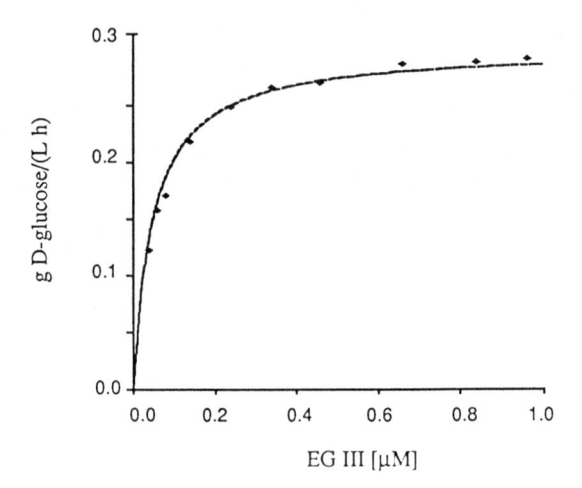

Figure 5. Enzyme-dependent rate of filter paper degradation by EG III from *T. reesei*: The broken line was calculated using the model in equation (1) and the constants in Table IVA.

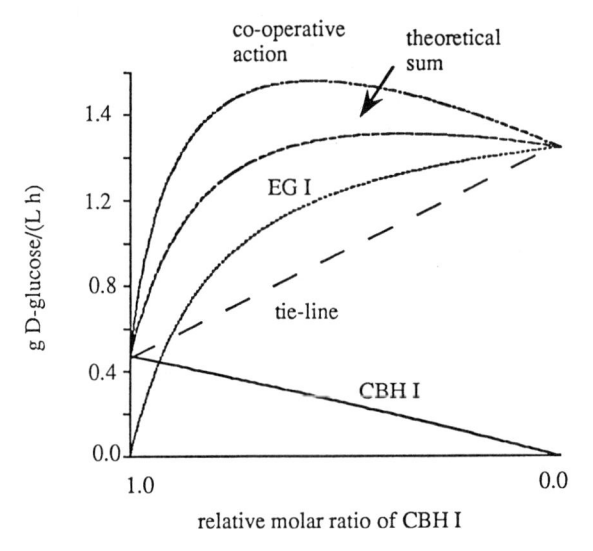

Figure 6. Theoretical analysis of the synergistic action of CBH I and EG I in varying mole fractions on filter paper (20 g/L; total enzyme is 0.5 µM). The lines were calculated according to the model in equations (2) and (3) using the constants in Table IVA and B.

Table IV. Kinetic Constants for the Action of Individual Cellulases from *Trichoderma Reesei* and Their Binary, Synergistic Combinations Acting on Filter Paper (S.D. is Given)

A.

enzyme	range [μM]	V_{max} [mg/(mL h)]	K_E [μM]	V_{max}/K_E	ratio
CBH I	0.20 - 8.00	0.448±0.072	4.243±1.285	0.106±0.036	
CBH I core	0.40 - 15.00	0.471±0.101	9.663±2.548	0.049±0.016	2.17
CBH II	0.10 - 7.00	0.518±0.052	3.301±0.644	0.157±0.034	
EG I	0.01 - 0.50	0.172±0.021	0.095±0.034	1.810±0.503	--
EG III	0.02 - 0.90	0.299±0.010	0.047±0.007	6.361±0.971	6.65
EG III core	0.02 - 0.90	0.271±0.012	0.241±0.031	1.125±0.153	

B.

enzyme1 / enzyme 2	$V_{syn,max}$ [mg/(mL h)]	$V(E_1)+V(E_2)$ [mg/(mL h)]	K_1 [μM]	K_2 [μM]
CBH I / CBH II	2.183±0.141	0.966±0.089	4.243[a]	0.297±0.063
CBH I / EG I	0.494±0.053	0.620±0.075	4.243[a]	0.022±0.010
CBH I / EG III	1.277±0.080	0.747±0.073	1.488±0.193	0.008±0.002
CBH II / EG III	0.635±0.069	0.817±0.053	0.803±0.216	0.042±0.009

C.

	calculated optimal ratios (total enzyme concentration)	
	(1 μM)	(10 μM)
CBH I / CBH II	62/38	70/30
CBH I / EG I	74/26	90/10
CBH I / EG III	82/18	92/8
CBH II / EG III	74/26	86/14

[a] Set constant during regression analysis because of highly linear correlation between $V_{syn,max}$ and K_1; reproduced with permission from reference 50.

V(E2) the hydrolysis rate of the individual cellulases (cf Table IV), V_{syn}(E1, E2) the synergistic hydrolysis rate [mg glucose/(mL h)], $V_{syn,max}$ the maximal synergistic hydrolysis rate [mg glucose/(mL h)] and K_1 and K_2 the half saturation constant corresponding to enzyme 1 and 2 in a binary combination [µM].

Estimates for the kinetic constants in equation (3) are given in Table IV. For CBH I/CBH II synergism the model predictions in comparison with the experimental data are graphically depicted in Figure 7. A good agreement of theory with the experimental data can be seen and no trend deviation detected (Figure 7 and constant of determination r^{2}). The highest synergistic rates were found with the combinations CBH I/CBH II and CBH I/EG III (Table IVB) and, in comparison to K_E-values for the individual actions, the half saturation constant of at least one enzyme in a synergistic combination (K_1 or K_2 in Table IVB) is reduced drastically.

For a given total enzyme concentration the synergy model in equations (2) and (3) allows to calculate the optimal ratio of the individual enzyme components in a binary combination to obtain (i) maximal hydrolysis rates or (ii) maximal degrees of synergistic action. The optimal ratio of enzymes is quite strongly shifted when the total protein concentration is changed (Table IV C). At a total protein concentration of 10 µM, the optimal ratios for combinations of EG with CBH components are in good agreement with those previously reported to be typical for endo-exo synergism (2,55).

Combinations of CBH I/CBH II from *T. reesei* on Avicel exhibited an optimal ratio of 4:1 (2,14). For CBH I and II from *P. pinophilum* this ratio was 1:1 using cotton fibers. Interestingly, at least traces of endoglucanase activity were required for significant hydroly-sis (12). A ratio of 7:3 was calculated to be optimal for exo-exo synergism from the data in Table IV using eqns. (2) and (3) at 10 µM total protein concentration. The different optimal molar ratios reported for mixtures of the CBH components may stem from varying experimental conditions such as (i) the type of substrate used, (ii) the total enzyme concentration and, probably most important, (iii) the addition of ßG which in cleaving cellobiose removes inhibition of e.g. *T. reesei* CBH I.

No synergism (based on release of soluble sugars) was found when CBH I core and its CBD were admixed in different concentrations and different molar ratios (not shown). In contrast, Kilburn et al. (32) detected a synergistic action between *Cellulomonas fimi* endoglucanase (CenA) and its binding domain on dewaxed cotton with an optimal molar ratio of both components of 1:4.

Effect of Enzymic Pretreatment of Filter Paper. The activities of individual cellulase components on filter paper either enzymatically pretreated or non-treated (control in step 1) were compared. In all cases, no hydrolysis was observed when only buffer and ß-glucosidase were added to the pretreated filter paper (control in step 2) thus proving complete enzyme inactivation after heat treatment. The rate enhancing effect of substrate pretreatment by EG I on CBH I activity is illustrated in Figure 8 and the results of all experimental observations are summarized in Figure 9.

The activities of EG I and EG III on substrate pretreated by one of these two components were lower than in the controls (Figure 9). This result corroborates the interpretation of non-linear time courses of hydrolysis in Figure 1 as only limited substrate sites in filter paper would be available for hydrolysis by the endocellulases.

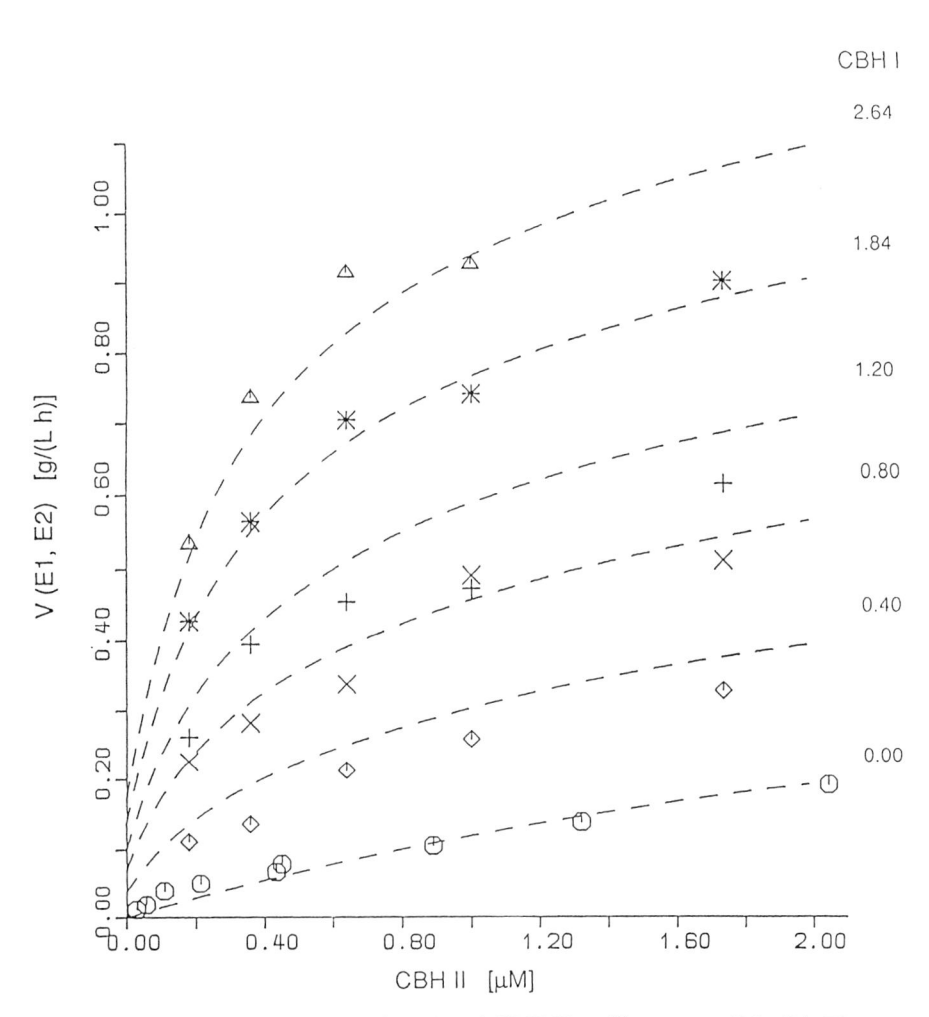

Figure 7. Synergism between CBH I and CBH II on filter paper (20 g/L). The lines were calculated according to the model in equations (2) and (3) using the constants in Table IVA and B (reproduced with permission from reference *50*).

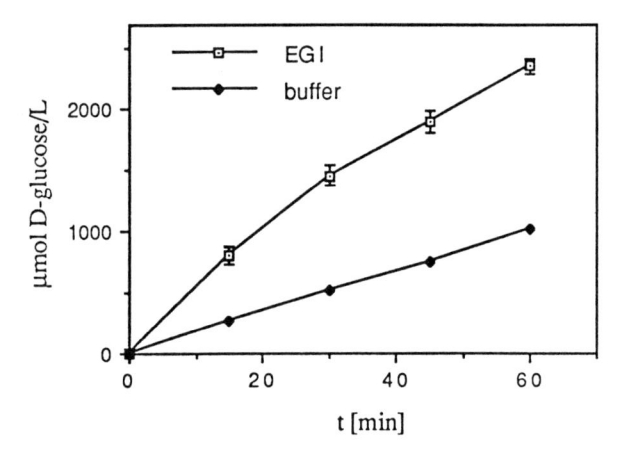

Figure 8. Effect of enzymic pretreatment of filter paper by EG I on the hydrolysis time course catalyzed by subsequently acting CBH I.

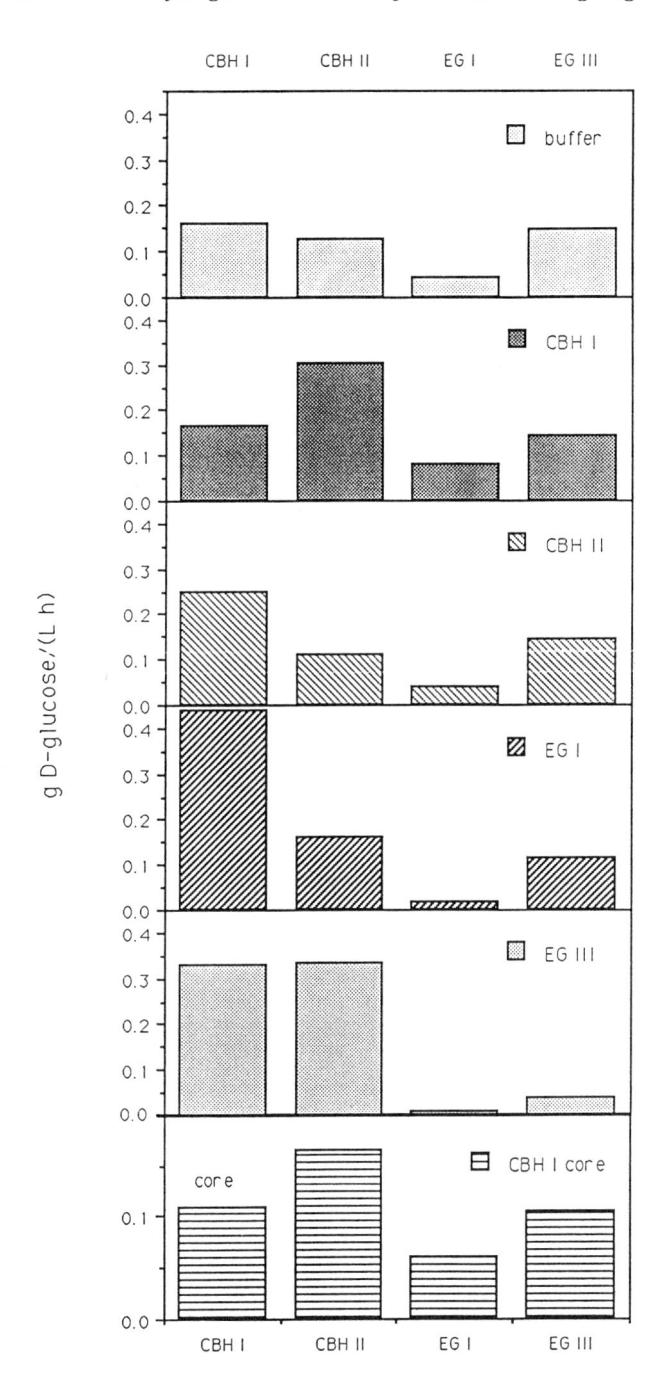

Figure 9. Influence of enzymic substrate pretreatment on the activities of individual cellulase components from *T. reesei*.

On the contrary, no indication for a depletion of substrate sites in an initial phase of hydrolysis was observed in case of CBH I, CBH I core and CBH II because reaction rates on untreated filter paper and on filter paper pretreated by these enzymes were identical (Figure 9). For the synergistic enzyme combinations (CBH I and CBH I core with CBH II, EG I or EG III, CBH II with EG III and EG I), pretreatment of filter paper by at least one enzyme component of the binary combination could significantly increase the activity of the other component (Figure 9). While a direct comparison of the results of a subsequent enzyme action with those of a simultaneous one may be difficult, because other effects such as competition for substrate sites could be important in the latter case, their significance in relation to the results in Table IV for synergism are discussed:

Endo-Exo-Synergism. For synergism between endo- and exo-components the model assuming a sequential mechanism of enzyme action initiated by an attack of the endocellulases (5) may in part may be valid: EG III prepares a more easily hydrolysable substrate for CBH I, its core and CBH II, but, after substrate pretreatment by the CBH components, its own activity is not significantly influenced (or even slightly decreased). A mutual influence of substrate pretreatment, however, was detected for combinations of EG I with each cellobiohydrolase. EG I, which synergistically interacts with CBH I, CBH I core and to a lesser extent with CBH II, shows higher activity on filter paper pretreated by either cellobiohydrolase (Figure 9). This effect was reciprocal, and on substrate pretreated by EG I the activities of all CBH components were significantly and up to twofold increased.

The hydrolysis rates seen after substrate pretreatment reflect and corroborate the degrees of synergism observed when the binary enzyme combinations were acting on untreated filter paper (Figure 9 and Table IV). Consequently, pretreatment of the substrate by EG I increased the activity of CBH II only slightly (about 20 %), whereas a hydrolysis rate 2.6 times higher than the control was monitored following a pretreatment of filter paper with EG III. Preincubation of the substrate with EG I affected the hydrolytic activities of CBH I and its core (Figure 9) and resulted in a 2.76 and 2.45 fold increase of the respective reaction rates. Pretreatment of filter paper with EG III increased the activity of CBH I 2.1 fold, whereas the effect on CBH I core was not as significant and only a 5 % increase of the activity was observed.

Exo-Exo-Synergism. CBH I is preparing a more easily hydrolysable substrate for CBH II and vice versa. The simultaneous concerted action of both enzymes is not strictly required to observe the synergistic effect. While this does not rule out that CBH-enzyme-complexes do exist (14), their formation is not intrinsically necessary for synergism. Compared with intact CBH I, substrate pretreatment by CBH I core was less effective to increase CBH II activity (Figure 9). Preincubation of filter paper with CBH II resulted in a 1.57 fold increase of CBH I activity whereas CBH I core activity was enhanced only by about 9 %. Quite similarly, during simultaneous action on filter paper the synergism between intact CBH I and CBH II was about 1.5 times higher than for a combination of CBH I core and CBH II (Table IV). From Figure 9 it is evident that the rate enhancing effect of substrate pretreatment was much more expressed when CBH II was acting on filter paper preincubated with

CBH I than vice versa. This result seems to corroborate the finding in Table IV (total protein 1 or 10 µM) that the optimal molar ratio in binary CBH combinations is a 1.5 - 2-fold excess of CBH I over CBH II.

Adsorption on to Pretreated Filter Paper. Within the estimated experimental error of 3-5 %, adsorption of the individual cellulase components onto enzymatically pretreated filter paper remained unchanged (Table V) when compared to the control binding experiments (i.e., substrate without enzyme treatment).

Table V. Binding of Individual Cellulase Components From *Trichoderma reesei* on to Filter Paper After Enzymatic Pretreatment

Enzyme	Untreated	Pretreated [a]
2.6 µM CBH I	84	85±2
2.6 µM CBH I core	31	32±2
2.6 µM CBH II	65	60±9
0.4 µM EG I	68	64±4
0.7 µM EG III	90	89±3

[a] Percentage of enzyme bound on to filter paper (20 g/L) after pretreatment with either 2.6 µM CBH I or its core, 2.6 µM CBH II, 0.4 µM EG I, or 0.4 µM EG III.

Hence, synergism cannot be interpreted as a consequence of increased binding onto adsorption sites newly created by enzyme action. It appears that after enzyme treatment the substrate is rendered more accessible to enzymatic attack and that consequently the specific activities of the individual cellulases are increased. A clear and distinctly different structural transformation of cellulose after treatment with either EG or CBH component has been recently visualized by electron microscopy. Both submicrofibril formation by EG component and conversion into (amorphous) cellulose plaques by CBH component were observed (*15,56,57*).

Conclusions

The present results demonstrate that the synergistic interaction between cellulolytic components of *T. reesei* (i) is maximal on crystalline cellulose whereas the individual activities are highest on hydrated and swollen cellulose types and that (ii) it decreases with increasing substrate concentration. The synergistic action between a core protein and an intact cellulase is always lower than the action between two intact components (except on an amorphous cellulosic substrate). Tentatively, endo-exo synergism is maximal on substrates with a high degree of polymerization whereas an opposite correlation seems to be valid for exo-exo synergism.

During filter paper degradation, non linear relations of hydrolysis rate and concentration of the individual cellulolytic components are found and the resulting half saturation constants point to a manyfold higher 'substrate affinity' of the endocellulases. The calculated catalytic efficiencies are up to 60 times higher for the endo-components. The fungal CBD does not appear to have an active role in the hydrolysis reaction since (i) the half-saturation constant but not the maximal hydrolysis rate are affected by proteolytic modification and (ii) no synergism in mixtures of CBH I core and its binding domain is detected. On filter paper, hydrolysis rates and synergistic rates observed for binary combinations are maximal at substrate saturating concentrations of the cellulases whereas the degrees of synergistic action (or else the synergistic factors) are maximal at non saturating enzyme concentrations as evident from the decrease of K_E-values in binary combinations. For each enzyme combination an optimal ratio of the two components exists, which is determined by the total enzyme concentration thus reflecting the degree of substrate saturation.

The present results obtained in two step experiments with substrate pretreatment by one enzyme component in the first step followed by complete inactivation and subsequent hydrolysis of filter paper by the same or by another cellulase component in the second step indicate (i) a sequential attack of the cellulases on to cellulose and point (ii) to an enzymatic transformation of the substrate, which would consequently lead to an increase in the specific activity of another simultaneously or subsequently acting cellulase component. While cellulase-cellulase complexes in *T. reesei* culture fluids may exist, a sequential action of the enzyme components seems to be as effective for the occurrence of a synergistic effect. Synergistic adsorption as a consequence of de novo created adsorption sites on the cellulose can thus be ruled out.

Acknowledgements

The authors are grateful to Dietmar Haltrich (Institute of Food Technology, Vienna, Austria) for his critical revision of the manuscript.

Literature cited

1. Woodward, J. *Bioresource Technol.* **1991**, *36*, 67.
2. Henrissat, B.; Driguez, H.; Viet, C.; Schülein, M. *Bio/Technology* **1985**, *3*, 722.
3. Van Tilbeurgh, H.; Bhikhabhai, R.; Pettersson, L.G.; de Boeck, H.; Claeyssens, M. *Eur. J. Biochem.* **1985**, *148*, 329.
4. Vrsanska, M.; Biely, P. *Carbohydr. Res.* **1992**, *227*, 19.
5. Wood, T.M.; McCrae, S.I. *Adv. Chem. Ser.* **1979**, *181*, 181.
6. Beldman, G.; Voragen, A.G.J.; Rombouts, F.M.; Pilnik, W., *Biotechnol. Bioeng.* **1988**, *31*, 173.
7. Wood, T.M.; McCrae, S.I.; Bhat, K.M. *Biochem. J.* **1989**, *260*, 37.
8. Ryu, D.D.Y.; Kim, C.; Mandels, M. *Biotechnol. Bioeng.* **1984**, *26*, 488.
9. Kyriacou, A.; Neufeld, R.J.; MacKenzie, C.R. *Biotechnol. Bioeng.* **1989**, *33*, 631.
10. Walker, L.P.; Belair, C.D.; Wilson, D.B.; Irwin, D.C. *Biotechnol. Bioeng.* **1993**, *42*, 1019.
11. Enari, T.M.; Niku-Paavola, M.L. *CRC Crit. Rvs. Biotechnol.* **1987**, *5*, 67.
12. Wood, T.M.; McCrae, S.I. *Biochem. J.* **1986**, *234*, 93.

13. Tomme, P.; van Tilbeurgh, H.; Pettersson, G.; van Damme, J.; Vandkerckhove, J.; Knowles, J.; Teeri, T.; Claeyssens, M. *Eur. J. Biochem.* **1988**, *170*, 575.
14. Tomme, P.; Heriban, V.; Claeyssens, M. *Biotechnol. Lett.* **1990**, *121*, 525.
15. Sprey, B.; Bochem, H.-P. *FEMS Microbiol. Lett.* **1993**, *106*, 239.
16. Nidetzky, B.; Zachariae, W.; Gercken, G.; Hayn, M.; Steiner, W. *Enzyme Microb. Technol.* **1994**, *14*, 43.
17. Rouvinen, J.; Bergfors, T.; Teeri, T.; Knowles, J.K.C.; Jones, T.A. *Science* **1990**, *249*, 380.
18. Teleman, A.; Valkeajärvi, A.; Koivula, A.; Ruohonen, L.; Drakenberg, T. **1993**, Paper presented at Tricel 93, Kirkkonummi, Finland, June 2-5 (1993).
19. Irwin, D.C.; Spezio, M.; Walker, L.P.; Wilson, D.B. *Biotechnol. Bioeng.* **1993**, *42*, 1002.
20. Ståhlberg, J.; Johansson, G.; Pettersson, G. *Biochim. Biophys. Acta* **1993**, *1157*, 107.
21. Wood, T.M.; McCrae, S.I.; McFarlane, C.C. *Biochem. J.* **1980**, *189*, 51.
22. Walker, L.P.; Wilson, D.B.; Irwin, D.C.; McQuire, C.; Price, M. *Biotechnol. Bioeng.* **1992**, *40*, 1019.
23. Van Tilbeurgh, H.; Loontiens, F.G.; de Bruyne, C.K.; Claeyssens, M. *Meth. Enzymol.* **1988**, *160*, 45.
24. Claeyssens, M.; Aerts, G. *Bioresource Technol.* **1992**, *39*, 143.
25. Converse, A.O.; Matsuno, R.; Tanaka, M.; Taniguchi, M. *Biotechnol. Bioeng.* **1988**, *32*, 38.
26. Tanaka, M.; Ikesaka, M.; Matsuno, R.; Converse, A.O. *Biotechnol. Bioeng.* **1988**, *32*, 698.
27. Teeri, T.T.; Reinikainen, T.; Ruohonen, L.; Jones, T.A.; Knowles, J.K.C. *J. Biotechnol.* **1992**, *24*, 169.
28. Ståhlberg, J.; Johansson, G.; Pettersson, G. *Bio/Technology* **1991**, *9*, 286.
29. Knowles, J.; Lehtovaara, P.; Teeri, T. *Trends Biotechnol.* **1987**, *5*, 255.
30. Kellett, L.E.; Poole, D.M.; Ferreira, L.M.A.; Durrant, A.; Hazelwood, G.P.; Gilbert, H.J. *Biochem. J.* **1990**, *272*, 369.
31. Din, N.; Gilkes, N.R.; Tekant, B.; Miller, R.C.; Warren, A.J.; Kilburn, D.G. *Bio/Technology* **1991**, *9*, 1096.
32. Kilburn, D.G.; Assouline, Z.; Din, N.; Gilkes, N.R.; Tomme, P.; Warren, A.J. In *Trichoderma reesei Cellulases and Other Hydrolases*; Suominen, P.; Reinikainen, T., Eds.; Foundation for Biotechnical and Industrial Fermentation Research: Helsinki, Finland, 1993, Vol. 8; pp. 281-290.
33. Suga, K.; van Dedem, G.; Moo-Young, M. *Biotechnol. Bioeng.* **1974**, *17*, 433.
34. Okazaki, M.; Moo-Young, M. *Biotechnol. Bioeng.* **1978**, *2*, 637.
35. Fujii, M.; Homma, T.; Ooshima, K.; Taniguchi, M. *Appl. Biochem. Biotechnol.* **1991**, *28/29*, 145.
36. Fujii, M.; Shimizu, M. *Biotechnol. Bioeng.* **1986**, *28*, 878.
37. Dean III, S.W.; Rollings, J.E. *Biotechnol. Bioeng.* **1992**, *39*, 968.
38. Converse, A.O.; Optekar, J.D. *Biotechnol. Bioeng.* **1993**, *42*, 145.
39. Converse, A.O.; Girard, D.J. *Biotechnol. Prog.* **1992**, *8*, 587.
40. Woodward, J.; Lima, M.; Lee, N.L. *Biochem. J.* **1988**, *255*:, 895.
41. Wood, T.M.; Garcia-Campayo, V. *Biodegradation* **1990**, *1*, 147 (1990).
42. Woodward, J.; Hayes, M.K.; Lee, N.E. *Bio/Technology* **1988**, *6*, 301.
43. Wood, T.M. *Meth. Enzymol.* **1988**, *160*, 19.
44. Nidetzky, B.; Macarron, R.; Hayn, M.; Steiner, W. *Biotechnol. Lett.* **1993**, *16*, 71.
45. Van Tilbeurgh, H.; Bhikhabhai, R.; Pettersson, L.G.; Claeyssens, M. *FEBS Lett.* **1984**, *169*, 215.

46. Van Tilbeurgh, H., Tomme, P., Claeyssens, M., Bhikhabhai, R.; Pettersson, G. *FEBS Lett.* **1986**, *204*, 223.
47. Macarron, R.; Acebal, C.; Castillon, P.; Dominguez, J.M.; de la Mata, I.; Pettersson, G.; Tomme, P.; Claeyssens, M. *Biochem. J.* **1993**, *289*, 867.
48. Ståhlberg, J.; Johansson, G.; Pettersson, G. *Eur. J. Biochem.* **1988**, *173*, 179.
49. Tomme, P.; McCrae, S.I.; Wood, T.M.; Claeyssens, M. *Meth. Enzymol.* **1988**, *160*, 187.
50. Nidetzky, B., Steiner, W., Hayn, M.; Claeyssens, M. *Biochem. J.* **1994**, *298*, 705.
51. Saloheimo, M.; Lehtovaara, P.; Penttilä, M.; Teeri, T.T.; Ståhlberg, J.; Johansson, G.; Pettersson, G.; Claeyssens, M.; Tomme, P.; Knowles, J.K.C. *Gene* **1988**, *63*, 11.
52. Grethlein, H.E. *Bio/Technology* **1985**, *2*, 155.
53. Henrissat, B.; Vigny, B.; Buleon, A.; Perez, S. *FEBS Lett.* **1988**, *231*, 177.
54. Gilkes, N.R.; Warren, R.A.J.; Miller, R.C.; Kilburn, D. *J. Biol. Chem.* **1988**, *263*, 10401.
55. Wood, T.M. In *Enzyme Systems for Lignocellulose Degradation*; Coughlan, M.P., Ed.; Elsevier Applied Science, London, UK, 1989, pp. 19-35.
56. Sprey, B.; Bochem, H.-P. *FEMS Microbiol. Lett.* **1992**, *97*, 113.
57. Sprey, B.; Bochem, H.-P. *FEMS Microbiol. Lett.* **1991**, *78*, 183.

RECEIVED September 18, 1995

Chapter 9

Synergism Between Purified Bacterial and Fungal Cellulases

John O. Baker, William S. Adney, Steven R. Thomas, Rafael A. Nieves, Yat-Chen Chou, Todd B. Vinzant, Melvin P. Tucker, Robert A. Laymon, and Michael E. Himmel

Alternative Fuels Division, Applied Biological Sciences Branch, National Renewable Energy Laboratory, 1617 Cole Boulevard, Golden, CO 80401–3393

A standardized comparative study measured glucose release and synergistic effects in the solubilization of microcrystalline cellulose by binary mixtures of 11 fungal and bacterial cellulases (eight endoglucanases and three exoglucanases). Evaluation of 16 endo/exo pairs revealed that bacterial/fungal hybrid pairs are very effective in solubilizing microcrystalline cellulose. Of nine bacterial/fungal hybrid pairs studied, six were ranked among the nine most synergistic combinations, and six bacterial/fungal pairs were also among the top nine pairs in terms of soluble-sugar release. One hybrid pair (*Acidothermus cellulolyticus* E1 and *Trichoderma reesei* CBH I) was ranked first in both synergism and sugar-release. In exo/exo synergism experiments, the performance of *Thermomonospora fusca* E_3 confirmed its mode of action as "CBH II-like" (i.e., E_3 is synergistic with *T. reesei* CBH I but not with *T. reesei* CBH II). Studies of endo/endo interactions suggested a possible means of categorizing endoglucanases in terms of substrate specificity.

The fact that cellulose is the most abundant organic polymer on earth confers tremendous ecological, industrial, and economic importance upon the enzymatic depolymerization of cellulosic materials. Every year, largely through the action of cellulases, some 2×10^{11} tons of cellulose is recycled through the carbon cycle (*1*). The potential of cellulose to provide fermentable sugars as a carbon source in the production of fuels and chemical feedstocks is now well appreciated (*2,3*).

It has long been recognized that "cellulase" activity is not the activity of a single enzyme, but the result of multiple activities working cooperatively, or "synergistically," to achieve the efficient solubilization of crystalline cellulose (*4,5*). There are three categories of cellulolytic enzymes required for this process: (1)

"endo-1,4-β-D-glucanases" or "endoglucanases" (1,4-β-D-glucan-4-glucanohydrolases; EC 3.2.1.74), which cleave glycosidic bonds randomly in the interior of the cellulose polymer chain, (2) the "exo-1,4-β-glucosidases," including both the 1,4-β-D-glucan glucohydrolases (EC 3.2.1.74), which liberate D-glucose from 1,4-β-D-glucans and hydrolyze cellobiose slowly, and 1,4-β-D-glucan cellobiohydrolases (EC 3.2.1.91) which liberate D-cellobiose from 1,4-β-glucans, and (3) "β-D-glucosidases," or β-D-glucoside glucohydrolases (EC 3.2.1.91), which relieve end-product inhibition of the endoglucanases and cellobiohydrolases by hydrolyzing the penultimate product, cellobiose, to the less-inhibitory final product, glucose.

The development during the past two decades of powerful molecular-biological methods for identifying, transferring and/or modifying, and overexpressing the genetic material encoding specific proteins, combined with the understanding that cellulase action is a multi-activity process, suggests that cellulase mixtures might be engineered by combining enzymes that originate from different organisms to form effective, and possibly improved, hybrid cellulase systems. Among the potential benefits offered by such heterologous systems are reductions in the production cost of a given quantity of overall activity, higher specific activity of the protein expressed, and the ability to "tailor" enzyme systems for use in unusual applications (6).

A step toward making possible the rational selection of enzymes for inclusion in engineered mixtures is to obtain consistent, uniform and quantitative measurements of the interactions of individual pairs of cellulase components, drawn from as large and diverse an inventory of enzymes as practical. The two data sets we have chosen to collect are the actual quantity of reducing sugar (as glucose) released from microcrystalline cellulose by different enzymes and binary enzyme mixtures under standardized conditions, and the degree of synergistic effect (DSE), which is defined as the ratio of the reducing-sugar release by a given mixture to the sum of the experimentally measured release of reducing sugar by equivalent, but separate, loadings of the individual components.

Historical Background. The concept of synergism in enzymic depolymerization of cellulose was introduced as early as 1950 by Reese et al. (4) These researchers proposed the cooperative action of a C_1 factor, acting in an unspecified way to disrupt the crystalline structure of cellulose, and a C_x activity, which encompasses all β-glucanase activity, including the endoglucanases and exoglucanases. The C_x factor was considered to act upon the cellulose made more accessible by the prior action of the C_1 factor. In 1954 Gilligan and Reese (5) provided quantitative evidence for synergism by showing that the amount of reducing sugar released from cellulose by the combined fractions of fungal culture filtrate was greater than the sum of the amounts released by the individual fractions. Over the next three decades, investigators demonstrated synergistic interactions between homologous (i.e., produced by the same organism) exo- and endo-acting cellulase components (7-18). "Cross-synergism" between purified heterologous endo- and exo-acting enzymes produced by different species of aerobic fungi is well known (9-11,13).

Exo/exo synergism between two cellobiohydrolases of *T. reesei*, CBH I and CBH II, was first reported by Fägerstam and Pettersson in 1980 (19), and has now been rationalized in terms of attack at either the reducing or the non-reducing terminus of the cellulose chain. Kinetic data have confirmed that *T. reesei* CBH II preferentially attacks the non-reducing end of the cellulose chain (*20*), and recent X-ray crystallographic studies of have shown a cellodextrin bound to CBH I catalytic domain with the reducing terminus toward the active-site tunnel (*21*).

Synergism between enzymes from different phylogenetic domains (i.e., between fungal and bacterial endo- and exoglucanases) was first proposed by Eveleigh in 1987 (*6*) and confirmed a year later by Wood, et al. (*22*). The concept of cross-synergy has most recently been investigated Wilson and colleagues (*23,24*) and in the authors' laboratory (*25*).

Present Experimental Approach. In the present study, we have assembled an array (Figure 1) of eight endoglucanases (two fungal and six bacterial) from five different organisms, plus three exoglucanases (two fungal and one bacterial), and have measured soluble-sugar production and synergism for a large fraction of the total number of binary pairs that could be formed from this inventory of enzymes. Despite the relative paucity of exoglucanases, this series represents one of the larger, and probably the most diverse, of the collections of cellulase pairs yet subjected to a study of this type.

Materials and Methods

Enzyme Purifications. Only highly purified enzymes were used to construct all "cellulase cocktails" for evaluation in this study. Enzyme purification was done either at NREL or by DOE subcontractors and provided to NREL under the terms of those subcontracts. The *T. fusca* enzymes were purified in the laboratory of D. Wilson at Cornell University, E_3 from the supernatant of *T. fusca* strain ER1 (*26*), and rE_5 (the prefix "r" denoting a recombinant enzyme) from cell lysates of *Streptomyces lividans* TK24 containing a plasmid carrying the *T. fusca* E_5 gene(*27*). Endoglucanase A (EndoA) from *Microbispora bispora*, expressed from a cloned genomic fragment in *Escherichia coli*, was purified in the laboratory of D. Eveleigh at Rutgers University (*28*). A xylanase/cellulase (Xyl/Cel) from *M. bispora* (*29*), and endoglucanase B from *Thermotoga neapolitana* (*30*) were both purified from the culture broths of the native organisms in D. Eveleigh's laboratory.

Enzymes purified at NREL included *Acidothermus cellulolyticus* endoglucanase 1 (E1; *31*), a truncated form (rEC-E1CAT) of *A. cellulolyticus* E1 that lacks the cellulose-binding domain (CBD), endoglucanases EG I and EG II, and exoglucanases CBH I and CBH II, all from *T. reesei*, and *Aspergillus niger* β-glucosidase. *A. niger* β-glucosidase was purified from the commercial preparation Novozym 188 Cellobiase by a procedure described earlier (*25*). The *T. reesei* and *A. cellulolyticus* enzymes were prepared according to the procedures detailed later in this section.

The purity of each enzyme preparation used in this study was verified by silver-stained sodium-dodecylsulfate gel electrophoresis. In addition, the purity of the *T. reesei* cellulases was checked by means of Western blots using monoclonal antibodies (*32*). The preparations were found to have compositions of at least 98% of the protein to be evaluated.

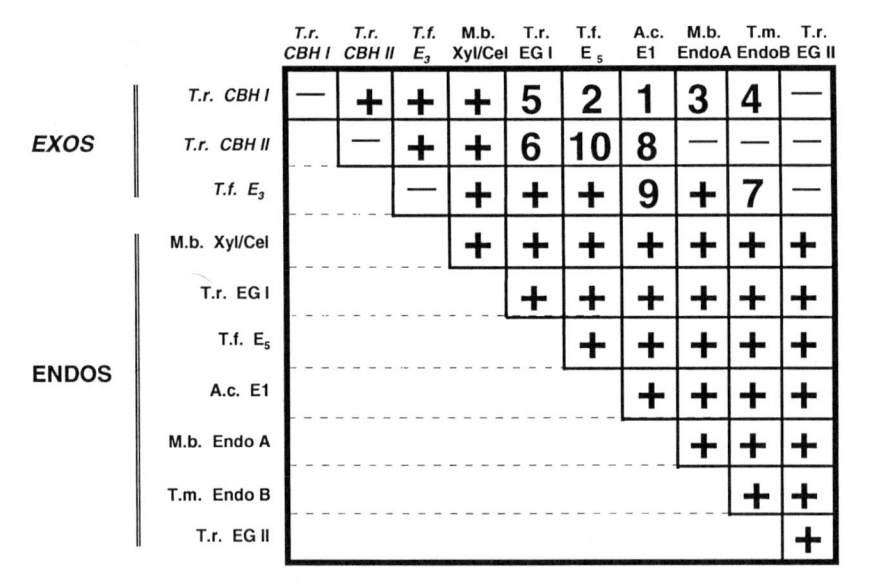

	T.r. CBH I	T.r. CBH II	T.f. E_3	M.b. Xyl/Cel	T.r. EG I	T.f. E_5	A.c. E1	M.b. EndoA	T.m. EndoB	T.r. EG II
T.r. CBH I	—	+	+	+	5	2	1	3	4	—
T.r. CBH II		—	+	+	6	10	8	—	—	—
T.f. E_3			—	+	+	+	9	+	7	—
M.b. Xyl/Cel				+	+	+	+	+	+	+
T.r. EG I					+	+	+	+	+	+
T.f. E_5						+	+	+	+	+
A.c. E1							+	+	+	+
M.b. Endo A								+	+	+
T.m. Endo B									+	+
T.r. EG II										+

EXOS: T.r. CBH I, T.r. CBH II, T.f. E_3

ENDOS: M.b. Xyl/Cel, T.r. EG I, T.f. E_5, A.c. E1, M.b. Endo A, T.m. Endo B, T.r. EG II

Figure 1. Grid pattern showing the organization of the overall experimental series of enzyme pairs evaluated for synergism and release of soluble sugar. "Plus" (+) signs indicate data reported later in the paper; "minus" (−) signs denote pairs not yet evaluated. The numerals appearing in some cells of the 3x7 "endo/exo' rectangle are the rankings of the ten most synergistic endo/exo pairs, as will be discussed under Results and Discussion in connection with Table I.

Purification of *T. reesei* EG I and CBH I. The purification procedure used in this study was that developed earlier (*25*), following the general size-exclusion-chromatography (SEC)/anion-exchange chromatography (AEC) protocol described by Shoemaker et al. (*33*). The initial SEC step utilized a Pharmacia Superose-12 prep grade 35/60 column loaded with 6 mL of Genencor Laminex (Lot 13-90091-01, code 6-5960). The AEC step (Pharmacia Q-Sepharose) was followed by an additional size-exclusion step using a Pharmacia Superdex 200 16/60 HiLoad column in 20 mM acetate, pH 5.0. In all cases, "heart cuts" of chromatographic peaks were harvested in order to minimize contamination by other proteins.

Purification of *T.reesei* CBH II and EG II. CBH II and EG II were purified from Genencor Laminex by using the thiocellobioside affinity chromatography approach of van Tilbeurgh et al. (*34*) following the SEC/AEC procedure described above for EG I and CBH I.

Purification of *A. cellulolyticus* E1 expressed in *S. lividans* (rS-E1). The recombinant enzyme was produced using *S. lividans* strain TK24 expressing E1 pIJ702 (Thomas, S. R., et al. in this volume) grown in tryptic soy broth (30 g/L) with 5 µg/mL thiostrepton in a New Brunswick 14-L fermentor. The fermentation broth was harvested using a CEPA continuous-flow centrifuge, and the supernatant was concentrated and diafiltered in an Amicon CH2 concentrator with 10,000 MW cutoff hollow-fiber cartridges, to a final volume of 300 mL in 20 mM Bis-Tris, pH 5.8. The recombinant enzyme was purified using a three-step process consisting of hydrophobic interaction chromatography (HIC) using a 250-mL column packed with Pharmacia Fast-Flow Phenyl Sepharose. The second step employed AEC using a Pharmacia Resource Q 6-mL column, and was followed by a final SEC step using a Pharmacia HiLoad 16/60 Superdex 200 prep grade column. For the hydrophobic interaction step, ammonium sulfate was added (final concentration 0.5 M) to the concentrate of one 10-L fermentation. A total volume of 300 mL of this concentrate was loaded onto the PhenylSepharose column using 20 mM Tris, pH 8.0, with 0.5 M ammonium sulfate. The recombinant enzyme was eluted with a decreasing salt gradient. Fractions containing the recombinant enzyme were identified by means of assay against 4-methylumbelliferyl-β-D-cellobioside. Fractions with activity were combined, concentrated, and diafiltered into 20 mM Tris, pH 8.0, after which they were loaded onto the anion-exchange column and eluted with an increasing NaCl gradient (0-1 M). The final purification step was SEC using a Pharmacia HiLoad 16/60 Superdex 200 prep grade column in 20 mM acetate, pH 5.0.

Production and purification of *A. cellulolyticus* E1 expressed in *E. coli* (rEC-E1CAT). Cells from a 10-L fermentation of *E. coli* expressing the cloned *A. cellulolyticus* E1 gene were suspended in 50 mL of 20 mM acetate, pH 5.0, containing 100 mM NaCl. The cells were lysed by a French-Press treatment employing two successive treatments at 12,000 psi in a 30-mL pressure cell with flow rate at a fast drip, with both supply and receiver vessels on ice. The cell lysate was heat-treated at 65°C for 60 min, then clarified by centrifugation to remove cell debris and coagulated *E. coli* proteins. The purification protocol outlined above for native E1 was then followed.

Cellulose Digestions. Enzyme digestions of crystalline cellulose were carried out at 50°C in 50 mM acetate buffer, pH 5.0, to which 0.004% (w/v) sodium azide had been added to prevent microbial growth. Total cellulase loadings were held constant at 0.36 µM in the 1.0-mL reaction mixtures, which contained 5% (w/v) Sigmacel, Type 20, as substrate. The total weight of protein added therefore varied with the molecular weights of the components involved, but total cellulase loadings (endoglucanase plus exoglucanase) were typically close to 20 µg protein per mL of digestion mixture, or 0.4 mg cellulase per g cellulose. Sufficient purified *A. niger* β-glucosidase (4.17 µg/mL of digestion mixture, equal to 0.61 units/mL or 12.2 units/mg of cellulose) was added to eliminate the problem of cellobiose inhibition. As described previously (25), the "sufficiency" of the β-glucosidase loading was established (for the individual enzymes and for representative mixtures) by means of experiments in which reducing-sugar output was measured in the presence of different loadings of β-glucosidase, to determine loadings above which further increases in β-glucosidase loading produced no further increases in yield of reducing sugar. The standard loading described above, equal to 83.33 µg β-glucosidase per g cellulose, was found to be approximately 10 times the minimum loading required to render the reducing-sugar output independent of the β-glucosidase loading (25).

To economize on purified enzyme, a miniaturized digestion apparatus was devised, using 1.5-mL Wheaton autoinjector vials as reaction vessels. The enzyme "cocktails" (0.3 mL total for each digestion of endoglucanase, exoglucanase, and β-glucosidase) for the different digestion mixtures were first placed in the vials, then substrate was added (as 0.7 mL of a 7.15% stirred slurry) to initiate the reaction. The vials were sealed with aluminum crimp-caps (PTFE-faced silicone septa, Kimble Glass, Inc., Vineland, NJ), placed in a custom-built rotator head immersed in a 50°C water bath and continuously mixed by inversion at 10 rpm. After a standard 120-h digestion period, representative 0.04-mL aliquots of each (well-dispersed) digestion mixture were withdrawn, diluted to 2.0 mL with deionized water, and centrifuged to remove all solid substrate. Reducing-sugar content of the supernatant was then determined (as glucose) using the bicinchoninic-acid method of Doner and Irwin (35).

Synergism ratios (abbreviated as "DSE" for "degree of synergistic effect") were calculated by dividing the reducing-sugar production of endoglucanase/exoglucanase mixtures by the sum of the productions in parallel digestion mixtures containing equivalent amounts of the endoglucanase and exoglucanase separately.

Results and Discussion

Given the variety of experimental approaches to synergism measurements found in the literature, and the complex, heterogeneous physical nature of the substrates employed, some comments on both of these matters should be made as background for the description and interpretation of the present results.

Physical Nature of the Substrate. Commercial "microcrystalline celluloses" such as Avicel or Sigmacell are, because of their ready availability in standardized form, widely used as model substrates to estimate the ability of cellulases to hydrolyze highly-ordered forms of cellulose. Superimposed on the "microcrystalline" structure of their basic building blocks, such cellulose preparations have, however, complex ultrastructures that interfere with enzymic attack on the microcrystals (*36*).

The "microcrystalline" substrates are produced from native cellulose by limited acid hydrolysis (*37,38*) to remove most of the "amorphous" regions that are seen as connecting the crystallites in the native material (*39*), followed by washing, caking, milling, and a size-fractionation step to yield the commercial product. The particles in the specified size range for a given commercial product are therefore aggregates of large numbers of "microcrystallites." The geometrical fashion in which the individual particles are aggregated is not clear, but a consideration of relative sizes can shed some light on the possible complexity of the particles as supplied. Electron-microscopic studies of acid-hydrolyzed materials have shown that the residual microcrystallites are 0.03-0.08 µm in length (depending on the conditions of hydrolysis and the cellulose starting material) and roughly 0.01 µm in width, or approximately the width of the original cellulose microfibrils (*38*). Frey-Wyssling has proposed a picture in which the microcrystallites of native cellulose are shaped like long, narrow tiles, with rectangular cross-sections measuring approximately 0.003 µm by 0.007 µm (*40*). These cross-sectional dimensions are of the same order of magnitude as the 0.006-µm and 0.0065-µm sizes reported by Abuja et al. (*41,42*) for the catalytic domains of *T. reesei* CBH II and CBH II. The length of the entire tadpole-shaped *T. reesei* CBH molecules, as measured by these workers, is on the order of 0.020 µm. Working with a similar picture of the elementary fibrils, Sagar (*43*) concluded that only one face of the fibril, that containing the accessible 2,6 and 2,3,6 hydroxyl groups, is attacked by *T. reesei* and *T. koningii* cellulases. The degree of compactness with which the residual microcrystallites are aggregated into the observable particles (inversely related to effective pore size) will affect both the accessibility of the cellulose to enzymes (*44*) and other reagents, and the number of microcrystallites making up each particle. Rough estimates based on 0.08µm x 0.007µm x 0.003µm microcrystallite dimensions would indicate that the number of individual crystallites in an "average" 20-µm-diameter particle should range from some hundreds of millions of individual crystallites for a quite loose, open structure, to a few billion for the tightest possible physical packing.

Choice of Experimental Design and Method of Calculating "Synergism". Despite general agreement on the definition of "synergism" or the "degree of synergistic effect" (DSE) as "the ratio of the activity of a given mixture of enzymes to the sum of the activities of the enzymes acting alone" (*23,45-49*), there are two different approaches in common use for estimating "activities of the enzymes acting alone." For the same enzyme combinations, the two computational approaches will yield different synergism values; the reader should therefore exercise caution in comparing numbers from one study with those from another.

In one approach, the "solo" activity of each component in the mixture is actually measured in assays identical to the assays used to measure the activity of the mixture, except that only one of the cellulase components is present, at a loading identical to the loading of that component in the mixture (*14,25,27,46*).

In other studies of binary synergism, the "sum of the separate activities" appears to have been estimated by linear interpolation between the sugar-release values for the samples containing 100% of the loading as one component or the other (*36,50*). In these latter studies, a curve is plotted to connect (or to fit) the data points for sugar release for each of the mixtures, including the "100%-either" samples at the ends, and a straight (usually broken) line without data points is drawn between the points for the two "100%" compositions as an estimator of the summed activities of the two enzymes acting separately.

Synergism studies also differ from one another with respect to whether the total cellulase loading is kept constant (*25,36,50*) or whether the total cellulase loading varies as the composition of the mixture is varied (as is the case for studies in which the composition is varied by holding the loading of one enzyme component fixed while the other is varied)(*45,46*). The magnitudes of the (fixed or varied) total cellulose loadings used in a variety of studies of synergism in the saccharification of microcrystalline cellulose (*24,25,36,45-48*) may be seen to range from as low as 0.4 µg/mg cellulose to as much as 100 µg/mg cellulose. The ratios of components in enzyme mixtures may be expressed either as mass ratios or as molar ratios.

Given the variety of experimental approaches used by different researchers, careful consideration is advisable in comparing results from different groups, with the most useful comparisons being those between studies that employ the most similar reaction conditions and computational methods. For this reason, we designed a survey of binary mixtures of as numerous and diverse a selection of enzymes as we could obtain in adequate quantities and purity, using identical experimental conditions and computational approaches throughout. For our standard approach we have chosen 5% (w/v) microcrystalline cellulose (Sigmacell, type 20, 20-µm average particle diameter) as substrate, enzyme mixtures expressed as molar percentages with a constant total cellulase loading of 7.2 pmol/mg cellulose, a long (120-h) digestion period with released (solubilized) glucose measured at the end, and calculation of DSE using the sum of actual activity measurements of individual activities in the denominator.

Our choices of 50°C and pH 5.0 as digestion conditions were based on the fact that these are optimal conditions (*51*) for the *T. reesei* enzymes, which constitute the reference system for our comparisons.

Endo/Exo Synergism Experiments. Figure 2 presents results from the 16 experiments carried out to evaluate and rank pairs of endoglucanases in terms of synergism and glucose release. The rankings obtained from these studies are presented in Table I. The millimolar values for soluble-sugar release shown in Figure 2 and Table I represent relatively small extents of conversion of the substrate; the highest glucose concentration, produced by a 20:80 mixture of *A. cellulolyticus* E1, corresponds to solubilization of approximately 13% of the substrate, and the endo/exo pair ranked 16th in terms of soluble-sugar release converted approximately 3.8% of the substrate when present in its optimal ratio.

Of the 16 endo/exo pairs presented in Figure 2 and Table I, nine are bacterial/fungal hybrids (one bacterial and one fungal enzyme). Seven pairs are formed within phylogenetic domains (five bacterial/bacterial pairs and two fungal/fungal pairs); of these, three pairs are homologous; i.e., formed using

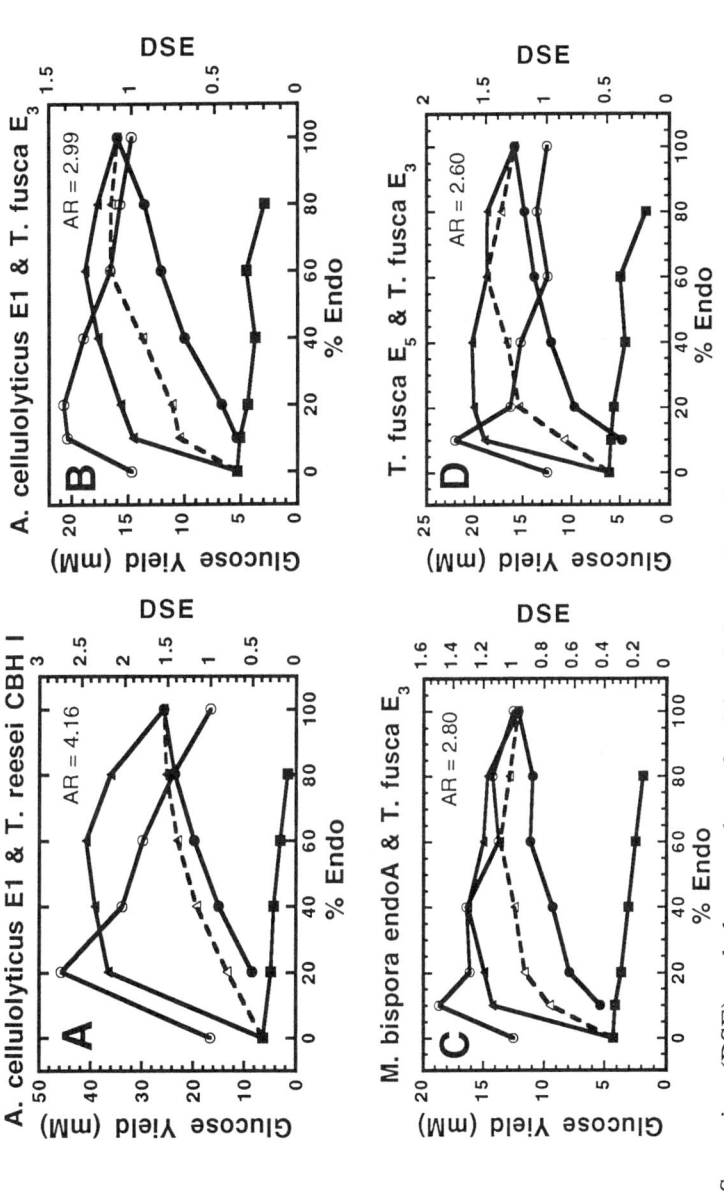

9. BAKER ET AL. *Synergism Between Purified Bacterial & Fungal Cellulases* 121

Figure 2. Synergism (DSE) and glucose-release for 16 endo/exo pairs, plotted as a function of cellulase mixture composition at a constant total cellulase loading of 7.2 nmol/g cellulose. Filled squares, glucose release by exoglucanase alone; filled circles, glucose release by endoglucanase alone; open triangles, sum of glucose-release values for endoglucanase and exoglucanase acting alone; filled triangles, actual glucose release by the mixture of endoglucanase and exoglucanase; open circles, degree of synergistic effect (DSE). Digestions were carried out at 50°C in 50 mM acetate, pH 5.0, for 120 h. The numerals shown in bold-face inside the plots are the ratios of the glucose release by a given endoglucanase, at "100%" loading, to the glucose release by the paired exoglucanase, also measured at "100%" loading. *Continued on next page*

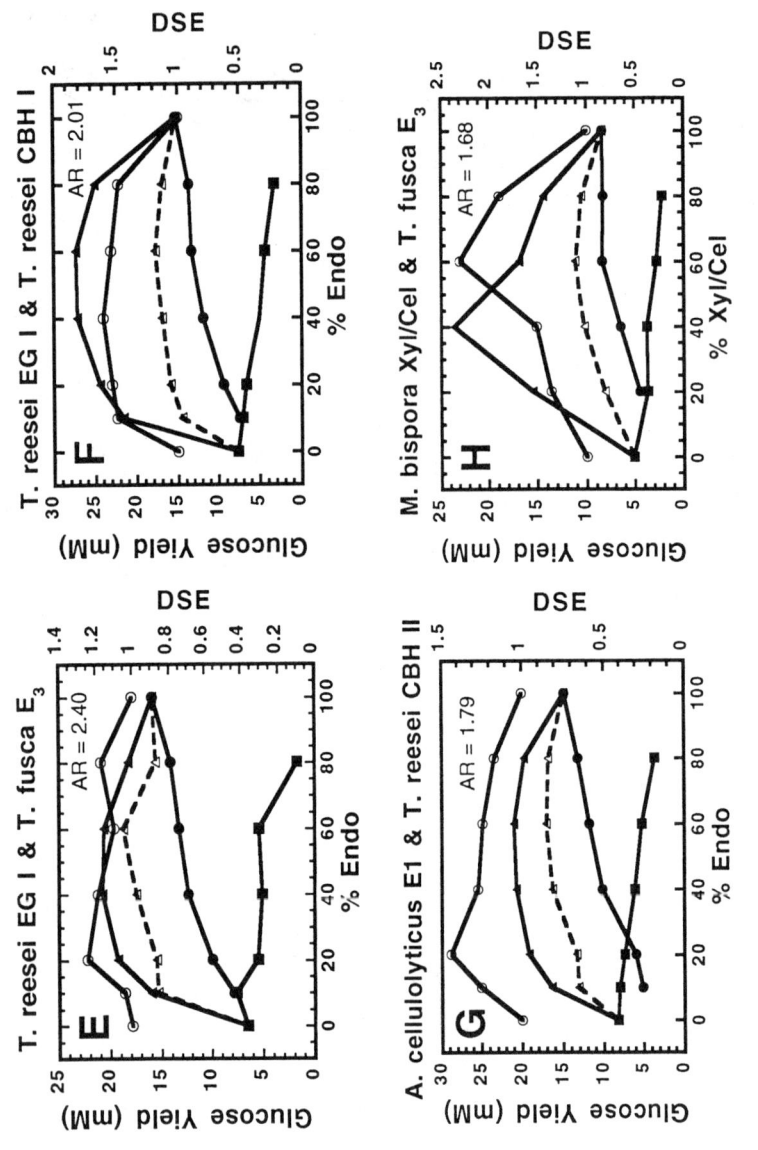

Figure 2. Continued.

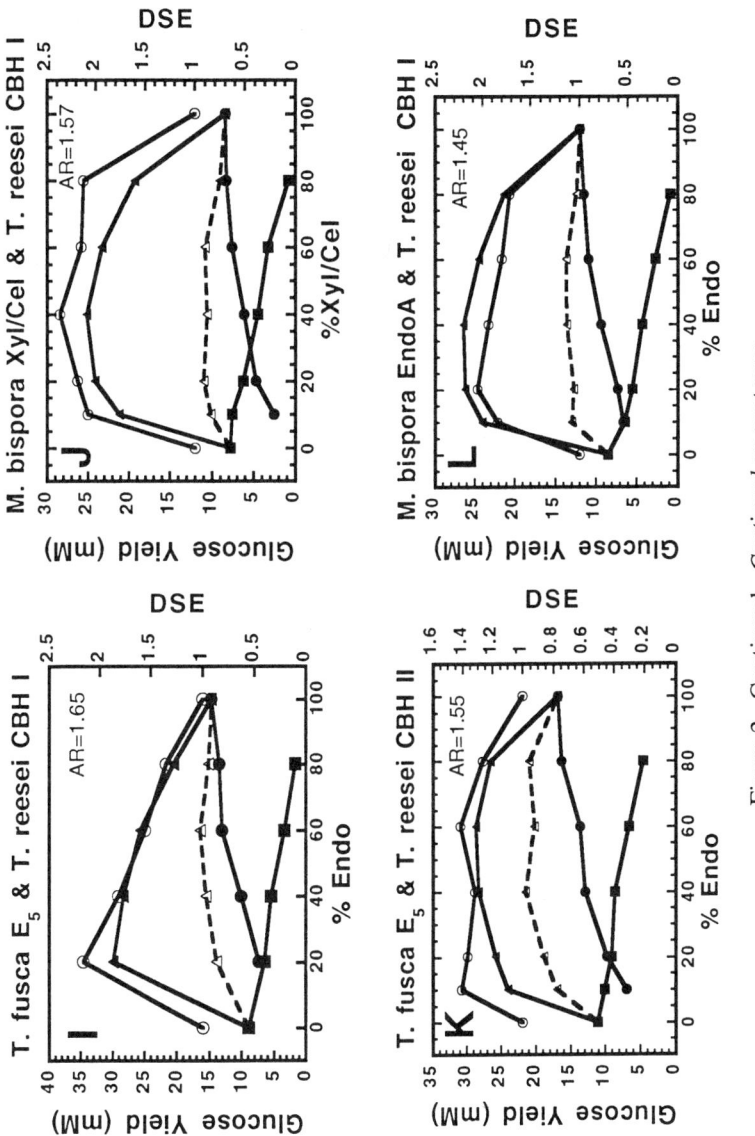

Figure 2. Continued. *Continued on next page.*

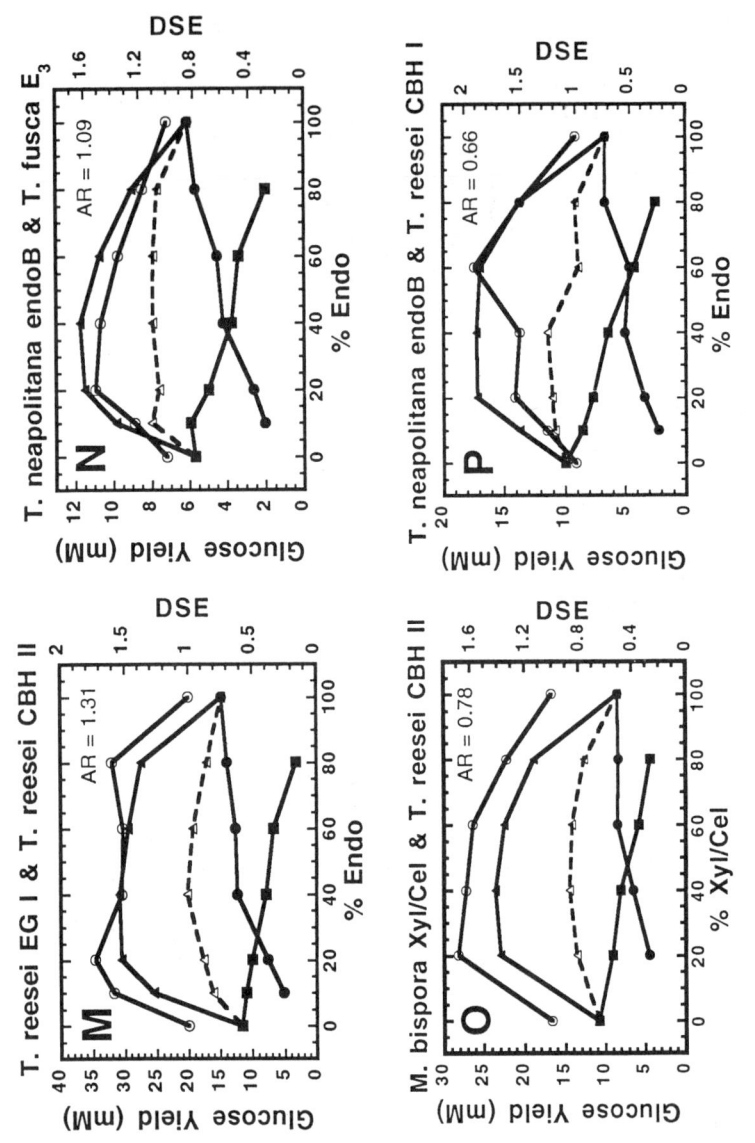

Figure 2. Continued.

Table I. Endo/Exo Cellulase Pairs Ranked in Order of Decreasing Synergism

	Enzyme Pair (Endo/Exo)	DSE	@ % Endo	Max. R.S.[a] (mM glucose)	@ % Endo	R.S. Ranking
1.	A. cellulolyticus E1 & T. reesei CBH I	2.75	20	40	60	1
2.	M. bispora Xyl/Cel & T.reesei CBH I	2.35	40	25	40	7
3.	M. bispora Xyl/Cel & T. fusca E$_3$	2.28	60	23	40	8
4.	T. fusca E$_5$ & T. reesei CBH I	2.16	60	30	20	3
5.	M. bispora Endo A & T. reesei CBH I	2.04	20	26	20-40	6
6.	T. neapolitana Endo B & T.reesei CBH I	1.91	20	17	20-60	14
7.	T. fusca E$_5$ & T. fusca E$_3$	1.76	10	20	20-40	12
8.	T. reesei EG I & T. reesei CBH II	1.72	20	31	20-40	2
9.	M. bispora Xyl/Cel & T. reesei CBH II	1.69	20	23	20-40	9
10.	T. reesei EG I & T. reesei CBH I	1.58	40	27	20-40	5
11.	T. neapolitana Endo B & T. fusca E$_3$	1.52	20	11	20-40	16
12.	M. bispora Endo A & T. fusca E$_3$	1.48	10	16	40	15
13.	A. cellulolyticus E1 & T.reesei CBH II	1.43	20	21	40-60	10
14.	A. cellulolyticus E1 & T. fusca E$_3$	1.38	10-20	18	60	13
15.	T. fusca E$_5$ & T. reesei CBH II	1.36	10-60	28	40-60	4
16.	T. reesei EG I & T. fusca E$_3$	1.25	20	20	40-60	11

[a] Reducing sugar (as glucose) measured after 120 h at pH 5.0, 50°C.

enzymes originating from the same organism. The intradomain and intraspecies pairs thus provide useful reference points for assessing the activity of the interdomain hybrids.

The performances of the bacterial/fungal hybrid pairs meet or exceed those of the intradomain and intraspecies pairs. Of the nine bacterial/fungal hybrid pairs (out of 16 pairs studied, overall), six were ranked among the nine most synergistic combinations, and six were among the top nine pairs in terms of soluble-sugar release. One of the more interesting observations in Table I is that in terms of both synergism and glucose release, the best-performing endo/exo pair is a hybrid composed of a bacterial endoglucanase, *A. cellulolyticus* E1, and a fungal exoglucanase, *T. reesei* CBH I. In terms of glucose release, two other bacterial/fungal hybrids, the third-ranked E_5/CBH I and the fourth-ranked E_5/CBH II, are essentially equivalent to the combinations of *T. reesei* EG I with its co-evolved exoglucanases.

The dominance of *T. reesei* CBH I in forming highly synergistic pairs is shown in Table I by the fact that five of the six most synergistic pairs contain CBH I, as do five of the top seven glucose-releasing pairs.

The individual data frames in Figure 2 are arranged in order of the descending ratio of the activity of the endoglucanase stock solution to that of the equimolar exoglucanase stock solution (activity ratio, AR), with the 120-h glucose release by the 100% endoglucanase and 100% exoglucanase loadings taken as the measure of the respective activities. (As discussed above, the cellulase loadings in all of these experiments were made to a constant *molarity*, not to a constant activity.) With the frames arranged in this manner, the only semblance of a trend observed is that enzyme pairs with high ratios of endoglucanase activity to exoglucanase (on the first page of the multi-page figure) tend to have reasonably sharp synergism maxima at low endoglucanase percentages, with broader glucose-release maxima significantly offset toward higher endoglucanase percentages, whereas for smaller ratios of endoglucanase to exoglucanase activity, the synergism and glucose-release maxima are closer to being coincident. Even this rather weak trend has its exceptions; frame 2H, *M. bispora* Xyl/Cel and *T. fusca* E_3, is an egregious pattern-breaker, and frame 2G may be as well. At the lower right corner of the fourth page of Figure 2, the rEndoB/CBH I pair, with an endo/exo activity ratio substantially smaller than 1.0, appears to reverse the pattern shown to by the enzyme pairs with high activity ratios, in that the synergism peak is found at the right-hand (high-endoglucanase) edge of the broad sugar-release maximum.

On consideration of the entire data set, it appears most likely that the differences noted are a function of the relative amounts of endoglucanase and exoglucanase activities loaded, rather than fundamental differences in mechanism. (In two alternative orderings (not shown) of the frames in Figure 2, the frames were arranged in order of either maximum DSE value or maximum sugar release; in neither case were any clear patterns discernible.)

The ranking of endo/exo pairs in terms of soluble-sugar release (Table I) is inevitably biased by the choice of a single temperature (50°C) for this study. This temperature was chosen to accommodate the less thermostable enzymes in the array, and therefore results in decreased activity measurements for enzymes with optima at higher temperatures, relative to the activities these enzymes would display at their optimal temperatures. Two of the endoglucanases in this study, *A. cellulolyticus* E1 and *M. bispora* rEndoB, have activity optima above 70°C (Himmel, M. E., et al., in this volume). E1 has a temperature/activity profile sufficiently broad that it retains one of the higher activities measured in this series, even when used some 25-30°C below its optimum, but the activity of rEndoB at 50°C is significantly depressed with respect to its optimal activity near 75°C. Both of these enzymes will have applications at temperatures at which the other enzymes in Table I cannot survive.

Exo/Exo Synergism Experiments. Figure 3(A-C) shows results for all possible binary combinations of the three exoglucanases used in this study. Here the exoglucanases are mixed according to the same protocol used for the endo/exo binary pairs, but without any endoglucanase being present.

Figure 3A demonstrates, as has been reported before (*19,23,48*), that CBH I and CBH II are strongly synergistic in the solubilization of crystalline cellulose. The maximum synergism ratio of 1.6 is observed at 40% CBH II. At the resolution of the array of compositions used in these experiments, this agrees well with the earlier finding that a 2:1 mixture of CBH I and CBH II produced maximum synergism, and serves to relate the present work to the earlier reports. E_3 displays (Figure 3B) even stronger synergism with CBH I than does CBH II, with a maximum synergism ratio of 2.16 at a composition of 60% E_3 and 40% CBH I.

In contrast to the results shown in the first two frames of Figure 3, mixtures of CBH II and E_3 (Figure 3C) do not show synergism; i.e., the synergism-ratio values are not greater than 1.0. The synergism ratios appear, in fact, to be somewhat less than 1.0 across the composition range, although the lowest values shown (0.92-0.93 in range of 40-80% CBH II) are close enough to 1.0 that the difference may not be significant given the noise levels in the data.

The fact that E_3 is synergistic with CBH I, but not with CBH II, is consistent with the conclusion by Irwin, et al., based on filter-paper synergism studies (*23*), that E_3 is "CBH II-like" in its specificity against cellulosic substrates; i.e., E_3 and CBH II both cleave cellobiosyl units from the non-reducing end of the cellulose chain, whereas CBH I attacks the reducing end of the chain. A reasonable interpretation of this phenomenon is that E_3 and CBH II compete for the same population of potential binding/hydrolysis sites, and therefore do not act synergistically. The DSE values for E_3 and CBH II do not fall much below 1.0, which may indicate that, under the conditions of these measurements, the substrate is not saturated with either exoglucanase or by the mixture. This, in turn, may be understandable in view of the physical nature of the substrate and the processive mode of attack proposed for cellobiohydrolases in general (*2*). The 20-μM cellulose particles, being aggregates of much smaller microcrystallites, may provide an abundance of chain ends, and the processive mode of attack would imply that once a cellobiohydrolase molecule has initiated attack at a given chain end, it will be occupied with that chain for a number of cleavage events before rejoining the pool of enzyme molecules diffusing to new initiation sites.

Endo/Endo Synergism Studies. Wood and co-workers have suggested that there may be two distinct classes of endoglucanases, each class (*14*) specific for one of two stereochemically-distinct types of potential cleavage sites presented on the

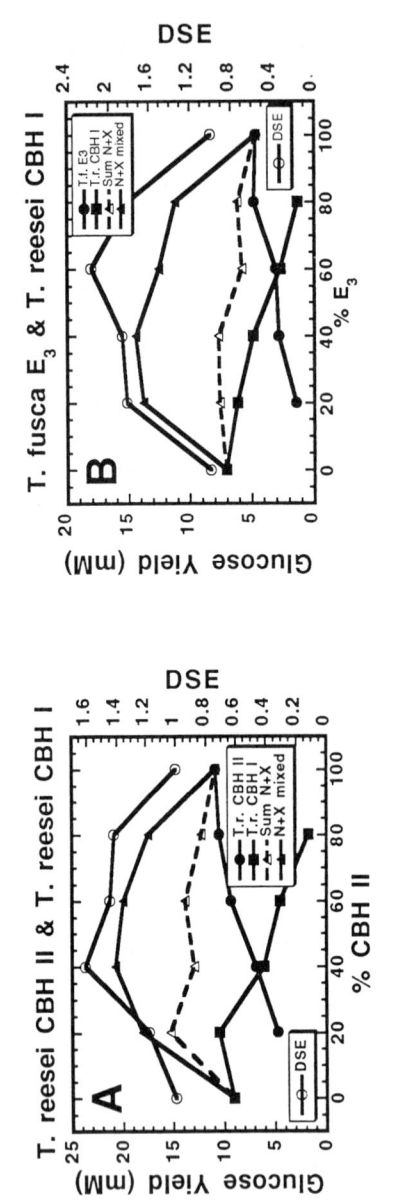

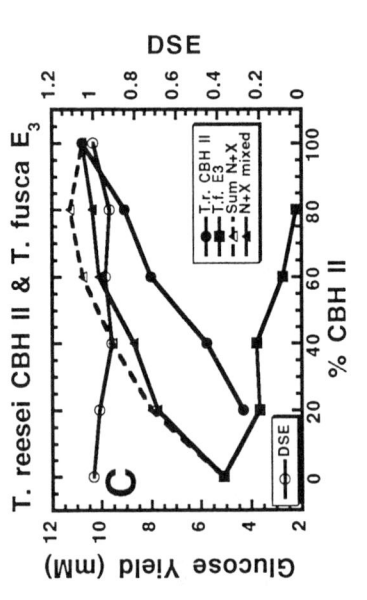

Figure 3. Synergism (DSE) and glucose-release for three pairs of exoglucanases. Digestion conditions and symbols as described for Figure 2.

surface of the cellulose crystal. In the present study, we sought to gather evidence for or against this hypothesis.

Figure 4(A-C) illustrates, in "round-robin" fashion, the performances of binary mixtures drawn from three endoglucanases: *A. cellulolyticus* E1, *T. fusca* E_5, and *T. reesei* EG I. In sharp contrast to all of the results for endo/exo pairs, and similar to the results for the E_3/CBH II exo/exo pair, the soluble-sugar productions for the binary mixtures of endoglucanases are seen to be *less* than the sums of the equivalent loadings of the two enzymes acting separately. The endo/endo pairs thus display values less than 1.0 for the synergism ratio (DSE). The interactions between the different pairs of endoglucanases are not identical, as may be seen in Figure 4(D), in which the synergism-ratio plots for the three pairs have been overlaid for direct comparison. For the mixture of E1 and EG I, the DSE values are only slightly below 1.0 (0.90 or higher across the composition range). For E1 and E_5, the value approaches 0.70 for compositions in the range of 60-70% E1. For the E_5/EG I pair, the values are still lower (near 0.60 across the middle composition range). Figure 5(A-C) shows the results of similar experiments pairing *M. bispora* xylanase/cellulase in turn with E1, E_5, and EG I, with the synergism ratios in the middle of the composition range being 0.90, 0.70, and 0.63, respectively.

A particularly striking feature of the results in Figure 4A, and to a somewhat lesser extent in Figure 5A, is that although the synergism ratios for the mixtures of endoglucanases are not greater than 1.0, the actual reducing-sugar release values for some mixtures are significantly greater than for equivalent molar loadings of either endoglucanase alone. In these two cases, the mathematical approach used to calculate synergism may mask a very real cooperativity.

Endoglucanase "Self-Synergism" as a Model for Enzymes with Identical Specificities. Some light can be cast on the results shown in Figures 4 and 5 by considering the results shown in Figure 6. Here *T. fusca* E_5 plays the role of both components of a "binary mixture," thus providing a model for the performance of mixtures of two hypothetical endoglucanases that have identical substrate specificities and kinetics. Whereas for the mixtures of different endoglucanases in Figures 4 and 5 results are shown only for a long (120-h) digestion time, for the E_5 "self-interaction" model system results are presented for successively longer digestion times (from 6.5 to 120 h). Even at the shortest digestion time (6.5 h), the synergism ratio is seen to be substantially below 1.0 over the entire range measured, decreasing to a value near 0.76 as the composition moves toward the middle of the range. At longer digestion times, the minimum values for the curves decrease to 0.65-0.70, and the curves flatten out, with the mid-range values becoming more typical of the values across the range of 20-80% "component 1."

One relatively simplistic explanation for the behavior of the "E_5 with E_5" model system, as shown in Figure 6, is that the two "components" of the mixture, being in fact exactly the same enzyme, are competing for exactly the same array of potential binding/hydrolysis sites on the substrate. If one assumes that the population of potential binding/hydrolysis sites is heterogeneous, consisting of sites differing widely in accessibility and/or reactivity, and that the enzyme/substrate ratio is not so high that the sites available at the beginning of the digestion are near

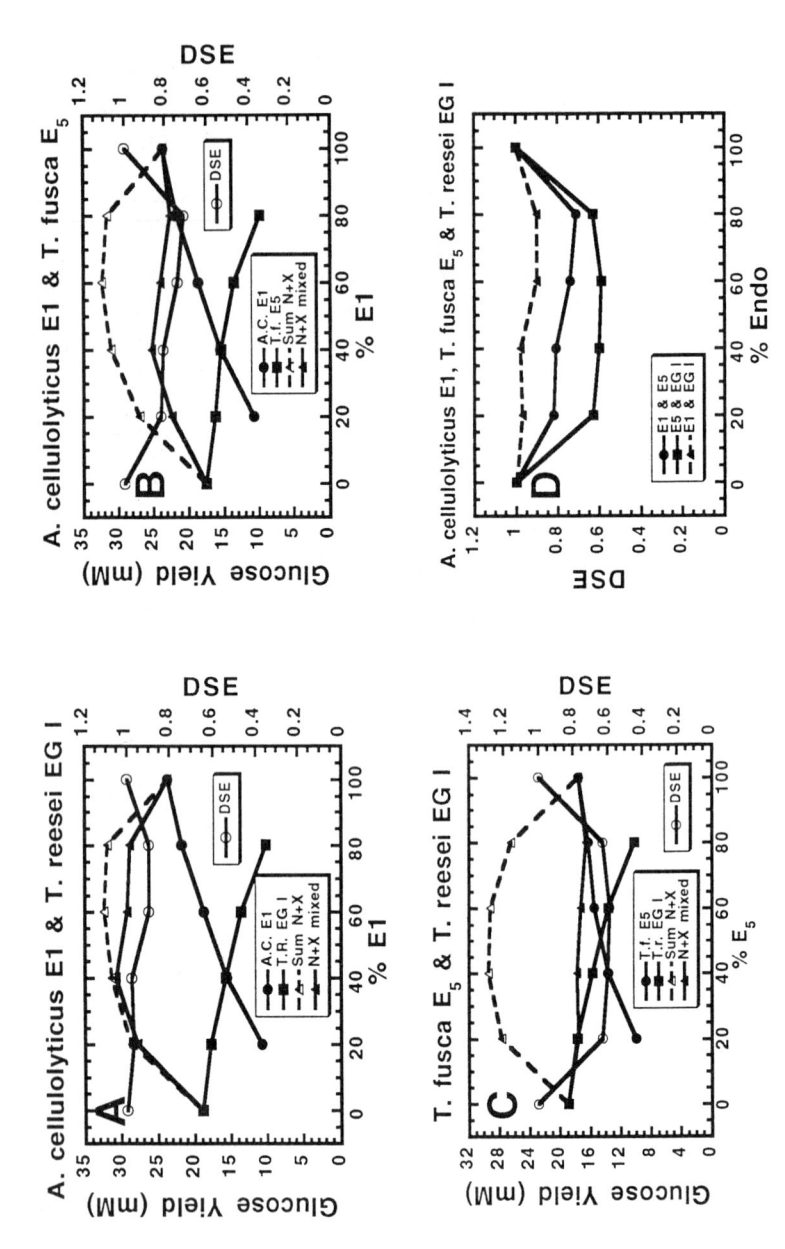

Figure 4. Frames (A-C): Synergism (DSE) and glucose release for three pairs of endoglucanases. Frame (D): Comparison of the DSE plots from frames a-c. Digestion conditions and symbols as described for Figure 2.

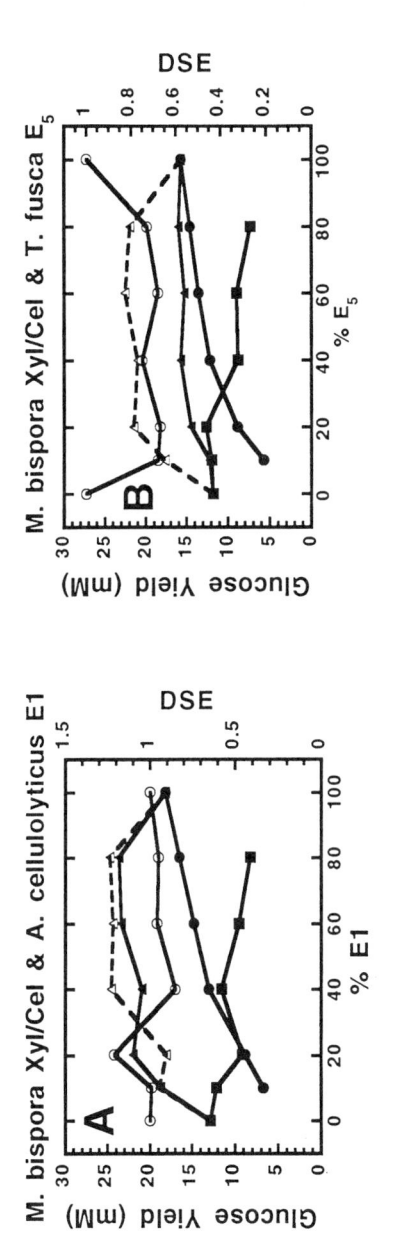

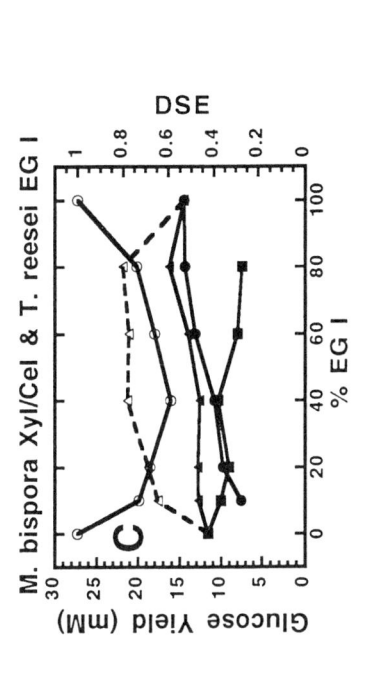

Figure 5. Synergism (DSE) and glucose release for three additional pairs of endoglucanases. Digestion conditions and symbols as described for Figure 2.

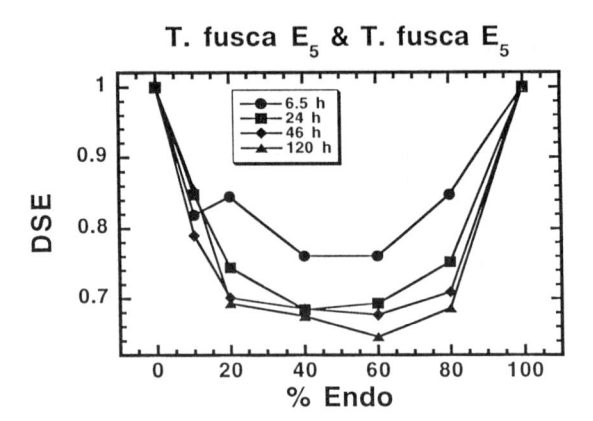

Figure 6. Synergism between *T. fusca* E$_5$ and itself. (Model system for synergism between identical endoglucanases.) The curves show DSE as a function of "mixture composition," for successively longer digestion times.

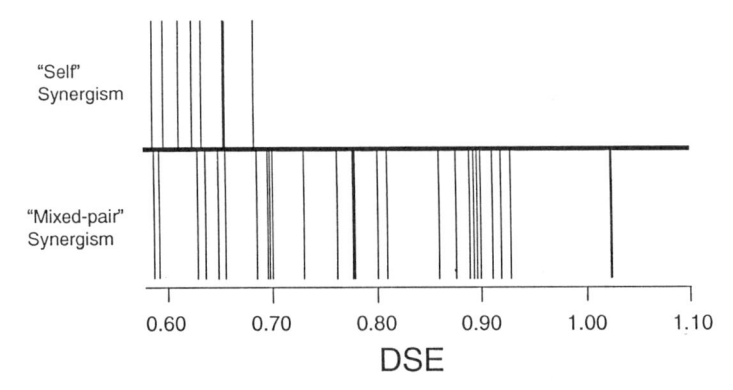

Figure 7. Graphical comparison between the ranges of DSE values found for pairing of eight endoglucanases with each other, and the range of values found for pairing each endoglucanase with itself. The positions of the vertical bars represent individual DSE values found for 1:1 mixtures of endoglucanases after 120-h digestions. (Values from Table II.)

saturation with enzyme, then one would expect to observe results of the type shown in Figure 6.

In the very early stages of the digestion, a large number of readily accessible, easily hydrolyzed sites are available to the (non-saturating) enzyme loading in each of the three digestion samples. In these early stages the identical enzyme molecules would not interfere with each other to any great extent, the soluble-sugar outputs of the "two components" operating together in the "mixture" sample would be nearly additive, and the synergism ratio observed would not be much less than 1.0. As the digestion progresses, and the abundance of the more readily accessible, easily hydrolyzed sites decreases (i.e., as the substrate becomes more recalcitrant), competition ("interference") between the enzyme molecules would be expected to increase. The observed synergism ratios would be expected to decrease further below 1.0, approaching a value of 0.5 for very long digestion times. For this model, therefore, a DSE value of 0.5 at long digestion times would be interpreted as indicating identical specificities for the two enzymes in a mixture. At the other extreme, a synergism ratio of 1.0 at long digestion times would indicate that the two enzymes in a mixture are specific for two completely distinct populations of binding/hydrolysis sites, and therefore operate completely independently.

Binary synergism values in the range (0.5 to 1.0) suggested by this model are not without precedent in the literature. Beldman, et al. (*45*) cited a DSE value of 0.97 for hydrolysis of Avicel by a 1:1 mass-ratio mixture of *T. reesei* Endo III with a known endoglucanase as conclusive evidence that Endo III was not an exoglucanase. Nidetzky, et al. (*48*) reported that no synergism was observed for combinations of *T. reesei* EG I and EG III (core) acting on a variety of cellulosic substrates, including microcrystalline cellulose, and noted that the activities of such combinations on the substrates were always less than the sums of the separate activities. Although Irwin et al. (*23*) used experimental and computational approaches somewhat different from those in the present study, it is worth noting that these workers, using filter paper as the substrate, found DSE values of 0.5, 0.7, and 0.5, respectively, for binary combinations of *T. fusca* endoglucanases ($E_1 + E_5$, $E_1 + E_2$, and $E_2 + E_5$).

Applying the above-described model to the results in Figures 4 and 5 would suggest that the four enzymes involved in these figures have differing specificities for the variety of stereochemically-distinguished potentially-hydrolyzable glycosidic bonds in the cellulose substrate, and that these different specificities overlap to different extents for the different binary combinations of enzymes. For example, the specificities of E1 and EG I (Figure 4A) might be seen as differing substantially, whereas the specificities of E_5 and EG I would be seen as being much more similar.

Other explanations that may invoked to explain the results of Figures 4 and 5. Two of these alternate explanations will be dealt with below, after consideration of some additional data (Table II and Figure 7) on interactions between endoglucanases in the solubilization of microcrystalline cellulose. Table II presents results for a series of experiments that extend the investigation of "endo-endo" interactions to all possible binary combinations that may be made of the eight endoglucanases dealt with in this study, including pairings of the enzymes with themselves, which will be shown to be very important as references in the interpretation of the other results. The enzyme pairs in Table II are ranked in

Table II. Synergism between Endoglucanases in 1:1 Mixtures

Enzyme Codes[a]	DSE	Mixture	Endo1	Endo2	Sum	Activity Ratio[c]
CG	1.02	5.2	2.4	2.7	5.1	0.89
DG	1.02	4.2	1.4	2.7	4.1	0.52
AC	0.92	5.4	3.1	2.8	5.9	0.90
AE	0.90	4.3	2.7	2.1	4.8	0.78
AH	0.90	4.3	2.8	2.0	4.8	0.71
FG	0.89	3.2	0.8	2.7	3.6	0.30
GH	0.89	4.1	2.7	2.0	4.6	0.74
EG	0.87	4.2	2.1	2.7	4.8	0.78
AF	0.86	3.0	2.7	0.8	3.5	0.30
EF	0.82	2.3	2.0	0.8	2.8	0.40
BF	0.81	2.5	2.3	0.8	3.1	0.35
CF	0.78	2.5	2.4	0.8	3.2	0.33
AB	0.78	4.4	3.1	2.6	5.7	0.84
BG	0.78	4.5	2.3	2.7	5.0	0.85
AD	0.78	3.1	2.5	1.9	4.4	0.76
DF	0.77	1.7	1.4	0.8	2.2	0.57
DE	0.77	2.7	1.4	2.1	3.5	0.67
BD	0.72	2.8	2.3	1.6	3.9	0.70
DH	0.70	2.3	1.4	2.0	3.3	0.70
AG	0.69	3.8	2.8	2.7	5.5	0.96
CE	0.69	3.1	2.4	2.1	4.5	0.88
CH	0.68	3.0	2.4	2.0	4.4	0.83
FH	0.64	1.8	0.8	2.0	2.8	0.40
BH	0.65	2.8	2.3	2.0	4.3	0.87
BE	0.64	2.8	2.3	2.1	4.4	0.91
CD	0.63	2.4	2.1	1.6	3.7	0.76
BC	0.60	3.1	2.6	2.6	5.2	1.00
EH	0.59	2.4	2.1	2.0	4.1	0.95
AA	0.67	3.0	2.2	2.3	4.5	0.96
BB	0.65	2.4	1.9	1.8	3.7	0.95
CC	0.62	2.3	1.8	1.9	3.7	0.95
DD	0.62	1.8	1.5	1.4	2.9	0.93
EE	0.59	2.0	1.7	1.7	3.4	1.00
FF	0.61	1.1	0.9	0.9	1.8	1.00
GG	0.65	2.8	2.2	2.1	4.3	0.95
HH	0.60	2.1	1.9	1.6	3.5	0.84

[a] A = *A. cellulolyticus* E1 fr. *S. lividans*; B = *T. fusca* E_5; C = *T. reesei* EG I; D = *M. bispora* Xyl/Cel; E = *T. reesei* EG II; F = *T. neapolitana* rEndoB; G = *A. cellulolyticus* E1 "CAT" fr. *E. coli*; H = *M. bispora* Endo-A.
[b] Averages of duplicate sugar determinations on each of duplicate digestion mixtures.
[c] Values shown are the ratios of the smaller of the two separate activities (as measured at "50%" loading) to the larger.

descending order of the calculated synergism ratios, upon which the discussion will focus. Table II also presents the data for sugar release from which the synergism ratios were calculated. This additional information is presented because (1) even though the relative productivities of the single enzymes and mixtures are not discussed in detail here, this information is useful in assessing the synergism numbers, and (2) presenting the data for soluble-sugar release by the individual enzymes at 50% loadings serves to eliminate the possibility that in some cases the activity of one member of the pair might be negligible with respect to the activity of the other, which would yield completely trivial values for the synergism ratio. The last column in the table displays the ratios between the separate, 50% loadings for the various pairings, demonstrating that, even though the ratios of the individual activities vary considerably from pair to pair, in no case is one activity negligible with respect to the other. (For the "self-interaction" pairings listed at the bottom of the table, the two "components" of the "mixture" are the same enzyme, so that if the data were perfect, for these pairs the activity ratios would all be 1.0.)

In Table II, several different values are given for the "separate" activities of some of the enzymes, depending on the pair in which the enzyme is participating. (The series from AF to BF is a case in point.) These differences are a result of the pairs making up the series being evaluated in different batches, at different times, some using enzyme preparations at different ages, and some using different preparations of the enzyme in question. Efforts were made to make the "solo" assays true replicates, in that the same protein concentrations were used throughout, but, as seen in the table, there were some differences in activity. Rather than attempting to average the "separate" activities for each enzyme for the sake of consistency in the table as a whole, the authors have chosen to compare the activities for mixtures with the activities measured for the separate enzymes in the same batch of digestions, run at the same time. *Some* of the separately-measured activities for a given enzyme are identical for different pairings. This is because the sugar-release values for all separate-enzyme measurements made *in a given batch* of experiments *were* averaged, and the results used for comparison with the activities of all mixtures that included that enzyme and were evaluated in the same batch of digestions. The "self-interaction" separate activities referred to above (Figure 2, lower section) provide an indication of the repeatability of digestions run in the same batch; with the exception of the HH pair, the replicates are repeatable within 5-7%.

The most significant observation from the synergism ratios shown in Table II is that the range of values covered by the data for mixtures of different endoglucanases is much wider than the range covered by the values for enzymes "paired" with themselves. The synergism ratios for equimolar mixtures of different endoglucanases range from a high of 1.02 ($\approx$1.0), which is interpreted as representing the behavior of a pair enzymes specific for completely separate, independent sets of binding/catalysis sites, to a low of 0.59, which suggests substrate specificities that overlap to a large extent and therefore approach the limiting case of enzymes with identical substrate specificities. When the same enzyme is used as both components of the mixture, however, the synergism ratios are all clustered between 0.59 and 0.67, or in the bottom one-third of the range covered by the pairs of different enzymes. This comparison is presented graphically in Figure 7.

Alternatives to the "Differing-Specificity" Model. It was noted above that factors other than simple differences in substrate specificity might contribute to the different synergism values measured for different endoglucanase pairs. Consideration of the observation described in the paragraph above would appear to dispose of one possible alternative explanation. The possibility in question is that the observed differences in synergism numbers do not reflect differing degrees of overlap of substrate specificity for the various endo/endo pairs, but instead are a function of differences in the magnitudes of kinetic parameters (binding constants and catalytic rate constants) for different enzymes. According to this picture, enzymes with low binding affinity, but relatively high catalytic rates once bound to the substrate surface, could produce substantial rates of sugar release while not physically saturating the substrate surface. In other words, the endoglucanases would spend most of the time in solution, rather than on the surface where they would sterically block the approach of other enzyme molecules to nearby cleavage sites. If one or more of the enzymes in a digestion mixture were of this type, then the two enzymes would be expected to interfere with each other much less than would a pair of enzymes, both of which had a relatively high binding affinity and low catalytic rate. For the same rate of sugar release, this latter hypothetical pair of enzymes would spend much more time on the substrate surface.

This kinetically-controlled saturation model could account for the range of synergism ratios shown in Table II for mixtures of different endoglucanases, and without invoking any differences in the populations of sites attacked by the two enzymes, were it not for the data for "pairs" composed of a given enzyme and itself. If the high (i.e., near 1.0) synergism ratios observed for pairs such as CG and DG were attributable to at least one of each pair being a weak-binding, rapidly-catalyzing enzyme that spends most of the time free in solution, that enzyme would be expected to be at least as "non-interfering" when paired with itself. This is not at all what is shown in the data for the enzymes paired with themselves. The GG, CC, and DD pairs have synergism ratios of 0.65, 0.62, and 0.62, respectively, all far below the value of 1.02 posted by both CG and DG.

Another possible alternative to a difference in substrate specificity as an explanation for the observed variability of endo/endo synergism cannot be so readily dismissed. This is the possibility that there may be direct physical interactions between enzyme molecules. It has been reported that *T. reesei* EG I can form dimers (52), and that complexes of as many as six glycohydrolases can be found in culture broths of *T. reesei* (53). Formation of homo- or heterodimers (or higher aggregates) from the endoglucanase mixtures would reduce the effective concentration of endoglucanase available for reaction with the substrate, and this effect would become more pronounced at higher enzyme concentrations. If the surface characteristics of some of the endoglucanases promote the formation in binary mixtures of heterodimers as well as homodimers, with the binding affinity for heterodimer formation varying with the identity of the endoglucanases involved, this might explain the variable (apparent) interference observed across the array of endoglucanase pairs. This alternate explanation does not lend itself either to proof or disproof by means of the approaches employed in this paper; more direct physical investigations (such as sedimentation-equilibrium studies by analytical centrifugation) are needed to address this question.

Influence of the Cellulose-Binding Domain on Endo/Endo Synergism.
Another comparison that is interesting in connection with the endo/endo "self-synergism" results is that between enzyme "A" (*A. cellulolyticus* E1 as expressed in *S. lividans*) and enzyme "G" (the product of the same gene expressed in *E. coli*). These two proteins are represented in the experimental series as distinct enzymes because, while *S. lividans* expresses the entire enzyme (catalytic domain, linker domain, and CBD), the *E. coli* product consists of the catalytic domain only, the CBD and most of the linker region apparently having been removed by post-translational proteolytic cleavage. This pair of enzymes therefore provides an opportunity to evaluate the effect of the CBD on the behavior being observed in this experimental series. The effect of the presence or absence of the CBD appears to be minimal in this context. Whereas both E1 versions, A and G, are relatively non-interfering enzymes, with G displaying an average synergism ratio of 0.93 when paired with all the other endoglucanases (except A), and A averaging 0.86 when paired with all the other endoglucanases (except G), the pair AG has a synergism ratio of 0.69, which is only slightly above the upper end of the range shown by the enzymes paired with themselves. (AA and GG have synergism ratios of 0.67 and 0.65, respectively.) In determining the degree of interference or non-interference between these two forms of E1, the fact that the two forms have identical *active sites* appears to be much more important than the fact that one has, and one completely lacks, a CBD.

Suggested Grouping of Endoglucanases Based on Endo/Endo Synergism Results. Figure 8 illustrates a method by which the relative synergism numbers for binary pairs amongst the eight endoglucanases might be used to divide the endoglucanases into functional categories. The dendrogram shown in this figure was constructed by applying hierarchical agglomerative cluster analysis (54-56) to the "synergism" data of Table II and Figure 7, with the controlling assumption being that lower "synergism" ratios are correlated with greater similarity of substrate specificity. At the level of greatest similarity, *M. bispora* EndoA would appear to be very similar in specificity to *T. reesei* EG II, and *T. fusca* E_5 to *T. reesei* EG I. At the other end of the similarity scale, the two *A. cellulolyticus* E1 expression products, taken together, would be seen as having far less overlap of substrate specificity with the other six enzymes than the other six have with each other. It will be noted that the groupings produced by the cluster analysis cut across the boundaries of cellulase "families" as determined from sequence analysis (57-59); this is not surprising, because the present clustering technique forms groups based on *function* and not on amino acid sequence.

The authors recommend caution with regard to taking the finer divisions of the cluster-analysis dendrogram too literally, because, as shown by the dashed vertical lines in Figure 8, the range of values for "synergism" between identical (i.e., the same) enzymes covers the values for linkage of B and C with E and H, as well as the linkage values between the individual members of these two groups. Thus, it is not clear whether B, C, E, and H should compose two groups (BC and EH), as shown, or a single group (BCEH) of four enzymes. In addition, the linkage level between the BCEH group and "D" (*M. bispora* Xyl/Cel) is not far removed from the high-value edge of the "same versus same" range. This analysis may actually

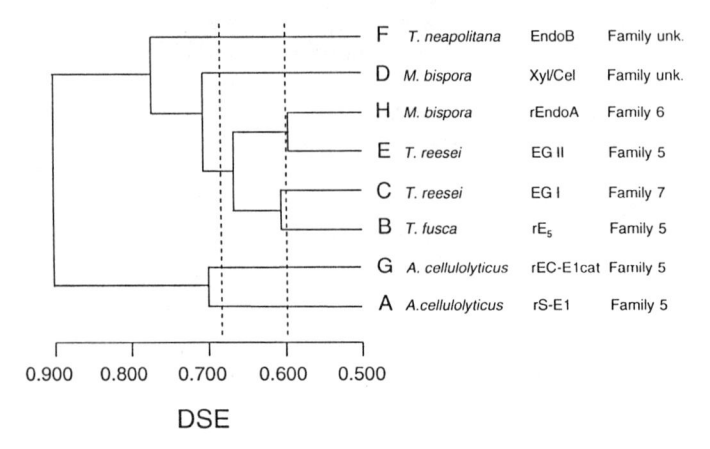

```
          F  T. neapolitana   EndoB       Family unk.
          D  M. bispora       Xyl/Cel     Family unk.
          H  M. bispora       rEndoA      Family 6
          E  T. reesei        EG II       Family 5
          C  T. reesei        EG I        Family 7
          B  T. fusca         rE₅         Family 5
          G  A. cellulolyticus rEC-E1cat  Family 5
          A  A. cellulolyticus rS-E1      Family 5
```

0.900   0.800   0.700   0.600   0.500
DSE

Figure 8. Groupings of endoglucanases based on possible similarities in substrate specificity. The dendrogram shown is the result of agglomerative hierarchical cluster analysis, based on the assumption that low values of synergism between pairs of endoglucanases are correlated with high degrees of similarity between their substrate specificities. The vertical dashed lines delimit the range of DSE values shown by combinations of the endoglucanases with themselves.

suggest only three reasonably robust clusters of enzymes: (1) *A. cellulolyticus* rS-E1/rEC-E1CAT, (2) *T. neapolitana* rEndoB, and (3) all remaining enzymes in the array. Interpretations are complicated by the fact that the substrate is physically heterogeneous and changes during the digestion, thus producing complex kinetics, and by the fact that, for the endoglucanases produced from native culture broths or commercial preparations, it is difficult to prove that the last trace of any contaminating cellulase has been removed. These groupings of endoglucanases, and the suggestion that the observed endo/endo synergism patterns may be ascribed to differences in substrate specificity, must therefore be regarded as preliminary. The discussion of these interactions is presented here to illustrate an approach to functional cellulase classification that may prove useful in future studies.

Conclusions

The high degrees of synergism observed for bacterial/fungal hybrid endo/exo pairs imply that a wide variety of choices are available to industrial researchers who wish to tailor enzyme mixtures to specific applications. The inclusion of a larger number of exoglucanases in comparisons such as that described here presents itself as the most effective and efficient route to expanding the list of useful possibilities even further; with *T. fusca* E_3 in the collection as a "CBH II-like" exoglucanase, the highest priority is to add a "CBH I-like" enzyme to the comparisons.

The proposal that endoglucanases may be classified in terms of the degree of overlap between their substrate specificities, and that this overlap may be evaluated by means of endo/endo synergism experiments of the type presented here, must be regarded as preliminary. The results and analyses used here to illustrate the concept do, however, leave open the possibility that endoglucanases may be classifiable into more than one group on a fundamental functional basis, perhaps in a manner analogous to that found useful for the exoglucanases.

Acknowledgments

This work was funded by the Ethanol from Biomass Program of the Biofuels System Division of the U.S. Department of Energy. The authors are indebted to R. Meglen for valuable advice concerning cluster analysis.

Literature Cited

1. Hall, D.O. *Solar Energy* **1979**, *22*, 307-328.
2. Coughlan, M.P. In *Microbial Enzymes and Biotechnology, 2nd Edition;* Fogarty, W.M.; Kelley, C.T., Eds.; Elsevier Applied Science: London, 1990; pp 1-36.
3. Grohmann, K.; Wyman, C.E.; Himmel, M.E. In *Emerging Technologies for Materials and Chemicals from Biomass*; Rowell, R.M.; Schultz, T.P.; Narayan, R., Eds.; ACS Symposium Series No. 476; American Chemical Society: Washington, DC, 1992; pp 354-392.
4. Reese, E.T.; Siu, R.G.H.; Levinson, H.S. *J. Bacteriol.* **1950**, *59*, 485-497.
5. Gilligan, W.; Reese, E.T. *Can. J. Microbiol.* **1954**, *1*, 90-107.
6. Eveleigh, D.E. *Phil. Trans. Roy. Soc. Lond.* **1987**, *A321*, 435-447.
7. Mandels, M.; Reese, E.T. *Devel. Indust. Microbiol.* **1964**, *5*, 5-20.
8. Li, H.; Flora, R.M.; King, K.W. *Arch. Biochem. Biophys.* **1965**, *111*, 439-447.
9. Selby, K. In *Cellulases and Their Actions*; Hajny, G.J.;Reese, E.T. Eds.; Advances in Chemistry Series 95, American Chemical Society: Washington, DC, 1969; pp 34-50.
10. Wood, T.M. *Biochem. J.* **1969**, *115*, 457-&&&.
11. Wood, T.M. *Biochem. Soc. Trans.* **1985**, *13*, 407-410.
12. Halliwell, G.; Riaz, M. *Biochem. J.* **1970**, *116*, 35-42.
13. Wood, T.M.; McCrae, S.I. *Carbohydr. Res.* **1977**, *57*, 117-133.
14. Wood, T.M.; McCrae, S.I. *Adv. Chem. Ser.* **1979**, *181*, 181-209.
15. Eriksson, K.E. In *Symposium on Enzymatic Hydrolysis of Cellulose*; Bailey, M.; Enari, T.-M.; Linko, M., Eds., SITRA: Helsinki, pp 263-280.
16. Pettersson, L.G. In *Symposium on Enzymatic Hydrolysis of Cellulose*; Bailey, M.; Enari, T.-M.; Linko, M., Eds., SITRA: Helsinki, pp 255-261.
17. McHale, A.; Coughlan, M.P. *FEBS Lett.* **1980**, *117*, 319-322.
18. Moloney, A.P.; McCrae, S.I.; Wood, T.M.; Coughlan, M.P. *Biochem. J.* **1985**, *225*, 365-374.
19. Fägerstam, L.G.; Pettersson, L.G. *FEBS Lett.* **1980**, *119*, 97-100.
20. Claeyssens, M.; van Tilbeurgh, H.; Tomme, P.; Wood, T.M.; McCrae, S.I. *Biochem. J.* **1989**, *261*, 819-826.

21. Teeri, T.T.; Koivula, A.; Reinikainen, T.; Ruohonen, L.; Srisodsuk, M. "Hydrolysis of Crystalline Cellulose by Native and Engineered *Trichoderma reesei* Cellulases," Presented at The Symposium on Enzymatic Degradation of Insoluble Polysaccharides, at the 1994 Annual American Chemical Society Meeting, San Diego, CA, March 13-17, 1994.
22. Wood, T.M.; McCrae, S.I.; Wilson, C.A.; Bhat, K.M.; Gow, L.A. In *Biochemistry and Genetics of Cellulose Degradation*; Aubert, J.-P.; Beguin, P.; Millet, J. Eds.; Academic Press: London, 1988; pp 31-52.
23. Irwin, D.C.; Spezio, M.; Walker, L.P.; Wilson, D.B. *Biotechnol. Bioeng.* **1993**, *42*, 1002-1013.
24. Walker, L.P.; Belair, C.D.; Wilson, D.B.; Irwin, D.C. *Biotechnol. Bioeng.* **1993**, *42*, 1019-1028.
25. Baker, J.O.; Adney, W.S.; Nieves, R.N.; Thomas, S.T.; Wilson, D.B.; Himmel, M.E. *Applied Biochem. Biotechnol.* **1994**, *45/46*, 245-256.
26. Wilson, D.B. *Methods Enzymol.* **1988**, *160A*, 314-323.
27. Walker, L.P.; Wilson, D.B.; Irwin, D.C.; McQuire, C.; Price, M. *Biotechnol. Bioeng.* **1992**, *40*, 1019-1026.
28. Yablonsky, M.D.; Bartley, T.; Elliston, K.O.; Kahrs, S.K.; Shalita, Z.P.; Eveleigh, D.E. In *Biochemistry and Genetics of Cellulose Degradation*; Aubert, J.-P.; Beguin, P.; Millet, J. Eds.; Academic Press: London, 1988; pp 249-266.
29. Wu, Y., M.S. Thesis, Rutgers University, May, 1994.
30. Bok, J.D.; Goers, S.K.; Eveleigh, D.E. in *Enzymatic Conversion of Biomass for Fuels Production*; Himmel, M.E.; Baker, J.O.; Overend, R.P. Eds.; ACS Symposium Series No. 566, American Chemical Society: Washington, DC, 1994; pp 54-65.
31. Himmel, M.E.; Adney, W.S.; Grohmann, K.; Tucker, M.P. United States patent No. 5,275,944; entitled "Thermostable Purified Endoglucanase from *Acidothermus cellulolyticus*"; **1994**.
32. Nieves, R.A.; Himmel, M.E.; Todd, R.J.; Ellis, R.P. *Appl. Environ. Microbiol.* **1990**, *56*, 1103-1108.
33. Shoemaker, S.; Watt, K.; Tsitowsky, G.; Cox, R. *Bio/Technology* **1983**, *1*, 687-690.
34. van Tilbeurgh, H.; Bhikhabhai, R.; Pettersson, L.G.; Claeyssens, M. *FEBS Lett.* **1984**, *169*, 215-218.
35. Doner, L.W.; Irwin, P.L. *Anal. Biochem.* **1992**, *202*, 50-53.
36. Henrissat, B; Driguez, H.; Viet, C.; Schulein, M. *Bio/Technology* **1985**, *3*, 722-726.
37. Immergut, E.A.; Ranby, B.G. *Indust. Engin. Chem.* **1956**, *48*, 1183-1189.
38. Immergut, E.A. In *The Chemistry of Wood*; Browning, B.L., Ed.; Wiley-Interscience : New York, 1963; pp 103-190.
39. Tarchevsky, I.A.; Marchenko, G.N. *Cellulose: Biosynthesis and Structure*; Springer-Verlag: Berlin, 1991; p 274.
40. Frey-Wyssling, A. *Science* **1954**, *119*, 80-82.
41. Abuja, P.M.; Pilz, I.; Claeyssens, M.; Tomme, P. *Biochem. Biophys. Res. Commun.* **1988**, *156*, 180-185.

42. Esterbauer, H.; Hayn, M.; Abuja, P.M.; Claeyssens, M. In *Enzymes in Biomass Conversion*; Leatham, G.F.; Himmel, M.E., Eds.; ACS Symposium Series 460, American Chemical Society: Washington, DC, 1991; pp 301-312.
43. Sagar, B.F. In *Cellulose and Its Derivatives: Chemistry, Biochemistry and Applications*; Kennedy, J.F.; Phillips, G.O.; Wedlock, D.J.; Williams, P.A., Eds.; John Wiley & Sons: New York, 1985, pp 199-207.
44. Grethlein, H.E. *Bio/Technol.* **1985**, *3*, 155-160.
45. Beldman, G.; Voragen, A.G.J.; Rombouts, F.M.; Pilnik, W. *Biotechnol. Bioengin.* **1988**, *31*, 173-178.
46. Woodward, J.; Lima, M.; Lee, N.E. *Biochem. J.* **1988**, *255*, 895-899.
47. Woodward, J.; Hayes, M.K.; Lee, N.E. *Bio/Technol.* **1988**, *6*, 301-304.
48. Nidetzky, B.; Hayn, M.; Macarron, R.; Steiner, W. *Biotechnol. Lett.* **1993**, *15*, 71-76.
49. Converse, A.L.; Optekar, J.D. *Biotechnol. Bioengin.* **1993**, *42*, 145-148.
50. Rouland, C.; Civas, A.; Renoux, J.; Petek, F. *Comp. Biochem. Physiol.* **1988**, *91B*, 459-465.
51. Schulein, M. *Methods Enzymol.* **1988**, *160*, 221-234.
52. Dominguez, J.M.; Pettersson, G.; Acebal, C.; Jimenez, J.; Macarron, R.; de la Mata, I.; Castillon, M.P. *Biotechnol. Appl. Biochem.* **1992**, *15*, 236-246.
53. Sprey, B.; Lambert, C. *FEMS Microbiol. Lett.* **1983**, *18*, 217-222.
54. Anderberg, M.R. *Cluster Analysis for Applications*; Academic Press: New York, 1973; pp 131-155.
55. Massart, D.L.; Kaufman, L; *The Interpretation of Analytical Chemical Data by the Use of Cluster Analysis*; Wiley-Interscience: New York, 1983; pp 75-99.
56. Davis, J.C. *Statistics and Data Analysis in Geology, 2nd Ed.*; John Wiley & Sons: New York, 1986; pp 502-514.
57. Henrissat, B.; Claeyssens, M.; Tomme, P., Lemesle, L.; Mornon, J.P. *Gene* **1989**, *81*, 83-95.
58. Gilkes, N.R.; Henrissat, B.; Kilburn, D.G.; Miller, R.C., jr.; Warren, R.A.J. *Microbiol. Rev.* **1991**, *55*, 303-315.
59. Henrissat, B. *Biochem. J.* **1991**, *280*, 309-316.

RECEIVED August 17, 1995

Chapter 10

Cellulose-Binding Domains: Classification and Properties

Peter Tomme, R. Antony J. Warren, Robert C. Miller, Jr., Douglas G. Kilburn, and Neil R. Gilkes[1]

Department of Microbiology and Immunology and Protein Engineering Network Centres of Excellence, University of British Columbia, 300–6174 University Boulevard, Vancouver, British Columbia V6T 1Z3, Canada

> More than one hundred and twenty cellulose-binding domains (CBDs) are classified into ten families (I-X). Three families (V, VII and VII) contain only single representatives. Two of the larger families (II and III) can be divided into distinct subfamilies. Most of the CBDs are from cellulases and xylanases but some are from other hydrolases or from non-catalytic proteins. Family I CBDs are all from fungi. With the exception of families VI and XIII, the remaining the families contain only representatives from bacteria. CBDs are found at the N or C termini of proteins or are internal. Representatives of certain families show a tendency to occur at terminal or internal positions and to be associated with representatives of particular catalytic domain families. Possible relationships between the the classification scheme and the functional properties of the CBDs are discussed.

The efficient enzymatic degradation of biomass appears to requires a tight interaction between the insoluble, highly structured cellulosic and hemicellulosic matrices and the various polysaccharidases involved in their hydrolysis. Many cellulases and hemicellulases (e.g., xylanases and mannanases) have special features which facilitate this intimate association. Typically, these enzymes are composed of a catalytic domain containing the active site for substrate hydrolysis, and a carbohydrate-binding domain or cellulose-binding domain (CBD) for binding to the insoluble matrix. This modular organization is shared by other polysaccharidases involved in the hydrolysis of insoluble, polymeric substrates, e.g., chitinases and amylases (1, 2)

The presence of two distinct domains in the cellulases from the fungus *Trichoderma reesei* and the bacterium *Cellulomonas fimi* was first demonstrated by limited proteolysis (3-5). The individual domains were released as discrete functional units because of the proteolytic susceptibility of the connecting linker sequence. Subsequently, the domain structures of many cellulases and xylanases have been probed by biochemical and genetic analyses. Domains in newly discovered hydrolases can usually now be recognized simply by comparison of their primary structures with those of enzymes whose structural and functional organization has been defined. Functional studies can then proceed by expression of appropriate gene fragments and biochemical analysis of the corresponding polypeptides. However, for many enzymes, functional characterization of individual domains, particularly CBDs, is still lacking.

[1]Corresponding author

The availability of a large number of cellulase and hemicellulase sequences, has enabled their classification into several families based on sequence similarities of the catalytic domain alone (*6*). Revisions of the original classification scheme have expanded the original six families of cellulase and xylanase catalytic domains to twelve (*7-10*). Functional properties, such as the stereoselectivity and substrate specificities of representative family members, have been correlated with this catalytic domain classification (*11, 12*).

A similar family classification for CBDs was originally proposed by Gilkes *et al.* (*7*) and several expanded classification schemes were subsequently published (*9, 13*). The relationship of these previous schemes to the current classification is summarized in Table I. At present, ten CBD families can be distinguished (Table II). An increasing awareness of the importance of these domains in the enzymatic degradation of biomass and of their potential applications in biotechnology and industry has provoked renewed interest in CBDs and it is anticipated that new family representatives and perhaps additional CBD families will appear in the near future.

Functions of Cellulose-Binding Domains

It is difficult to ascribe precise roles to CBDs for several reasons. These include our currently limited understanding of the mechanism of substrate binding, conflicting reports of the properties of CBDs and incomplete biochemical characterization. Furthermore, most studies have focused on the qualitative and quantitative analysis of binding to purified, insoluble celluloses and xylans and have overlooked other less obvious roles. Nevertheless, it is clear that removal of the CBD often reduces enzyme activity on insoluble substrates (*3, 5, 7, 14, 15*). CBDs could enhance activity by increasing the local enzyme concentration on the substrate surface, or by disrupting non-covalent interactions, thereby increasing substrate accessibility (*16, 17*). Either mechanism could involve the targeting of enzymes to distinct regions of the substrate by CBDs with different specificities. We cannot assume that all CBDs have the same function and the roles of CBDs in some families may prove to be different from those in others. The classification of CBDs into families is a first step towards discerning and understanding these possible differences and may provide new insights into the mechanism of biomass degradation.

Classification of Cellulose-Binding Domains

Representative sequences for each recognized CBD family were used to scan the GenBank, SWISS-PROT, EMBL and PIR sequence databases using the BLAST program with the PAM or BLOSUM weight matrices (*18*). Retrieved sequences were aligned using the multiple alignment program AllAll (National Center for Biotechnology Information) using a gap penalty of 5. Each alignment was then edited manually for maximal fit.

More than a hundred and twenty putative CBD sequences, many previously unrecogminzed, were identified. Nearly all could be classified into seven families (Table II). The three remaining domains, each with demonstrated affinity for cellulose, were placed in separate families. Families were named using Roman numerals, as proposed by Coutinho *et al.* (1992), to avoid confusion with the Arabic numbering scheme used for catalytic domain classification (*8*). The availability of large numbers of sequences in two families allowed subfamilies to be distinguished.

Each of the ten CBD families is discussed individually below and we attempt to correlate the structural classification with various known functional properties.

Family I. All CBDs in this family are from fungal β-1,4-glycanases with catalytic domains belonging to families 6 and 7 (Figure 1 and Table II). The CBDs are not restricted to cellulases but are also present in two xylanases and one mannanase. One

Table I. Relationship of the Current Classification to Previous Schemes

	CBD Family Nomenclature		
This Work	Gilkes et al. (7)	Coutinho et al. (13)	Béguin (9)
I	B	I	2
II	A	II	1
III	not listed	III	3
IV	not listed	not listed	4
V	not listed	V	unclassified
VI	not listed	not listed	not listed
VII	not listed	IV	unclassified
VIII	not listed	not listed	not listed
IX	not listed	not listed	not listed
X	not listed	not listed	not listed

TABLE II. Classification of Cellulose-Binding Domains (CBDs) According to Amino Acid Sequence Similarity

Family[a]	Organism	Enzyme	Location[b]	Residues[c]	Binding[d]	Catalytic Domain Family[g]	Accession No.[e]
I	Agaricus bisporus	Cel1	C	36		-	M86356
	Agaricus bisporus	Cel3a	N	36		6	L24519
	Agaricus bisporus	Cel3b	N	36		6	L24520
	Fusarium oxysporum	B homolog	N	33		6	L29377
	Fusarium oxysporum	C homolog 2	C	33		6	L29379
	Fusarium oxysporum	K homolog	C	37		7	L29381
	Fusarium oxysporum	Xyn	N	36		45	L29380
	Humicola grisea	CBHI	C	33		10	X17258
	Humicola insolens	EII	N	36		7	X76046
	Humicola insolens	CBHII	C	33		5(5)	(48)
	Humicola insolens	A-1	-	33		-	(48)
	Humicola insolens	A-5	-	33		-	(48)
	Humicola insolens	A-8	-	33		-	(48)
	Humicola insolens	A-9	-	33		-	(48)
	Humicola insolens	A-11	-	33		-	(48)
	Humicola insolens	A-19	-	33		-	(48)
	Humicola insolens	43kDa	-	33		-	(48)
	Neocallimastix patriciarum	XylB	C	33		10	S71569
	Neurospora crassa	CBHI	C	33		7	X77778
	Penicillium janthinellum	CBHI	C	33		7	X59054
	Phanerochaete chrysosporium	CBHI	C	34		7	M22220
	Phanerochaete chrysosporium	CBHI-1	C	34		7	Z22528
	Phanerochaete chrysosporium	CBHI-2	C	34		7	X54411/S40817
	Phanerochaete chrysosporium	CBHI-3	C	34		7	(49)
	Phanerochaete chrysosporium	CBHI-4	C	34		7	L22656
	Phanerochaete chrysosporium	CBHII	N	36		7	(50)
	Porphyra purpurea	PBP[f]	x4	33		n-h	U08843
	Trichoderma koningii	CBHI	C	33		7	X69976

Continued on next page

TABLE II. Continued

Family[a]	Organism	Enzyme	Location[b]	Residues[c]	Binding[d]	Catalytic Domain Family[g]	Accession No.[e]
I (contd.)	Trichoderma longibrachiatum	CBHI	C	33	+	7	X60652
	Trichoderma reesei	CBHI	C	36	+	7	P00725
	Trichoderma reesei	CBHII	N	36	+	6	M16190/M55080
	Trichoderma reesei	EGI	C	33	+	7	M15665
	Trichoderma reesei	EGII	N	36	+	5.5	M19373
	Trichoderma reesei	EGV	C	36		45	Z33381
	Trichoderma reesei	Man	C	34		5(1)	L25310
	Trichoderma viride	CBHI	C	33		7	X53931
IIa	Butyrivibrio fibrisolvens	End1	C	95		5(4)	X17538
	Cellulomonas fimi	CenA	N	106	+	6	M15823
	Cellulomonas fimi	CenB	C	103	+	9(2)	M64644
	Cellulomonas fimi	CenD	C	105	+	5(1)	L02544
	Cellulomonas fimi	CbhA	C	106	+	6	L25809
	Cellulomonas fimi	CbhB	C	104	+	48	L29042
	Cellulomonas fimi	Cex	C	106	+	10	L11080/M15824
	Cellulomonas flavigena	CflX	C	106		–	(51)
	Clostridium cellulovorans	EngD	C	108	+	5(4)	M37434
	Clostridium longisporum	CelA	C	97	+	5(4)	L02868
	Microbispora bispora	CelA	C	100		6	P26414
	Micromonospora cellulolyticum	McenA	C	100		6	(52)
	Pseudomonas fluorescens	EglA	C	100		9(1)	X12570
	Pseudomonas fluorescens	CelB	N	102	+	45	X52615
	Pseudomonas fluorescens	CelC	N	99	+	5(1)	X61299
	Pseudomonas fluorescens	CelE	C	100		5(2)	X86798
	Pseudomonas fluorescens	XynA	N	101	+	10	X15429
	Pseudomonas fluorescens	XynB/C	N	99	+	10	X54523
	Pseudomonas fluorescens	XynD	N	102		10	X58956

	Organism	Protein	Pos	Length	+	Residues	Accession
	Streptomyces lividans	CelA	N	108		5(2)	M82807
	Streptomyces lividans	CelB	C	106		12	U04629
	Streptomyces lividans	ChiC	N	105		-	D12647
	Streptomyces plicatus	Chi63	N	102		-	M82804
	Streptomyces rochei	EglS	C	103		12	X73953
	Thermomonospora fusca	E1	C	96	+	9(1)	L20094
	Thermomonospora fusca	E2	C	96	+	6	M73321
	Thermomonospora fusca	E3	N	103	+	6	U18978
	Thermomonospora fusca	E4	C	104	+	9(2)	L20093
	Thermomonospora fusca	E5	N	103			L01577
IIb	*Cellulomonas fimi*	XynD1	I	90	+	11	X76729
	Cellulomonas fimi	XynD2	C	90	+	11	X76729
	Streptomyces lividans	Axe	C	86	+	-	(27)
	Streptomyces lividans	XlnB	C	86	+	11	M64552/S68767
	Thermomonospora fusca	XynA	C	86		11	U01242
IIIa	*Bacillus lautus*	CelA	C	150	+	44	M76588
	Bacillus lautus	ORF	C	150		-	M76588
	Bacillus subtilis BSE616	End	C	133		5(2)	D01057
	Bacillus subtilis CK2	Cel	C	133		5(2)	X67044
	Bacillus subtilis DLG	End1	C	132		5(2)	M16185
	Bacillus subtilis N-24	End2	C	132		5(2)	M28332
	Bacillus subtilis PAP115	End3	C	132		5(2)	X04689/Z29076
	Caldocellum saccharolyticum	CelA	I x 2	172/172		9(2)/48	L32742
	Caldocellum saccharolyticum	CelB	I	172		10/5(1)	X13602
	Caldocellum saccharolyticum	CelC	I x 2	172/172		9(2)/5(4)	(53)
	Caldocellum saccharolyticum	ManA	I x 2	172		5(4)/44	L01257

TABLE II. Continued

Family[a]	Organism	Enzyme	Location[b]	Residues[c]	Binding[d]	Catalytic Domain Family[g]	Accession No.[e]
IIIa (contd.)	Clostridium cellulovorans	CbpA	N	161	+	n-h	M73817
	Clostridium stercorarium	CelZ	C	133	+	9(2)	X55299
	Clostridium thermocellum	Cbh3	C	132		9(1)	X80993
	Clostridium thermocellum	CelI	C	150		9(2)	L04735
	Clostridium thermocellum	CipA	I	156		n-h	L08665
	Clostridium thermocellum	CipB	I	167	+	n-h	X68233
	Erwinia carotovora	CelV	C	156		9(2)	X76000/X79241
IIIb	Cellulomonas fimi	CenB	I	131	+	9(2)	M64644
	Clostridium cellulolyticum	CelCCG	I	138		9(2)	M87018
	Clostridium stercorarium	CelZ	I	144		9(2)	X55299
	Clostridium thermocellum	CelF	I	142		9(2)	X60545
	Clostridium thermocellum	CelI	I	137		9(2)	L04735
IV	Cellulomonas fimi	CenC	N	148	+	9(1)	X57858
	Cellulomonas fimi	CenC	I	148	+	9(1)	X57858
	Clostridium cellulolyticum	CelCCE	N	168		9(1)	M87018
	Myxococcus xanthus	CelA	N	139		6	X76726
	Streptomyces reticuli	CelI	N	125		9(1)	L04735/X65616
	Thermomonospora fusca	E1	N	141		9(1)	L20094
V	Erwinia chrysanthemi	EgZ	C	63	+	5(2)	Y00540
VI	Bacillus polymyxa	XynD	C	90		-	X57094
	Clostridium stercorarium	XynA	C × 2	87/92	+	11	D13325
	Clostridium thermocellum	XynZ	I	92		10	M22624
	Limulus sp.	Factor Gα	C × 2	87		-	(54)
	Microspora bispora	BglA	C	85		-	L06134

VII	Clostridium thermocellum	CelE	I	240		5(4)	M22759
VIII	Dictyostelium discoideum	CelA	N	152	+	9(2)	M33861
IX	Clostridium thermocellum	XynX	C x 2	174/189	+	10	M67438
	Thermoanaerobacterium saccharolyticum	XynA	C x 2	174/187		10	M97882
	Thermotoga maritima	XynA	C x 2	170/180	+	10	Z46264
X	Cellvibrio mixtus	XynA	C	51		11	Z48925
	Pseudomonas fluorescens	EglA	I	55		9(1)	X12570
	Pseudomonas fluorescens	EglB	I	55		45	X52615
	Pseudomonas fluorescens	CelC	I	53		5(1)	X61299
	Pseudomonas fluorescens	CelE	I	53		11	X86798
	Pseudomonas fluorescens	XynA	I	53		10	X15429
	Pseudomonas fluorescens	XynE	C	55	+	5(2)	Z48927

[a] Roman numerals indicate families; lower case letters indicate subfamilies.
[b] "N", "C" and "I", indicate N-terminal, C-terminal or internal CBDs, respectively; "x2" and "x4" indicate two and four CBDs, respectively.
[c] The number of amino acid residues indicated is based on sequence alignment and is considered tentative.
[d] "+" indicates that a functional CBD has been demonstrated experimentally.
[e] GenBank, SWISS-PROT or EMBL data base accession numbers are given if available; literature references are cited in other cases.
[f] PBP: polysaccharide-binding protein consisting entirely of four type I CBDs.
[g] Catalytic domain families are numbered according to reference 8: subfamilies are shown in parenthesis. The alternative letter designation (6) is as follows : 5 (A), 6 (B), 7 (C), 8 (D), 9 (E), 10 (F), 11 (G), 12 (H), 26 (I), 44 (J), 45 (K) & 48 (L). A dash indicates that the catalytic domain does not belong to one of the above families. "n-h" indicates that protein is non-hydrolytic.

```
                                •            •                    •
AbiCell.......TIPQYGQCGGI--GW-TGGTG-CVAPYQCKVIND--YYSQCL*
AbiCel3a......QSPVWGQCGGN--GW-TGPTT-CASGSTCVKQND--FYSQCL
AbiCel3b......QSPVWGQCGGN--GW-TGPTT-CASGSTCVKQND--FYSQCL
FoxBhomolog...SNGVWAQCGGQ--NW-SGTPC-CTSGNKCVKLND--FYSQCQ
FoxChomolog2..SVDQWGQCGGQ--NY-SGPTT-CKSPFTCKKIND--FYSQCQ*
FoxKhomolog...VVPAYYQCGGSKSAYPNGNLA-CATGSKCVKQNE--YYSQCV
FoxXyn........QAPIWGQCGGN--GW-TGATT-CASGLKCEKIND--WYYQCV
HgrCBHI.......KHGRWQQCGGI--GF-TGPTQ-CEEPYICTKLND--WYSQCL*
HinEGII.......QGGAWQQCGGV--GF-SGSTS-CVSGYTCVYLND--WYSQCQ
HinCBHII......    WGQCGGI--GF-NGPTC-CQSGSTCVKQND--WYSQCL
HinA-1........    WGQCGGQ--GW-NGPTC-CEAGTTCRQQNQ--WYSQCL
HinA-5........    WGQCGGI--GW-NGPTT-CVSGATCTKIND--WYHQCL
HinA-8........    WGQCGGN--GY-SGPTT-CAEGT-CKKQND--WYSQCT
HinA-9........    WGQCGGQ--GW-QGPTC-CSQGT-CRAQNQ--WYSQCL
HinA-11.......    WGQCGGQ--GY-SGCTN-CEAGSTCRQQNA--YYSQCI
HinA-19.......    WGQCGGQ--GY-SGCRN-CESGSTCRAQND--WYSQCL
Hin43kDa......TAERWAQCGGN--GW-SGCTT-CVAGSTCTKIND--WYHQCL
NpaXylB.......CAAKWGQCGGN--GF-NGPTC-CQNGSRCQFVNE--WYSQCL*
NcrCBHI.......GAAHWAQCGGI--GF-SGPTT-CPEPYTCAKDHD--IYSQCV*
PjaCBHI.......GARDWAQCGGN--GW-TGPTT-CVSPYTCTKQND--WYSQCL
PcrCBHI.......TVPQWGQCGGI--GY-TGSTT-CASPYTCHVLNP--YYSQCY*
PcrCBHI-1.....TVPQWGQCGGI--GY-TGSTT-CASPYTCHVLNP--YYSQCY*
PcrCBH1-2.....APPGFSQCGGL--GY-AGPTGVCPSPYTCQALNI--YYSQCI*
PcrCBH1-3.....TVPQWGQCGGI--GY--GPT-VCASPYTCQVLNP--YYSQC-*
PcrCBH1-4.....TVPQWGQCGGI--GY-SGSTT-CASPYTCHVLNP--YYSQCY*
PcrCBHII......QASEWGQCGGI--GW-TGPTT-CVSGTTCTVLNP--YYSQCL
PpuPBP........CVGLYEQCGGI--GF-DGVTC-CSEGLMCMKMGP--YYSQCR
              QVKPYGQCGGM--NY-SGKTM-CSPGFKCVELNE--FFSQCD
              CGKEYAACGGE--MF-MGAKC-CKFGLVCYETS-GKWQSQCR
              EVGRYAQCGGM--GY-MGSTM-CVGGYKCMAISEGSMYKQCL
TkoCBHI.......TQSHYGQCGGI--GY-SGPTV-CASGTTCQVLNP--YYSQCL
TloCBHI.......TQTHWGQCGGI--GY-TGCKT-CTSGTTCQYGND--YYSQCL*
TreCBHI.......TQSHYGQCGGI--GY-SGPTV-CASGTTCQVLNP--YYSQCL*
TreCBHII......SSPVWGQCGGQ--NW-SGPTC-CASGSTCVYSND--YYSQCL
TreEGI........TQTHWGQCGGI--GY-SGCKT-CTSGTTCQYSND--YYSQCL*
TreEGII.......QQTVWGQCGGI--GW-SGPTN-CAPGSACSTLNP--YYAQCI
TreEGV........QQTLYGQCGGA--GW-TGPTT-CQAPGTCKVQNQ--WYSQCL*
TreMan........CSPLYGQCGGS  GY-TGPTC-CAQG-TCIYSNY--WYSQCL*
TviCBHI.......TQTHYGQCGGI--GY-IGPTV-CASGSTCQVLNP--YYSQCL*
```

Figure 1. Sequence alignment of family I CBDs. Conserved tryptophan residues are shown in bold; exposed tryptophan residues are marked (•). Organism and enzyme abbreviations are explained in Table II.

non-hydrolytic polysaccharide-binding protein (PBP), composed entirely of four family I CBDs and a short proline rich linker between the second and third domains, is produced by the red alga *Porphyra purpurea*. Family I CBDs are found at either the N or C terminus of the mature polypeptide (Table II and Figure 7).

Family I CBDs are small, compact domains, 32 to 36 amino acids long, with highly conserved sequences (Figure 1). There are four strictly conserved aromatic residues. In the CBD of *Trichoderma reesei* CBH I, these residues are implicated in the interaction with cellulose (*19, 20*). Four strictly conserved cysteine residues form two disulfide bridges in the *T. reesei* CBH I CBD (*21*). The CBDs from *T. reesei* cellulases have a high affinity for both amorphous and crystalline cellulose (*5, 20, 22*) but do not bind to chitin, a β-1,4-linked polymer of N-acetyl-glucosamine (*23*). In contrast to CBDs from some other families, family I CBDs do not appear to disrupt the structure of Ramie fibers, nor do they release "small particles" from cotton (*23*).

Family II. The CBDs in this family are found in bacterial hydrolases with catalytic domains belonging to families 5, 6, 9, 10 or 11 (Table II). The enzymes include β-1,4-glucanases, several family 10 and 11 xylanases, one arabinofuranosidase, two acetyl esterases and two chitinases (Figure 2 and Table II). There are two family II CBDs in a spore germination-specific protein from the slime mold *Dictyostelium* discoideum (not included in Figure 2 and Table II) and at least one of these binds to cellulose (*24* and Innis, H., personal communication).

Family II CBDs are about a 100 amino acid residues long and usually contain a cysteine residue near the N and C termini (Figure 2) The latter form a disulfide bridge in the CBD from *Cellulomonas fimi* CenA and Cex (*25*). These residues are not conserved in the CBDs of End1 from *Butyrivibrio fibrisolvens*, EngD from *Clostridium cellulovorans*, CelA from *Clostridium longisporum* and CenA from *Micromonospora cellulolyticum* (Figure 2). Family II CBDs also contain four highly conserved tryptophan residues. Sequence alignment reveals two subfamilies, a and b (Figure 2). A stretch of eight amino acid residues, including the conserved C-proximal tryptophan residues, is deleted in the IIb subfamily (Figure 2). The IIb subfamily is found predominantly in xylanases with a family 11 catalytic domain. Several CBDs in subfamily IIb bind to xylan but some also bind to cellulose (*26-28*). In contrast, subfamily IIa CBDs appear to bind only to cellulose (Table III). The subfamily IIb CBD from *C. fimi* XynD (i.e., CBD2) is unusual because it binds only to crystalline cellulose (*28*), in contrast to other family II CBDs which bind to both crystalline and "amorphous" (i.e., regenerated or acid-swollen) cellulose (*29, 30*). Some of the subfamily IIa CBDs from *C. fimi* enzymes also bind tightly to chitin, (*30, 31*). Binding of other subfamily IIa CBDs to chitin has not been reported.

The family II CBDs occur mainly at the N- or C-termini of enzymes but a few are internal (Table II and Figure 7). Some enzymes contain a family II CBD and another CBD from a different family. For example, *C. fimi* CenB contains a family II and a family III CBD, and *Thermomonospora fusca* E2 contains a family II and a family IV CBD. In both these enzymes, the family II CBD is at the C-terminus (Figure 6).

Three of the four conserved tryptophan residues in the family II CBD of *C. fimi* Cex are located on the surface of the polypeptide and potentially available for interaction with cellulosic substrates (see below and Figure 8). Single mutations of two corresponding residues (W14A and W68A) greatly reduce the affinity of the family II CBD of *C. fimi* CenA for cellulose (*32*). Furthermore, gene fusion studies show that the C-terminus of the CenA CBD (including W68) is important for binding (*33*). In contrast, mutations of the conserved aromatic residues in the CBD from *Pseudomonas fluorescens* XynA suggest that W13 and W34 (corresponding to W14, W34 in the CenA CBD) are involved in binding, while W66 (corresponding to W68 in the CenA CBD) is not (*34*). These data indicate structural and functional differences between CBDs in family II. A further example of possible differences

Subfamily a

```
BfiEnd1.......GALK-AEYTI-NNWGSGYQVLIKVKNDSASRVDGWTLKISKSE--VKIDSSWCVN-
CfiCenA.......APGCRVDYAVTNQWPGGFGANVTITNLG-DPVSSWKLDWTYTA-GQRIQQLWNGT-
CfiCenB.......TPSCTVVYS-TNSWNVGFTGSVKITNTGTTPL-TWTLGFAFPS-GQQVTQGWSAT-
CfiCenD.......TGSCAVTYT-ANGWSGGFTAAVTLTNTGTTALSGWTLGFAFPS-GQTLTQGWSAR-
CfiCbhA.......SGGCTVKYSASS-WNTGFTGTVEVKNNGTAALNGWTLGFSFAD-GQKVSQGWSAE-
CfiCbhB.......GGSCSVAYNASS-WNSGFTASVRITNTGTTTINGWSLGFDLTA-GQKVQQGWSAT-
CfiCex........PAGCQVLWGV-NQWNTGFTANVTVKNTSSAPVDGWTLTFSFPS-GQQVTQAWSST-
CflX..........TGSCKVEYNASS-WNTGFTASVRVTNTGTTALNGWTLTFPFAN-GQTVQQGWSAD-
CcvEngD.......QSAVEVTYAITNSWGSGASVNVTIKNNGTTPINGWTLKWTMPI-NQTITNMWSAS-
CloCelA.......DNEKISITSKINDWGGAYQADFTLKNNTSSDINNWSFKIKKND--IVFTNYWDVK-
MbiCelA.......GRACEATYALVNQWPGGFQAEVTVKNTGSSPINGWTVQWTLPS-GQSITQLWNGD-
MceMcenA......GNGLSASVAIT-QWNGGFTAS--VNVTAGSAINGWTVTVALPG-GAAITGTWNAQ-
PflEglA.......GGNC--QYVVTNQWNNGFTAVIRVRNNGSSAINRWSVNWSYSD-GSRITNSWNAN-
PflCelB.......AAVC--EYRVTNEWGSGFTASIRITNNGSSTINGWSVSWNYTD-GSRVTSSWNAG-
PflCelC.......AAGC--EYVVTNSWGSGFTAAIRITNSTSSVINGWNVSWQYN--SNRVTNLWNPN-
PflXynA.......TATC--SYNITNEWNTGYTGDITITNRGSSAINGWSVNWQYAT--NRLSSSWNAN-
PflXynB/C/D...--AC--TYTIDSEWSTGFTANITLKNDTGAAINNWNVNWQYSS--NRMTSGWNAN-
SliCelA.......ATGCKAEYTITSQWEGGFQAGVKITNLG-DPVSGWTLGFTMPDAGQRLVQGWNAT-
SliCelB.......PSACAVSY-GTNVWQDGFTADVTVTNTGTAPVDGWQLAFTLPS-GQRITNAWNAS-
SliChiC.......TSATA-TFAKTSDWGTGFGGSWTVKNTGTTSLSSWTVEWDFPT-GTKVTSAWDAT-
SplChi63......ATSATATFQKTSDWGTFGGKWTVKNTGTTSLSSWTVEWDFPS-GTKVTSAWDAT-
SroEglS.......PAACTVSYAT-NVWPGGFTANVTVTNNGSAPVDGWRLAFTLPS-GQSVVHAWNAS-
TfuE1.........SASCAVTYQT-NDWPGGFTASVTLTNTGSTPWDSWELRFTFPS-GQTVSHGWSAN-
TfuE2.........SGACTATYTIANEWNDGFQATVTVTANQN--ITGWTVTWTFTD-GQTITNAWNAD-
TfuE4.........DASCTVGYST-NDWDSGFTASIRITYHGTAPLSSWELSFTTFPA-GQQVTHGWNAT-
TfuE5.........AGLTATVTKESS-WDNGYSASVTVRNDTSSTVSQWEVVLTLPG-GTTVAQVWNAQ-
```

Subfamily b

```
CfiXynD1......STGGCSVTATRAEEWSDRFNVTYSVS--GSS---AWTVNLALNG-SQTIQASWNAN-
CfiXynD2......TGSCSVSAVRGEEWADRFNVTYSVS--GSS---SWVVTLGLNG-GQSVQSSWNAA-
SliAxeA.......GGGCTATLSAGQRWGDRYNLNVSVS--GAS---DWTVTMNVPS-PAKVLSTWNVNA
SliXlnB.......GGGCTATVSAGQKWGDRYNLDVSVS--GAS---DWTVTMNVPS-PAKVLSNWNVNA
TfuXynA.......GGGCTATLSAGQQWNDRYNLNVNVS--GSN---NWTVTVNVPW-PARIIATWNIHA
```

Subfamily a

```
BfiEnd1.......IAEEGGYYVITPMSWNSSLEP-SASVDFGIQGS---GS-IGTS---VNISVQ*
CfiCenA.......ASTNGGQVSVTSLPWNGSIPT-GGTASFGFNGSWA-GSNPTPASFSLNGTTCTGT
CfiCenB.......WSQTGTTVTATGLSWNATLQP-GQSTDIGFNGSHP-GTNTNPASFTVNGEVCG*
CfiCenD.......WAQGSGSSVTATNEAWNAVLAP-GASVEIGFSGTHT-GTNTAPATFTVGGATCTTR*
CfiCbhA.......WSQSGTAVTAKNAPWNGTLAA-GSSVSIGFNGTHN-GTNTAPTAFTLNGVACTLG*
CfiCbhB.......WTQSGSTVTATNAPWNGTLAP-GQTVDVGFNGSHT-GQNPNPASFTLNGASCT*
CfiCex........VTQSGSAVTVRNAPWNGSIPA-GGTAQFGFNGSHT-GTNAAPTAFSLNGTPCTVG*
CflX..........WSQSGTTVTAKNAAWNGSLAA-GQTVDIGFNGAHN-GTNNKPASFTLNGATCTVG*
CcvEngD.......FVASGTTLSVTNAGYNGTIAANGGTQSFGFNINYS-GVLSKPTGFTVNGTECTVK*
CloCelA.......ITEENGYYVVTPQAWKTTILA-NSSIVISIQGT---GKVISNFEYKFD*
MbiCelA.......LSTSGSNVTVRNVSWNGNVPA-GGSTSFGFLGS---GTGQL-----SSSITCSAS*
MceMcenA......ASGTSGTVRFTNVGYNGQVGA-GQTTNFGFQGT---GTGQG------ATATCAA*
PflEglA.......VTGNNPY-AASALGWNANIQP-GQTAEFGFQGTKGAGSRQVPA---VTGSVCQ*
PflCelB.......LSGANPY-SATPVGWNTSIPI-GSSVEFGVQGNN GSSRAQVPA-VTGAICGGQ
PflCelC.......LSGSNPY-SASNLSWNGTIQP-GQTVEFGFQGVTNSGTVESPT---VNGAACTGG
PflXynA.......VSGSNPY-SASNLSWNGNIQP-GQSVSFGFQVNKNGGSAERPS---VGGSICSGS
PflXynB/C/D...FSGTNPY-NATNMSWNGSIAP-GQSISFGLQGEKN-GSTAERPT--VTGAACNSA
SliCelA.......WSQGSAVTAGGVDWNRTLAT-GASADLGFVGSFT-GANPAPTSFTLNGATCSGS
SliCelB.......LTPSSGSVTATGASHNARIAP-GGSLSFGFQGTYG-GAFAEPTGFRLNGTACTTV*
SliChiC.......VTNSGDHWTAKNVGWNGTLAP-GASVSFGFNGSGP-GSPSNCKL---NGGSCDGT
SplChi63......VTNSADHWTAKNVGWNGTLAP-GASVSFGFNGSGP-GSPSGCKI---NGGSCDGS
SroEglS.......VSPSSGAVTATGPAESARIAAGGSQ-SFGFQGAYS-GSFAQPAAFQLNGTACSTV*
TfuE1.........WQQSGSDVTATSLPWNGSVPPGGGSVNIGFNGTWG-GSNTKPEKFTVNGAVCSIG*
TfuE2.........VSTSGSVSTARNVGHNGTLSQ-GASTEFGFVGSK- GNSNS-----VPTLTCAAS*
TfuE4.........WRQDGAAVTATPMSWNSSLAP-GATVEVGFNGSWS GSNTPPTDFTLNGEPCALA*
TfuE5.........HTSSGNSHTFTGVSWNSTIPP-GGTASSGFIASGS-GEPTHCT---INGAPCDEG
```

Subfamily b

```
CfiXynD1......VTGSGSTRTVTPN---------GSGNTFGVTVMKN-GSSTTP------AATCAGS
CfiXynD2......LTGSSGTVTARPN---------GSGNSFGVTFYKN-GSSATP------GATCATG*
SliAxeA.......SYPSAQTLTAKSN---------GSGGNWGATIQAN-GNWTWP------SVSCTAG*
SliXlnB.......SYPSAQTLTARLN---------GSGNNWGATIQAN-ANWTWP------SVSCSAG*
TfuXynA.......SYPDSQTLVARPN---------GNGNNWGMTIMHN-GNWTWP------TVSCSAN*
```

Figure 2. Sequence alignment of family II CBDs. The family is divided into two subfamilies, IIa and IIb. Conserved tryptophan residues are shown in bold; exposed tryptophan residues are marked (•). Organism and enzyme abbreviations are explained in Table II.

within family II is provided by XynD from *C. fimi*. This enzyme contains two very similar CBDs from subfamily IIb (Figure 2). However, the internal CBD (CBD1) binds to xylan but not cellulose, whereas the C-terminal CBD (CBD2) binds to both cellulose and xylan (Table III). In this context, it is interesting to note that the various family II CBDs have widely different net charges and charge distributions (Figure 2). These differences could influence the affinity, specificity or desorption of the various CBDs on different substrates.

Family II CBDs derived from *C. fimi* enzymes disrupt cotton fibers and release "small particles" (*23, 35*). They also prevent flocculation of bacterial microcrystalline cellulose (*36*). The significance of these properties in cellulose hydrolysis is presently unclear and it is not yet known whether all family II CBDs share these characteristics.

Family III. The majority of family III CBDs occur in β-1,4-glucanases from families 5 or 9 (Table II). Exceptions are the CBD of *Caldocellum saccharolyticum* ManA (*37*) and the CBDs of the non-hydrolytic scaffolding proteins (CbpA, CipA and CipB) from the cellulosomes of *Clostridium* spp. (Figure 3 and Table II). Family III CBDs are usually C-terminal or internal. Internal family III CBDs occur between the two catalytic domains in hydrolases from *Caldocellum saccharolyticum* (Figure 7). The only known N-terminal family III CBD occurs in *Clostridium cellulovorans* CbpA (*38*).

Two subfamilies in family III can be distinguished (Figure 3). The RY(Y/W)Y motif found near the N-terminus in the IIIa domains is only partially conserved (RYF) in the IIIb domains and several other aromatic residues are conserved in subfamily IIIa but not subfamily IIIb (Figure 3). All the subfamily IIIb CBDs are internal and all occur in association with family 9 (subfamily 2) catalytic domains. A CBD from a different family or subfamily is sometimes found with a family IIIb CBD in the same enzyme, e.g., *Clostridium stercorarium* CelZ and *Clostridium thermocellum* CelI (IIIa and IIIb CBDs) and *C. fimi* CenB (IIa and IIIb CBDs) (Table II and Figure 7).

The subfamily IIIa CBD from *C. cellulovorans* CbpA has a high affinity for cellulose (K_d = 0.8-1.4 µM), comparable with that determined for the subfamily IIa CBD from *C. fimi* Cex (K_d = 0.5 µM) (*40* and Jervis, E., unpublished data). The subfamily IIIb CBDs may have much lower affinities because a truncated form of *C. fimi* CenB, lacking the C-terminal IIa CBD but retaining the internal IIIb CBD, binds only weakly to cellulose (*39*). The subfamily IIIa CBD from *C. cellulovorans* CbpA also binds to chitin (*40*).

Family IV. The two tandem CBDs (N1 and N2) in CenC from *C. fimi*, were originally classified in family II, based on very weak sequence similarity with the other *C. fimi* CBDs (*13* and Table I). However, they are now classified in a separate family, family IV (Figure 4). N1 and N2 in tandem, and N1 alone, are unusual because they bind to "amorphous" cellulose but not to crystalline cellulose (*13*). The binding characteristics of the other family IV CBDs have not been reported. Most family IV CBDs are present at the N termini of family 9 (subfamily 1) enzymes; an exception is the CBD from *Myxococcus xanthus* CelA. CenC is the only known enzyme with two family IV CBDs in tandem (Table II), but E1 from *T. fusca* also contains two CBDs: a single, N-terminal family IV CBD and a C-terminal subfamily IIa CBD. Two cysteine residues are conserved in all family IV CBDs, with the exception of the CBD from of *C. cellulolyticum* CelCCE (Figure 4).

Family VI. Family VI (Figure 5) is a recently recognized family of CBDs first identified in *C. stercorarium* XynA (*41*). Most known family VI CBDs are from xylanases and nearly all are C-terminal. An exception is the internal family VI CBD of *C. thermocellum* XynZ (Table II and Figure 7).

At least one family VI CBD, that from *C. stercorarium* XynA, binds to both cellulose and xylan (*41*). *C. stercorarium* XynA is a xylanase with a family 11

Subfamily a

```
BlaCelA.....NQIKPHFNIQNKGTSPVDLSSLTLRYYFTKDSSAAMNGWIDWAKL---------
BlaORF......NQIKPSFNIKNNGTSAVDLSTLKIRYYFTKDGSAAVNGWIDWAQL---------
BsuEnd......NQIRPQLQIKNNGNTTVDLKDVTARYWYNAK-NKGQNVDCDYAQL---------
BsuCel......NQIRPQLQIKNNGNTTVDLKDVTARYWYNAK-NKGQNFDCDYAQI---------
BsuEnd1.....NQIRPQLQIKNNGNTTVDLKDVTARYWYKAK-NKGQNFKCDYAQI---------
BsuEnd2.....NQIRPQLHIKNNGNATVDLKDVTARYWYNAK-NKGQNFDCDYAQI---------
BsuEnd3.....NQIRPQLHIKNNGNATVDLKDVTARYWYNAK-NKGQNFDCDYAQI---------
CasCelA.....NTIRPWLKVVNSGSSSIDLSRVTIRYWYTVDGERAQSAVSDWAQI---------
CsaCelB.....NTIRPWLKVVNSGSSSIDLSRVTIRYWYTVDGERAQSAVSDWAQI---------
CsaCelC.....NTIRPWLKVVNSGSSSIDLSRVTIRYWYTVDGERAQSAISDWAQI---------
CsaManA.....NTIRPWLKVVNSGSSSIDLSRVTIRYWYTVDGERAQSAISDWAQI---------
CcvCbpA.....NSITPIIKITNISDSDLNLNDVKVRYYYTSDGTQGQIFWCDDHAGALLGNSYVD
CstCelZ2....NGIMPRYRLTNTGTTPIRLSDVKIRYYYTIDGEKDQNFWCDWSSV---------
CthCbh3.....QEIKGKFNIVNTGNRDYSLKDIVLRYYFTKEHNSQLQFICYYTPI---------
CthCelI2....NSINPRFKIINNGTKAINLSDVKIRYYYTKEGGASQNFWCDWSSA---------
CthCipA.....NSINPQFKVTNTGSSAIDLSKLTLRYYYTVDGQKDQTFWCDHAAIIGSNGSYNG
CthCipB.....NSINPQFKVTNTGSSAIDLSKLTLRYYYTVDGQKDQTFWCDHAAIIGSNGSYNG
EcaCelV.....DAIRMAVNIKNTGSTPIKLSDLQVRYYFHDDGKPGANLFVDWANV---------
```

Subfamily b

```
CfiCenB.....TEVKAMIR-NQSAFPARSLKNAKVRYWFTTD-----GFAASDVTL---------
CceCelCCG...TEIKAVVY-NQTGWPARVTDKISFKYFMDLSEIVAAGIDPLSLV---------
CstCelZ1....IEIKALLH-NQSGWPARVADKLSFRYFVDLTELIEAGYSASDVTI---------
CthCelF.....VNIKASII-NKSGWPARGSDKLSAKYFVDISEAVAKGITLDQITV--------Q
CthCelI1....IEIKAIVN-NKSGWPARVCENLSFRYFINIEEIVNAGKSASDLQV---------
```

Subfamily a

```
BlaCelA....GGSNIQISF-----GNHNGAD---SDTYAELGFSSGAGSIAEGGQS---GEIQLR
BlaORF.....GGSNIQISF-----GNHTGTN---SDTYVELSFSSEAGSIAAGGQS---GETQLR
BsuEnd.....GCGNVTYKF-VTLHKPKQGA-----DTYLELGFKN--GTLAPGAST---GNIQLR
BsuCel.....GCGNVTHKF-VTLHKPKQGA-----DTYLELGFKN--GTLAPGAST---GNIQLR
BsuEnd1....GCGNVTHKF-VTLHKPKQGA-----DTYLELGFKN--GTLAPGAST---GNIQLR
BsuEnd2....GCGNLTHKF-VTLHKPKQGA-----DTYLELGFKT--GTLSPGAST---GNIQLR
BsuEnd3....GCGNLTHKF-VTLHKPKQGA-----DTYLELGFKT--GTLSPGAST---GNIQLR
CsaCelA....GASNVTFKF-VKLSSSVSGA-----DYYLEIGFKSGAGQLQPGKDT---GEIQIR
CsaCelB....GASNVTFKF-VKLSSSVSGA-----DYYLEIGFKSGAGQLQPGKDT---GEIQIR
CsaCelC....GASNVTFKF-VKLSSSVSGA-----DYYLEIGFKSGAGQLQPGKDT---GEIQIR
CsaManA....GASNVTFKF-VKLSSSVSGA-----DYYLEIGFKSGAGQLQPGKDT---GEIQMR
CcvCbpA....NTSKVTANF-VKETASPTSTY----DIYVEFGFASGRATLKKGQFI---T-IQGR
CstCelZ2...GSNNITGTF-VKMAEPKEGA-----DYYLETGFTDGAYGLYQPNQSI----EVQNR
CthCbh3....GSGNLIPSF---GGSGDE--------HYLQLEFKD-V-KLPAGGQT---GEIQFV
CthCelI2...GNSNVTGNF-FNLSSPKEGA-----DTCLEVGFGSGAGTLDPGGS----VEVQIR
CthCipA....ITSNVKGTF-VKMSSSTNNA-----DTYLEISFTG--GTLEPGAHH---VQIQGR
CthCipB....ITSNVKGTF-VKMSSSTNNA-----DTYLEISFTG--GTLEPGAH----VQIQGR
EcaCelV....GPNNIVTST-GTPAASTDKA-----NRYVLVTFSSGAGSLQPGAET---GEVQVR
```

Subfamily b

```
CfiCenB....-SANYSEC--GAQSGKGVS--AGGTLGYVELSC-VGQDIH-PGGQSQHRREIQFR
CceCelCCG..TSSNYSEGKN-TKVSGVLPWDVSNNVYYVNVIL-TGENIY-PGGQSACRREVQFR
CstCelZ1...-TTNYNAG---AKVTGLHPWNEAENIYYVNVDF-TGTKIY-PGGQSAYRKEVQFR
CthCelF....STTN---GG--AKVSQLLPWDPDNHIYYVNIDF-TGINIF-PGGINEYKRDVYFT
CthCelI1...SSS-YNQG---AKLSDVKHY--KDNIYYVEVDL-SGTKIY-PGGQSAYKKEVQFR
```

Figure 3. Sequence alignment of family III CBDs. The family is divided into two subfamilies, IIIa and IIIb. Organism and enzyme abbreviations are explained in Table II.

Subfamily a

```
BlaCelA.....MSKADWSN-FNAENDYSFDG----AKTAYI-DWDRVTLYQDGQLVWGIEP
BlaORF......MSKTDWSN-FNEANDYSFDG----TKTAFA-DWDRVVLYQNGQIVWGTAP
BsuEnd......LHNDDWSN-YAQSGDYSFFK----SNTFKTTK--KITLYDQGKLIWGTEP
BsuCel......LHNDDWSN-YAQSGDYSFFK----SNTFKTTK--KITLYDQGKLIWGTEP
BsuEnd1.....LHNDDWSN-YAQSGDYSFFK----SNTFKTTK--KITLYDQGKLIWGTEP
BsuEnd2.....LHNDDWSN-YAQSGDYSFFQ----SNTFKTTK--KITLYHQGKLIWGTEP
BsuEnd3.....LHNDDWSN-YAQSGDYSFFQ----SNTFKTTK--KITLYHQGKLIWGTEP
CsaCelA.....FNKSDWSN-YNQGNDWSWIQ----SMTSYG-ENEKVTAYIDGVLVWGQEP
CsaCelB.....FNKSDWSN-YN-GNDWSWLQ----SMTSYG-ENEKVTAYIDGVLVWGQEP
CsaCelC.....FNKSDWSN-YNQGNDWSWLQ----SMTSYG-ENEKVTAYIDGVLVWGQEP
CsaMan......FNKDDWSN-YNQGNDWSWIQ----SMTSYG-ENEKVTAYIDGVLVWGQEP
CcvCbpA.....ITKSDWSN-YTQTNDYSFDA----SSSTPVVNPK-VTGYIGGAKVLGIAP
CstCelZ2....FSKADWTD-YIQTNDYSF------STNTSYGSNDRITVYISGVLVSGIEP
CthCbh3.....IRYADNSF-HDQSNDYSFDP----TIKAF-QDYGKVTLYKNGELVWGTPP
CthCelI2....FSKEDWSN-YNQSNDYSFKQ----ACLRQ-----RTLIYL--YATWLR
CthCipA.....FAKNDWSN-YTQSNDYSFK-----SASQFV-EWDQVTAYLNGVLVWGKEP
CthCipB.....FAKNDWSN-YTQSNDYSFK-----SRSQFV-EWDQVTAYLNGVLVWGKEP
EcaCelV.....IHAGDWSN-VNETNDYSYGA----NVTSYA-NWDKITVHDKGTLVWGVEP
```

Subfamily b

```
CfiCenB.....LTGPAG---WNPANDPSYTGLTQTALA---KASA-ITLYDGSTLVWGKEP
CceCelCCG...IAAPQGRRYWNPENDFSYDGLP-TTSTVNTVTN--IPVYDNGVKVFGNEP
CstCelZ1....IAAPQGRRYWNNDNDYSFRDIK-GVTSGNTVKTVYIPVYDDGVLVFGVEP
CthCelF.....ITAPYGEGNWDNTNDFSFQGLEQGFTS---KKTEYIPLYDGNVRVWGKVP
CthCelI1....ISAPEGTV-FNPENDYSYQGL----SAGTVVKSEYIPVYDAGVLVFGREP
```

Figure 3. Continued.

Table III. Binding Characteristics of Family II CBDs

Enzyme	Family Classification		Binding Specificity	
	CBD	Catalytic Domain	Cellulose	Xylan
CfiCex	IIa	10	+	-
PflXynA	IIa	10	+	-
PflXynB	IIa	10	+	-
SliXlnB	IIb	11	?	+
TfuXynA	IIb	11	+	+
CfiXynD1	IIb	11	-	+
CfiXynD2	IIb	11	+	+

```
CfiCenC(N1)...ASPIGEGTTFDDG-PEGWVAYGTDGPLDTSTGALCVAVPAGSAQ-YGVGVVL
CfiCenC(N2)...VELLPH-TSFAE-SLGPWSLYGTSEPVF-ADGRMCVDLPGGQGNPWDAGLVY
CceCelCCE.....V-GLPWHVVESYPAKASFEITSDGKYKITAQ-KI-GEA--GKGERWDIQFRH
MxaEGL........LTELVSNGTFNGGTVSPWWSGPNTQSRVENARLRVDVGGGTANP-WDALIGQ
SriCel1.......VEQVRNGT-FDT-TTDPWW-TS-NVTAGLSDGRLCADVPGGTTNRWDSAIGQ
TfuE1.........VNQIRNGD-FSSGT-APWWGTE-NIQLNVTDGMLCVDVPGGTVNPWDVIIGQ


CfiCenC(N1)...NGVAIEEGTTYTLRYTATAS-TDVTTVRALVGQNGAPYGTVLDTSPA-LTSE
CfiCenC(N2)...NGVPVGEGESYVLSFTASAT-PDMPPVRVLVGEGGGAYRTAFEQGSAPLTGE
CceCelCCE.....RGLALQQGHTYTVKFTVTASRACKIYPK--IGDQGDPYDEYWNMNQQWNFLE
MxaEGL........DDIPLVNGRAYTLSFTASAS--VSTTVRVTVQLESAPYTAPLDR-QITLDGT
SriCel1.......NDITLVKGETYRFSFHASG-IPEGHVVRAVVGLAVSPYDTWQEASPV-LTEA
TfuE1.........DDIPLIEGESYAFSFTASSTVPV--SIRALVQEPVEPWTTQMDERAL-LGPE


CfiCenC(N1)...-----PRQVTETFTASATYPATPAADDPEGQIAFQLGGF------------S
CfiCenC(N2)...-----PATREYAFTSNLT---FPPDGDAPGQVAFHLG--------------K
CceCELCCE.....LQANTPKTVTQTFTQTK-------GDKKNVEFAFHLAPDKTTSEAQNPASFQ
MxaEGL........-----SRRFTFPFTSTLA--------TQAGQVTFQMGGRA-----------T
SriCel1.......-----DGSYSYTFTAPV--------DTTQGQVAFQVGG-------------S
TfuE1.........-----AETYEFVFTSNV--------DWDDAQVAFQIGG-------------S


CfiCenC(N1)...ADAWTFCLDDVAL--
CfiCenC(N2)...AGAYEFCISQVSL--
CceCELCCE.....PITYTF--DEIYIQD
MxaEGL........-GFSAF-IDDISL--
SriCel1.......TDAWRFCVDDVSLLG
TfuE1.........DEPWTFCLDDVALLG
```

Figure 4. Sequence alignment of family IV CBDs. Organism and enzyme abbreviations are explained in Table II.

```
BpoXynD...--IHNGDWIAVGKADFGSAGAKTFKANVATNV-GGNIEVRLDSETGPLVGSLKVP
CstXynA...GYIENGYSTTYKNIDFGD-GATSVTARVATQN-ATTIQVRLGSPSGTLLGTIYVG
CstXynA...GYIENGNTVTYSNIDFGD-GATGFSATVATEV-NTSIQIRSDSPTGTLLGTLYVS
CthXynZ...GYITSGDYLVYKSIDFGN-GATSFKAKVANAN-TSNIELRLNGPNGTLIGTLSVK
LimGα.....---NEGAWMAYKDIDFPSSGNYRIEYRVASERAGGKLSLDLNAGS-IVLGMLDVP
LimGα.....---KEGAWMAYKDIDFPSSGSYRVEYRVASERAGGKLSLDLNAGS-IVLGMLDIP
MbiBglA...G--STGDWIVFKDVDLKR-RPSRVTAGVAST-SGGSIELRLGSPKGKLIATVPVA


BpoXynD...STGGMQTWREVETTINNATGVHNIYLVFTG-SGSGNLLNL
CstXynA...STGSFDTYRDVSATISNTAGVKDIVLVFSG-PVNVDWFVFM
CstXynA...STGSWNTYQPYLQTSAKLPAFMILY-WYS--QVQSMWCthX
CthXynZ...STGDWNTYEEQTCSISKVTGINDLYLVFKG-PVNIDWFTFC
LimGα.....STGGWQKWTTISHTVNVDSGTYNLGIYVQRASWNINWIKIB
LimGα.....STGGLQKWTTISHIVNVDLGTYNLGIYVQKASWNINWIRIL
MbiBglA...ATGDVYRYETAAARVTGPSGVKDLYLVFQG-DVRI
```

Figure 5. Sequence alignment of family VI CBDs. Organism and enzyme abbreviations are explained in Table II.

catalytic domain. Subfamily IIb CBDs are also found on xylanases with family 11 catalytic domains, although only in 4 out of 31 known enzymes (Table II). Interestingly, the CBDs found in association with family 11 catalytic domains appear to be the only CBDs that bind to purified xylans. Xylanases with family 11 catalytic domains are specific for xylan whereas several xylanases with family 10 catalytic domains, e.g., *C. fimi* Cex, hydrolyse both β-1,4 xylosidic and β-1,4 glucosidic bonds. Apparently, the CBDs from xylanases with family 10 catalytic domains do not bind isolated xylans. However, xylans may adopt different conformations when tightly associated with cellulose and the possiblity that the CBDs from family 10 xylanases bind to xylan under these conditions cannot be excluded.

Family IX. Family IX (Figure 6) is a second recently recognized CBD family first seen in XynA from *Thermotoga maritima* (*42*). All known family IX CBDs are present as tandem repeats at the C-termini of thermostable xylanases with family 10 catalytic domains but only three enzymes carrying this CBD have been identified so far (Table II).

Family X. Xylanase E from *Pseudomonas fluorescens* ssp. *cellulosa* contains a 55 amino acid residue domain at its C-terminus that was recently shown to function as a CBD (see entry under GenBank accession number Z48927). Five other *P. fluorescens* ssp. *cellulosa* enzymes and XynA from *Cellvibrio mixtus* contain related domains and comprise a third novel CBD family (Figure 6 and Table II).

Structures of Cellulose-Binding Domains

To date, the structures of only two CBDs have been solved in detail: the family I CBD from *T. reesei* CBH I (*21*) and the subfamily IIa CBD from *C. fimi* Cex (*45*). Both structures were solved by NMR spectroscopy. Structural analyses of the family V CBD from *Erwinia chrysanthemi* EGZ (Brun, E., personal communication) and the subfamily IIIa CBD from *C. cellulovorans* (*43*) are in progress.

The solution structure for the *T. reesei* CBH I CBD is wedge-shaped. It consists of a irregular triple-stranded antiparallel β-sheet with overall dimensions of 30 x 18 x 10 Å (Figure 8). Two flat faces can be distinguised: one predominantly hydrophobic and containing three exposed tyrosine residues, the other predominantly hydrophilic (*21* and Figure 8). The tyrosine residues are spaced at 10.4 Å, a distance comparable to the length of a cellobiose repeating unit in cellulose (*44*).

The *C. fimi* Cex CBD consists of ten antiparallel β-strands. Nine strands form a β-barrel with two faces (*45* and Figure 7). The overall topology is similar to tendamistat (*46*). The CBD is elongated with overall dimensions of 45 x 25 x 25 Å. The β-strands are oriented along the longest dimension (Figure 8). In contrast to the CBH I CBD, there is no obvious partitioning of hydrophobic or hydrophilic residues. The Cex CBD has a coned-shaped cavity, approximately 7Å deep and 5Å across. The function of this cavity is presently unclear. Two tryptophan residues (W54, and W72, corresponding to W50 and W68, respectively, in the CBD from *C. fimi* CenA) are on the surface and interact with cellohexaose (*45*). The exposed positions of aromatic residues in the *T. reesei* and *C. fimi* CBDs are consistent with a role in binding to cellulose. The functions of these residues have been probed by site-directed mutation, as described above.

Conclusions

We can now recognize ten families of CBDs and twelve families of catalytic domains amongst the known cellulases and xylanases. Although a large number of CBD and catalytic domain combinations is possible (*47*), a relatively small number is observed (Table II). Family 8 and family 26 catalytic domains, and family 9 catalytic

Family IX

```
CthXynX1....QSAKALEGSPTIGANVDSSWKLVKPLYANTYVEGTV--GATATVKSMW
CthXynX2....QVATAIYGTPVIDGKIDDIWNKVDAITTNTWVLGS--DGATATAKMMW
TsaXynA1....QSAKALEGSPTIGANVDSSWKLVKPLYVNTYVEGTV--GATATVKSMW
TsaXynA2....QIATAIYGTPVIDGKVDDIWNNVEPISTNTWILGS--NGATATQKMMW
TmaXynA1....KESRISEGEAVVVGMMDDSYLMSKPIEILDE-EGNV----KATIRAVW
TmaXynA2....MVATAKYGTPVIDGEIDEIWNTTEEIETKAVAMGSLDKNATAKVRVLW


CthXynX1....DTKNLYLLVQVSDNT-------PSSNDGIEIFVDKNDNKSTSYETDDE
CthXynX2....DDKYLYVLADVTDSNLNKSSVNPYEQDSVEVFVDQNNDKTSYYESDDG
TsaXynA1....DTKNLYLLVQVSDNT-------PSNNDGIEIFVDKNDDKSTSYETDDE
TsaXynA2....DDKYLYVLADVTDSNLNKSSINPYEQDSVEVFVDQNNDKTTYYENDDG
TmaXynA1....KDSTIYIYGEVQDKT-----KKPAE-DGVAIFINPNNERTPYLQPDDT
TmaXynA2....DENYLYVLAIVKDPVLNKDNSNPWEQDSVEIFIDENNHKTGYYEDDDA


CthXynX1....HYTIKSD-----GTGSSDIT--KYVTSNAD-GYIVQLAIPIEDISPTL
CthXynX2....QYRVNYDNEQSFGGSTNSNGF-KSATSLTQSGYIVEEAIPWTSITLLN
TsaXynA1....RYTIKRD-----GTGSSDIT--KYVTSNAD-GYVAQLAIPIEDISPAV
TsaXynA2....QYRVNYDNEQSFGGSTNSNGF-KSATSLTQSGYIVEEAIPWTSITPSN
TmaXynA1....-YAVLWTNWKT-EVNREDVQVKKFVGPGFRR-YSFEMSITIPGVEFKK
TmaXynA2....QFRVNYMNEQTFGTGGSPARF-KATVKLIEGGYIVEAAIKWKTIKPTP


CthXynX1....NDKIGLDVRLNDDKGSGSIDTVTVWNDYT-NSQDTNTSYFGDIVLS-K
CthXynX2....GTIIGFDLQVNDADENGKRTGIVTWCDPSGNSWQD-TSGFGNLLLTGK
TsaXynA1....NDKIGFDIRINDDKGNGKIDAITVWNDYT-NSQNTNTSYFGDIVLS-K
TsaXynA2....GTIIGFDLQVNNADENGKRTGIVTWCDPSGNSWQD-TSGFGNLLLTGK
TmaXynA1....DSYIGFDAAVIDD---GKW---YSWSDTT-NSQKTNTMNYGTLKLEGI
TmaXynA2....NTVIGFNIQVNDANEKGQRVGIISWSDPTNNSWRD-PSKFGNLRLI-K
```

Family X

```
CmiXynA...SSAVTGNACQCNWWGTRYPLCTNTASGWGWENNTSCITTSTCNSQGAGGGGVVCN
PflEglA...SSSSVSGGLRCNWYGTLYPLCVTTQSGWGWENSQSCISASTCSAQPAPYGIVGAA
PflEglB...SSSVLTGAQACNWYGTLTPLCNNTSNGWGYEDGRSCVARTTCSAQPAPYGIVSTS
PflCelC...SSSVVSGGGQCNWYGTLYPLCVSTTSGWGYENNRSCISPSTCSAQPAPYGIVGGS
PflCelE...VSSAVSG-QQCNWYGTLYPLCSTTTNGWGWENNASCIARATCSGQPAPWGIVGGS
PflXynA...TPGSSSGNQQCNWYGTLYPLCVTTTNGWGWEDQRSCIARSTCAAQPAPFGIVGSG
PflXynE...AGGNTGGNCQCNWWGTFYPLCQTQTSGWGWENSRSCISTSTCNSQGTGGGGVVCN
```

Figure 6. Sequence alignments of family IX and family X CBDs. Organism and enzyme abbreviations are explained in Table II.

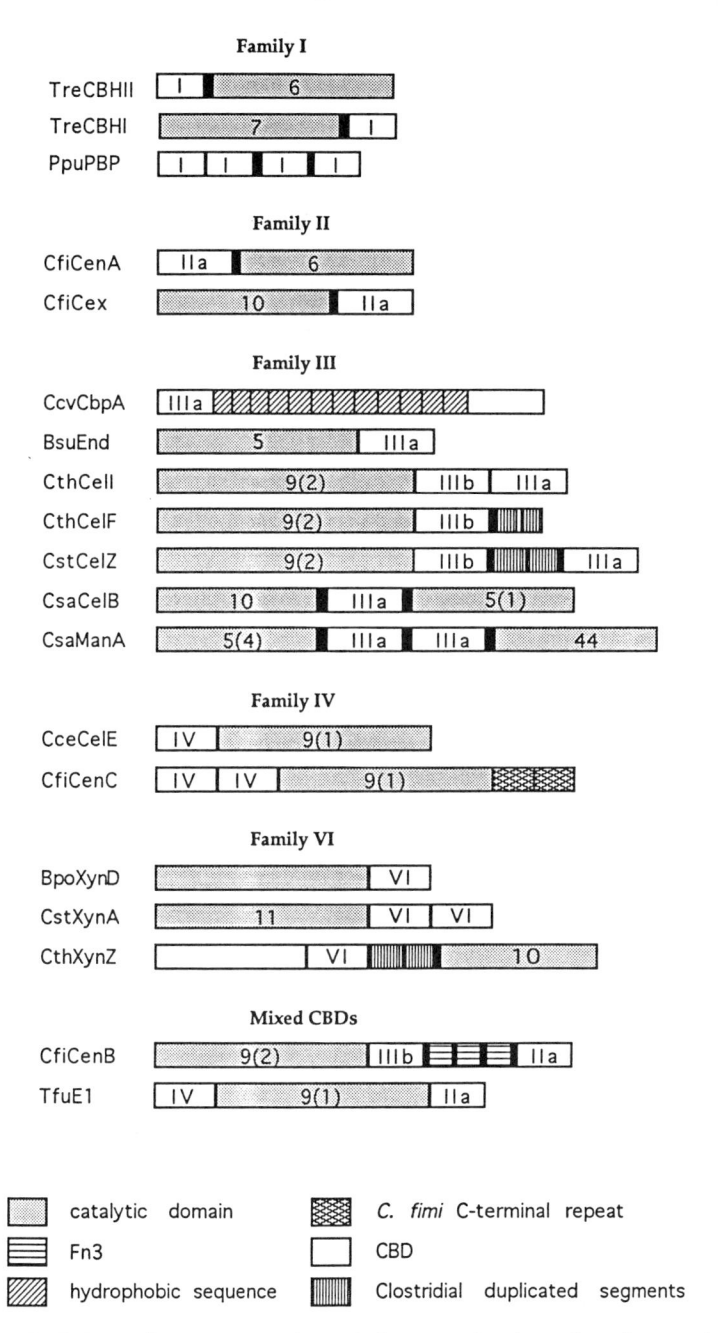

Figure 7. Schematic representation of the organization of various β-1,4-glycanases and non-catalytic proteins containing CBDs. Roman numerals denote CBD families; lower case letters denote CBD subfamilies. Arabic numerals denote catalytic domain families; the catalytic domain subfamily is shown in parentheses where appropriate.

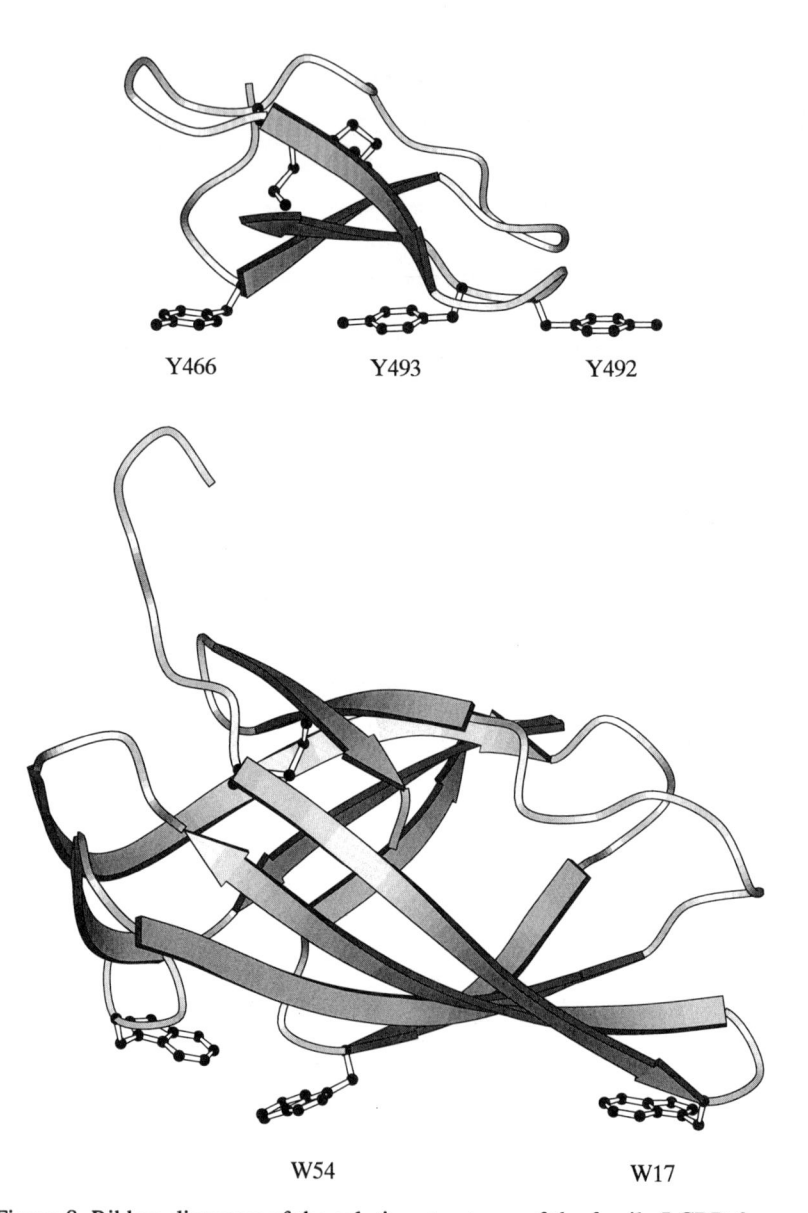

Figure 8. Ribbon diagrams of the solution structures of the family I CBD from *T. reesei* CBH I (top) and the subfamily IIa CBD from *C. fimi* Cex (bottom). β-sheets are represented as arrows. Aromatic residues implicated in binding to cellulose and disulfide bonds are shown in "ball and stick" form.

domains of plant origin do not appear to require CBDs at all. Other CBD families show a marked tendency to occur in combination with particular catalytic domain families. For example, most family IV CBDs occur in cellulases with family 9 (subfamily 1) catalytic domains. Similarly, all known family IX CBDs occur as tandem repeats joined to family 10 catalytic domains. Other, less obvious trends can be discerned from an examination of the data in Table II. A tendency for CBDs from some families to occur at particular positions in enzymes is also noticeable. Family I CBDs are found at the N or C terminus, but not internally. All known family IV CBDs are N-terminal whereas all known family IX CBDs are C-terminal. In contrast, all known subfamily IIIb are internal. It is anticipated that such observations will provide important clues about enzyme function and domain interactions and contribute to a more detailed understanding of the mechanisms of biomass degradation.

Acknowledgments

We thank Laurence Macintosh for providing Figure 8.

Literature Cited

1. Coutinho, P. M.; Reilly, P. J. *Protein Engineering* **1994**, *7*, 393-400.
2. Flach, J.; Pilet, P.- E.; Jollès, P. *Experimentia* **1992**, *48*, 701-716.
3. Van Tilbeurgh, H.; Tomme, P.; Claeyssens, M.; Bhikhabhai, R.; Pettersson, G. *FEBS Lett.* **1986**, *204*, 223-227.
4. Gilkes, N. R.; Warren, R. A. J.; Miller, R. C., Jr.; Kilburn, D. G. *J. Biol. Chem.* **1988**, *263*, 10401-10407.
5. Tomme, P.; Van Tilbeurgh, H.; Pettersson, G.; Van Damme, J.; Vandekerckhove, J.; Knowles, J.; Teeri, T.; Claeyssens, M. *Eur. J. Biochem.* **1988**, *170*, 575-581.
6. Henrissat, B.; Claeyssens, M.; Tomme, P.; Lemsle, L.; Mornon, J.-P. *Gene* **1989**, *81*, 83-95.
7. Gilkes, N. R.; Henrissat, B.; Kilburn, D. G.; Miller, R. C., Jr.; Warren, R. A. J. *Microbiol. Rev.* **1991**, *55*, 303-315.
8. Henrissat, B.; Bairoch, A. *Biochem. J.* **1993**, *293*, 781-788.
9. Béguin, P.; Aubert, J.-P. *FEMS Microbiol. Rev.* **1994**, *13*, 25-58.
10. Tomme, P.; Warren, R. A. J.; Gilkes, N. R. *Adv. Microbial Physiol.* **1995**, *37*, 1-81
11. Claeyssens, M.; Henrissat, B. *Prot. Sci.* **1992**, *1*, 1293-1297.
12. Gebler, J.; Gilkes, N. R.; Claeyssens, M.; Wilson, D. B.; Béguin, P.; Wakarchuk, W. W.; Kilburn, D. G.; Miller, R. C., Jr.; Warren, R. A. J.; Withers, S. G. *J. Biol. Chem.* **1992**, *267*, 12559-12561.
13. Coutinho, J. B.; Gilkes, N. R.; Warren, R. A. J.; Kilburn, D. G.; Miller, R. C., Jr. *Molec. Microbiol.* **1992**, *6*, 1243-1252.
14. Hefford, M. A.; Laderoute, K.; Willick, G. E.; Yaguchi, M.; Seligy, V. L. *Protein Eng.* **1992**, *5*, 433-439.
15. Coutinho, J. B.; Gilkes, N. R.; Kilburn, D. G.; Warren, R. A. J.; Miller, R. C., Jr. *FEMS Microbiol. Lett.* **1993**, *113*, 211-218.
16. Knowles, J.; Lehtovaara, P.; Teeri, T. *Tibtech* **1987**, *5*, 255-261.
17. Teeri, T.; Reinikainen, T.; Ruohonen, L.; Jones, T. A.; Knowles, J. K. C. *J. Biotechnol.* **1992**, *24*, 169-176.
18. Altschul, S. F.; Gish, W.; Miller, W.; Myers, E. W.; Lipman, D. J. *J. Mol. Biol.* **1990**, *215*, 403-410.
19. Claeyssens, M.; Tomme, P. *Enzyme Systems of Lignocellulose Degradation* **1989**, 37-49.
20. Reinikainen, T.; Ruohonen, L.; Nevanen, T.; Laaksonen, L.; Kraulis, P.; Jones, T. A.; Knowles, J. K. C.; Teeri, T. T. *Proteins* **1992**, *14*, 475-482.

21. Kraulis, P. J.; Clore, G. M.; Nilges, M.; Jones, T. A.; Petterson, G.; Knowles, J.; Gronenborn, A. M. *Biochemistry* **1989**, *28*, 7241-7257.
22. Srisodsuk, M.; Reinikainen, T.; Penttilä, M.; Teeri, T. T. *J. Biol. Chem.* **1993**, *268*, 20756-20761.
23. Tomme, P.; Driver, D. P.; Amadoron, E. A.; Miller, R. C., Jr.; Warren, R. A. J.; Kilburn, D. G. *J. Bacteriol.* **1995,** *In Press*
24. Meinke, A.; Gilkes, N. R.; Kilburn, D. G.; Miller, R. C. J.; Warren, R. A. J. *Protein Seq. Data Anal.* **1991,** *4*, 349-353.
25. Gilkes, N. R.; Claeyssens, M.; Aebersold, R.; Henrissat, B.; Meinke, A.; Morrison, H. D.; Kilburn, D. G.; Warren, R. A. J.; Miller, R. C., Jr. *Eur. J. Biochem.* **1991,** *202*, 367-377.
26. Irwin, D.; Jung, E. D.; Wilson, D. B. *Appl. Environ. Microbiol.* **1994,** *60*, 763-770.
27. Shareck, F.; Biely, P.; Morosoli, R.; Kleupfel, D. *Gene* **1995,** *153*, 105-109.
28. Millward-Sadler, S. J.; Poole, D. M.; Henrissat, B.; Hazlewood, G. P.; Clarke, J. H.; Gilbert, H. J. *Mol. Microbiol.* **1994,** *11*, 375-382.
29. Gilkes, N. R.; Jervis, E.; Henrissat, B.; Tekant, B.; Miller, R. C., Jr.; Warren, R. A. J.; Kilburn, D. G. *J. Biol. Chem.* **1992,** *267*, 6743-6749.
30. Ong, E.; Gilkes, N. R.; Miller, R. C., Jr.; Warren, R. A. J.; Kilburn, D. G. *Biotechnol. Bioeng.* **1993,** *42*, 401-409.
31. Tomme, P.; Gilkes, N. R.; Miller, R. C., Jr.; Warren, R. A. J.; Kilburn, D. G. *Protein Eng.* **1994,** *7*, 117-123.
32. Din, N.; Forsythe, I. J.; Burtnick, L. D.; Gilkes, N. R.; Miller, R. C. J.; Warren, R. A. J.; Kilburn, D. G. *Molec. Microbiol.* **1994,** *11*, 747-755.
33. Greenwood, J. M.; Gilkes, N. R.; Kilburn, D. G.; Miller, R. C., Jr.; Warren, R. A. J. *FEBS Lett.* **1989,** *244*, 127-131.
34. Poole, D. B.; Hazlewood, G. P.; Huskisson, N. S.; Virden, R.; Gilbert, H. J. *FEMS Microbiol. Lett.* **1993,** *106*, 77-84.
35. Din, N.; Gilkes, N. R.; Tekant, B.; Miller, R. C. J.; Warren, R. A. J.; Kilburn, D. G. *Bio/Technology* **1991,** *9*, 1096-1099.
36. Gilkes, N. R.; Chanzy, H.; Kilburn, D. G.; Miller, R. C., Jr.; Sugiyama, J.; Warren, R. A. J.; Henrissat, B. *Int. J. Biol. Macromol.* **1993,** *15*, 347-351.
37. Bergquist, P. L.; Gibbs, M. D.; Saul, D. J.; Te'o, V. S. J.; Dwivedi, P. P.; Morris, D.; Donald, A.; Kessler, A. *CTAPI Seventh International Symposium on Wood and Pulping Chemistry* **1993,** 276-285.
38. Shoseyov, O.; Tagaki, M.; Goldstein, M. A.; Doi, R. H. *Proc. Natl. Acad. Sci. USA* **1992,** *89*, 3483-3487.
39. Meinke, A.; Gilkes, N. R.; Kilburn, D. G.; Miller, R. C., Jr.; Warren, R. A. J. *J. Bacteriol.* **1991,** *173*, 7126-7135.
40. Goldstein, M. A.; Takagi, M.; Hashida, S.; Shoyeyov, O.; Doi, R. H.; Segel, I. H. *J. Bacteriol.* **1993,** *175*, 5762-5768.
41. Sakka, K.; Kojima, Y.; Kondo, T.; Karita, S.; Ohmiya, K.; Shimada, K. *Biosci. Biotech. Biochem.* **1993,** *57*, 273-277.
42. Winterhalter, C.; Heinrich, P.; Candussio, A.; Wich, G.; Liebl, W. *Mol. Microbiol.* **1995,** *15*, 431-444.
43. Lamed, R.; Tormo, J.; Chirino, A. J.; Morag, E.; Bayer, E. A. *J. Mol. Biol.* **1994,** *244*, 236-237.
44. Béguin, P and Aubert, J.P. *Ann. Rev. Microbiol.* **1990,** *44*, 219-248.
45. Xu, G.-Y.; Ong, E.; Gilkes, N. R.; Kilburn, D. G.; Muhandiran, D. R.; Brandeis, M.; Carver, J. P.; Kay, L. E.; Harvey, T. S. *Biochemistry* **1995,** *34*, 6993-7009.
46. Pflugrath, J. W.; Wiegrand, G.; Huber, R. *J. Mol. Biol.* **1986,** *189*, 383-386.
47. Henrissat, B. *Cellulose* **1994,** *1*, 169-196.
48. Saloheimo, A.; Henrissat, B.; Penttilä, M. *Trichoderma reesei Cellulases and Other Hydrolases* **1993,** 139-146.

49. Covert, S. F.; Vanden Wymeleberg, A.; Cullen, D. *Appl. Environ. Microbiol.* **1992**, *58*, 2168-2175.
50. Tempelaars, C.; Birch, P. R. J.; Sims, P. F. G.; Broda, P. *Appl. Environ. Microbiol.* **1994**, *60*, 4387-4393.
51. Al-Tawheed, A. R. **1988,** *Ph.D. thesis*.
52. Lin, F.; Marchenko, G.; Cheng, Y.-R. *J. Ind. Microbiol.* **1994**, *13*, 344-350.
53. Bergquist, P. L.; Gibbs, M. D.; Saul, D. J.; Te'O, V. S. J.; Dwivedi, P. P.; Morris, D. *Genetics, Biochemistry and Ecology of Lignocellulose Degradation* **1993,** 276-285.
54. Seki, N.; Muta, T.; Oda, T.; Iwaki, D.; Kuma, K.; Miyata, T.; Iwanaga, S. *J. Biol. Chem.* **1994,** *269*, 1370-1374.
55. Kraulis, J.P. *J. Appl. Crystallogr.* **1991**, *5*, 802-810.

RECEIVED September 12, 1995

Chapter 11

Identification of Two Tryptophan Residues in Endoglucanase III from *Trichoderma reesei* Essential for Cellulose Binding and Catalytic Activity

Ricardo Macarrón[1,4], Bernard Henrissat[2], Jozef van Beeuman[3], Juan Manuel Dominguez[1,5], and Marc Claeyssens[3]

[1]Departamento de Bioquimica y Biologia Molecular I, Facultad de Cienceas Quimicas, Universidad Complutense, Madrid, Spain
[2]Centre de Recherches sur les Macromolécules Végétales, Centre National de la Recherche Scientifique, B.P. 53X, F-38041, Grenoble, France
[3]Laboratory of Biochemistry, Department of Physiology, Microbiology, and Biochemistry, University of Ghent, K. L. Ledeganckstraat 35, B-9000 Ghent, Belgium

Three tryptophan residues are readily oxydized by N-bromosuccinimide in endoglucanase III from *Trichoderma reesei*. Evidence was obtained that the residue first modified (Trp[5]) is situated in the cellulose-binding domain and the second (Trp[255]) in the enzyme's catalytic site. The latter influences the binding and hydrolysis of soluble substrates and more specifically that of chromophoric cellotriosides. The modification of a third residue does not further affect the catalytic properties.

Van der Waals contacts and hydrogen bonds are dominant forces in carbohydrate-binding proteins. Aromatic amino acid residues (tryptophan, tyrosine) are thought to pack onto the sugar rings, conferring additional specificity and stability to the ligand-protein complexes (*1*).

Specific chemical modification, e.g. oxidation with N-bromosuccinimide, can be used to evaluate the importance of tryptophan residues in biologically active proteins (*2,3*). This has been demonstrated for many carbohydrate-binding proteins such as xylanases (*4*), galactosidases (*5*), glucoamylases (*6,7*), lectins (*8,9*) and cellulases (*10-12*).

The intact form of endoglucanase III (EG III) from *Trichoderma reesei*, presently studied, contains a N-terminal cellulose-binding domain (CBD) linked by a Pro-Ser-Thr rich sequence to the catalytic domain (*13*). The latter has been classified as a family A cellulase (*14*) and this core-protein can be isolated from *T. reesei* culture broths (*15*). The CBD, a homologous region in all *Trichoderma* cellulases (*16*), contains four conserved tryptophan and/or tyrosine residues. The three-dimensional structure of the CBD from cellobiohydrolase I (CBH I) has been elucidated (*17*). It is a wedge-like structure containing a hydrophobic and a hydrophylic surface. The importance of aromatic amino acid residues for adsorption has been demonstrated by site-directed

[4]Current address: SmithKline Beecham S. A., Ctiago Grisolia 4, E-28760 Madrid, Spain
[5]Current address: Glaxo, C/Severo Ochoa 2, E-28760 Madrid, Spain

mutagenesis (*18*). In the catalytic domain several tryptophan residues are present, some possibly involved in substrate binding.

The present study deals with the influence of NBS oxidation of tryptophan residues on cellulose binding and catalytic activity of EG III. Relevant data from peptide mapping and sequence analysis are given. These results supplement published data obtained during modification and site-directed mutation studies with this and other family A cellulases.

Experimental

Materials. NBS (Aldrich, Belgium) was recrystallized twice from distilled water. Trypsin (Ref.nr. 37260) was from Serva (Germany) and microcrystalline cellulose (Avicel pH 101) from FMC Corp. (USA). Cellooligosaccharides, their chromophoric and methyl glycoside derivatives were prepared as described (*19*). All other reagents and chemicals were analytical grade.

Enzyme Purification and Properties. EG III was purified from *T. reesei* QM 9414, grown on wheat straw (*20*), as previously described (*19*). EG III core protein was a kind gift of Dr. Pettersson (Uppsala). The homogeneity of the preparations was proven by SDS-PAGE and analytical isoelectric focusing (*19*). Concentrations were determined from the absorption coefficients at 280 nm: 77 000 $M^{-1}cm^{-1}$ for the intact protein and 63 600 $M^{-1}cm^{-1}$ for the core (*13,15*).

Adsorption Experiments and Activity Measurements. 200 µl Samples containing 0.6 µM EG III and freshly swollen Avicel (1 % w/v in 0.1 M sodium acetate, pH 5.0) were rotated in 1.5 ml Eppendorf centrifuge tubes for 30 min at 4°C (hydrolysis of Avicel during adsorption could be neglected under these conditions). After centrifugation, the concentration of enzyme remaining in the supernatant was determined by assaying activity towards 1 mM $CNP(Glc)_3$. The release of 2-chloro,4-nitrophenol was monitored continuously at 405 nm (*21*). With $MeUmb(Glc)_3$ (100 µM in 0.1 M sodium acetate pH 5.0, at 30°C) liberated 4-methylumbelliferone was measured fluorimetrically (*21*). The hydrolysis patterns and rates for cellooligosaccharides and both chromophoric trisaccharides were determined by HPLC as described (*22*); the reactions were performed in 10 mM pyridine/acetic acid buffer, pH 5.0, at 30°C with EG III 50 nM.

Chemical Modifications. Oxidation of tryptophan residues with NBS was performed according to Spande and Witkop (*2*). Modification was followed by difference absorption spectrophotometry (Uvikon 810, Kontron, Switzerland) using two matched and thermostatted (15°C) 1 ml quartz cuvettes, containing equal volumes of 9.1 µM EG III in 50 mM sodium acetate buffer (pH 5.0), one as reference, the other as sample cell. 20 µl Aliquots were taken from each cell for activity assays and 20 µl 0.15 mM NBS (in bidistilled water) was added to the sample solution (20 µl of bidistilled water to the reference cell) with rapid mixing. After 2 min a difference spectrum was again recorded. To 20 µl aliquots subsequently withdrawn an adequate amount of 0.1 mM tryptophan was added to quench further oxidation of the protein. Stepwise addition of the oxydizing reagent to the sample cell (water for the reference solution) was continued until the protein solution became turbid. The number of oxydized Trp was calculated from the spectral characteristics and residual activities (against $CNP(Glc)_3$ and $MeUmb(Glc)_3$) and adsorption (onto Avicel) determined for each 20 µl aliquot taken. The same procedure was employed to follow the oxidation of EG III core and in the experiments using protecting ligands. Larger quantities of enzyme with different degrees of Trp modification (one or two residues oxydized) were prepared by adding 10 µl NBS (respectively, 1.5 mM and 3 mM) to 90 µl 100 µM EG III (50 mM sodium acetate, pH 5.0). EG III core with one Trp residue oxydized was prepared similarly. L-Cysteinyl group modification was performed with Ellman's reagent [5,5'-dithiobis(2-nitrobenzoic acid)], according to the method described by Habeeb (23). Nitration of tyrosine residues was according to Sokolovsky et al. (24).

Thermal Stability and Circular Dichroism Spectra. 150 µl Aliquots from 3.4 µM stock solutions of unmodified EG III and modified EG III (respectively one and two Trp oxydized) were incubated for 30 min at several temperatures (40-75°C) in 0.1 M sodium acetate, pH 5.0. After cooling the samples (4°C) the remaining activity was measured with CNP(Glc)$_3$ as substrate. Circular dichroism spectra of intact and oxydized EG III were recorded with a Jobin Yvon Mark III dichrograph (France). Samples were analyzed in 0.1 cm optical path cells, in the far ultraviolet region (< 250 nm). Molar ellipticities were determined using 106.2 Da as the average molecular mass per residue for EG III.

Isolation and Analysis of Peptides from Unmodified and Oxydized EG III. 300 µl Samples of intact and modified EG III or EG III core (0.5 mg in 0.4 M ammonium bicarbonate, pH 7.8) were reduced (8 M urea, 1.7 mM dithiothreitol, 10 min at 50°C) and carboxymethylated (3.7 mM iodoacetamide, 15 min at 37°C). After 1:4 dilution with water, 45 µg trypsin was added and the mixture was incubated for 10 h at 37°C. The resulting peptides were analyzed by reversed-phase liquid chromatography (Waters chromatographic system, USA) on a C4-Vydac 214 TP-54 column (0.46 x 25 cm) using a 1 ml min^{-1} linear gradient of 5-50 % acetonitrile in 0.1 % trifluoroacetic acid and monitoring 190 to 320 nm absorbancies of the eluates with a photo-diode array detector (Waters, USA). Selected peptides were concentrated (30 µl) and stored at -20°C. Amino-terminal sequence analysis was carried out on a 477A pulsed-liquid sequenator with on-line analysis of the PTH-aminoacids on a 120A PTH-analyzer (Applied Biosystems, USA), following the protocol suggested by the manufacturer. Plasma desorption mass spectrometry (PDMS) was performed on a BIO ION 20 Biopolymer Analyzer (Applied Biosystems, USA). Spectra were recorded for 1 million fission counts of the Californium^{252} source. Desorbed analyte ions were detected with a channel plate detector operated at -2 kV (for the first plate), the measurements being carried out in the 'positive mode'.

Results

N-Bromosuccinimide Oxidation of EG III and its Core Protein. As shown in Figure 1, NBS added in equimolar or small excess quantities effectively oxidises tryptophan residues in EG III: absorbance decrease at 280 nm and the concomitant increase at 250 nm are consistent with the formation of oxindolealanine derivatives (2). The degree of oxidation is linear with reagent concentration (Figure 1, inset). After modification of 3 Trp residues (4.5 molar excess of NBS) further oxidation leads to the appearance of turbidity as often seen with other proteins, e.g. α-chymotrypsin (25), wheat-germ agglutinin (8), and lysozyme (26).

Although under the conditions used NBS mainly oxidizes Trp residues, modification (and peptide cleavage) at Tyr, Cys, Met, and His can occur (2-3). In the present study this possibility can be discarded considering the specific spectrophotometric changes observed (Figure 1) and the low excess of reagent needed (1.5 mol NBS per Trp). This is almost identical to that found for the oxidation of the model compound N-acetyl-L-tryptophan and, compared with other proteins, one of the lowest values reported (2). The appearance of a single isosbestic point at 264 nm throughout modification (Figure 1) further indicates the absence of Tyr oxidation (27). Reaction with Ellman's reagent under denaturing conditions (6 M guanidine

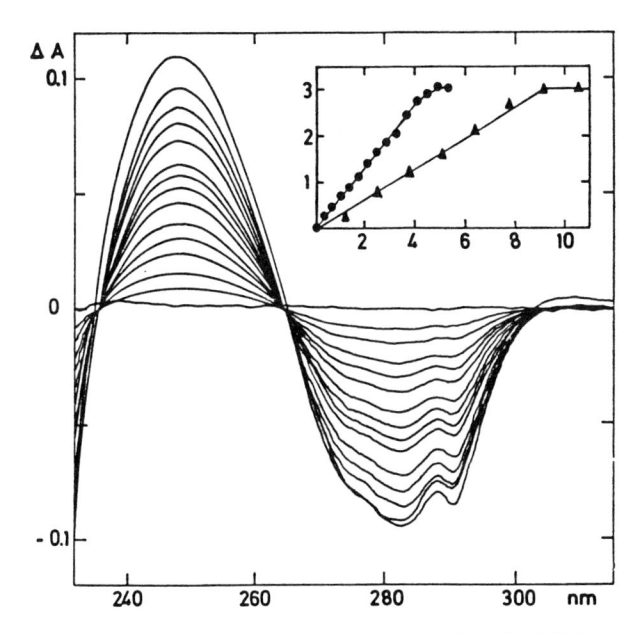

*Figure 1. Difference absorption spectra registered during NBS oxidation of EG III. **Inset**: NBS oxidation of EG III as a function of reagent concentration. Molar ratios of NBS to EG III on abscissas, number of Trp oxydized on ordinates. The modification was performed in the absence of ligands (circles), or in the presence of 10 mM methyl-β-cellotetraoside (triangles).*

hydrochloride) fails to reveal the presence of free sulphydryl groups in EG III. Nitration of Tyr with tetranitromethane (24) leads to partial inactivation which cannot be inhibited by specific ligands (results not shown). In this case the decrease (approx.50 %) of the activity towards the chloronitrophenyl cellotrioside (hydrolysis of chromophoric bond) is in contrast to the apparent increase (approx.150 %) in case of Trp modified EG III (see below)

Upon SDS-PAGE analysis of NBS oxydized EG III samples (1, 2 or 3 Trp modified) a single protein band with the same apparent molecular weight as that for the intact protein is observed (results not shown). Also no gross molecular changes can be observed by far UV circular dicroism spectroscopy (not shown). Thermal stability of EG III with two Trp oxydized is slightly decreased as compared to that of unmodified enzyme.

Protection by ligands. NBS modification of EG III is significantly retarded in the presence of cellooligosaccharides (10 mM methyl cellotetraoside or cellotriose; 100 mM cellobiose) (Figure 1, inset). The amount of reagent needed increases from 1.5 (control) to 2.9 mol NBS per mol oxidized Trp. At the same concentrations these ligands have no effect on the oxidation of N-acetyl-L-tryptophan-methyl ester. D-Glucose (100 mM) shows no significant binding affinity for the enzyme (19) and does not influence EG III oxidation by NBS. These results suggest that the oxidizable Trp residues are located in or near the binding site of one or both EG III functional domains (binding and catalytic) (15).

Influence of oxidation on binding and catalytic properties of EG III. Modification of the first Trp reduces EG III adsorption on Avicel by almost 50 % (Figure 2A), whereas only upon oxidation of a second Trp rates of the cellotrioside hydrolysis (cleavage at the chromoforic aglycon) are affected (Figure 2C). As described also in Table I, this is clearly dependent on the nature of the chromophore, modification influencing differently the ratios for cleavage at the holosidic vs heterosidic bonds. More extensive modification (3 Trp) does not further affect binding or catalytic properties.

Thus, in the initial stage of Trp oxidation only the adsorption of EG III on Avicel seems to be affected (Figure 2A). Since this binding is governed by the N-terminal cellulose binding domain (15,28), the first modified residue is probably located in this conserved peptide sequence. On the other hand, the effects on the hydrolysis of the small substrates appear only upon modification of a second Trp, probably at or near the enzyme's active site. To corroborate these observations the NBS oxidation of the catalytic core of EG III was examined separately (Figure 2D). The effects evident when the first Trp is oxidized in the core protein correspond with the changes in activity observed upon modification of the second Trp in the intact enzyme (Figure 2C). These findings confirm the location of the first modifiable residue in the binding domain and the presence of a second oxydizable Trp in the catalytic core domain.

The oxidation of the second Trp (the first residue in the core) causes a totally different effect on the hydrolysis of two chromophoric substrates (Table I). The hydrolysis patterns, analyzed by HPLC, reveal a shift in cleavage preference which indicates an apparent increase in chromophoric product formation in the case of the chloronitrophenyl derivative but a decrease in the case of the umbelliferyl cellotrioside. When natural substrates were used (cellotriose, cellotetraose and cellopentaose) the pattern of hydrolysis did not change upon modification of the enzyme but the activity was reduced to approximately 35 % of the initial. In the presence of specific ligands modification was inhibited, but, eventually, the same changes in adsorption and catalytic properties, with respect to the degree of Trp oxidation, were observed.

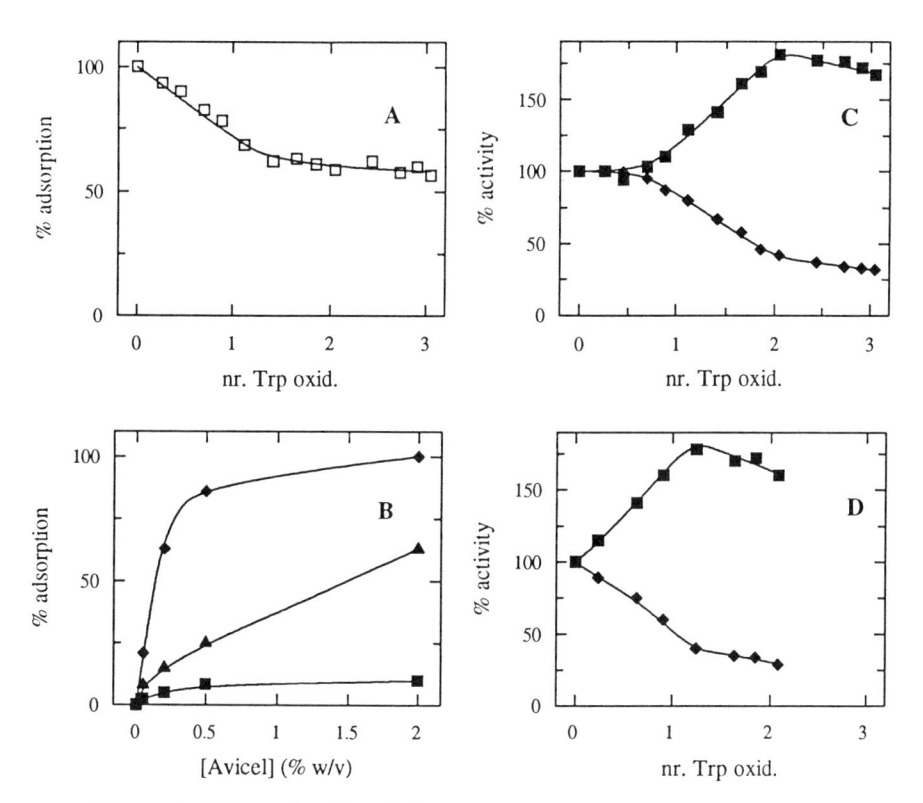

Figure 2. Effects of NBS modification on: A. relative adsorption of EG III on 1 % Avicel. B. adsorption isotherms for intact EG III (diamonds), oxydized (1 Trp) (triangles), and core protein (squares). C. residual activity against CNP(Glc)$_3$ (squares) and MeUmb(Glc)$_3$ (diamonds) as substrates (release of chromophore). D. residual activity (EG III core); symbols as for C.

Table I. Comparison of the hydrolytic patterns of chromophoric substrates by intact and oxydized (2 Trp) EG III. Ratios of hydrolysis rates (v_1 / v_2) determined with 1 mM CNP(Glc)$_3$ or 100 μM MeUmb(Glc)$_3$. (v_1, ß-D-glucopyranosyl ; v_2, chromophoric group)

Enzyme	CNP(Glc)$_3$	MeUmb(Glc)$_3$
Intact	1.3 ± 0.1	0.5 ± 0.05
Oxydized (2 Trp)	0.3 ± 0.05	0.8 ± 0.1

Identification of two Functional Trp Residues in EG III. Tryptic proteolysates of (i) intact, (ii) modified EG III (two Trp oxydized) or (iii) core (one Trp oxydized) were compared by reverse-phase HPLC. In the resulting peptide maps two regions were specifically affected by NBS oxidation (Figure 3A). In region "1" two fractions were isolated: a and b. Upon PDMS analysis peak "a" proved to be particularly heterogeneous and was not further investigated, whereas peak b contained three peptides: one main fraction (absent in the oxydized EG III and core protein) and two peptides consistently found in the three cases. They proved to be autolytic tryptic fragments (N-terminal sequences not shown). Cumulative evidence from PDMS and sequence analysis of the main peptide in b(i) (Table II) proved it to be a chymotryptic fragment from the N-terminal sequence in intact EG III (chymotryptic contamination of the commercial trypsin preparation). Thus, Trp^5, present in the cellulose binding domain, is probably oxydized by NBS. This peptide is not present in b(ii), probably since modified Trp is no longer recognized as a chymotryptic cleavage site, resulting in a larger peptide fragment (not detected). It is absent also in fraction b(iii), since, of course, the core enzyme lacks the N-terminal peptide.

Table II. Peptide analyses. Peptide digests were analysed by HPLC and fractions b, c, and d (Figure 3) examined by PDMS. N-terminal sequences for selected peptides are given. Data for (i) intact EG III, (ii) oxydized EG III (2 Trp) and (iii) core protein.

Fraction	PDMS M found (Da)	N-terminal sequence	Identification (M theoretical, Da)
b (i)	835.0 *	G N X G G I G W	Gly^6-Trp^{13} (834.7)
b (ii)	n. d. **	---	---
b(iii)	n. d. **	---	---
c (i)	1464.3	I S L P G N D W Q--	Ile^{248}-Phe^{263} (1462.6)
d (ii)	1478.7	I S L P G N D Wox Q--	Ile^{248}-Phe^{263} (1478.6)

* Main peptide detected
** Main 835 Da peptide non detected

The appearance of chymotryptic cleavage sites are also evident when peptides in region "2" are analysed (fraction c, Table II). Fraction d, specifically absorbing at 250 nm, appears in oxydized samples concommitant with a decrease in intensity of fraction c (Figure 3C). PDMS reveals the presence of 2-3 peptides in both c and d. Molecular

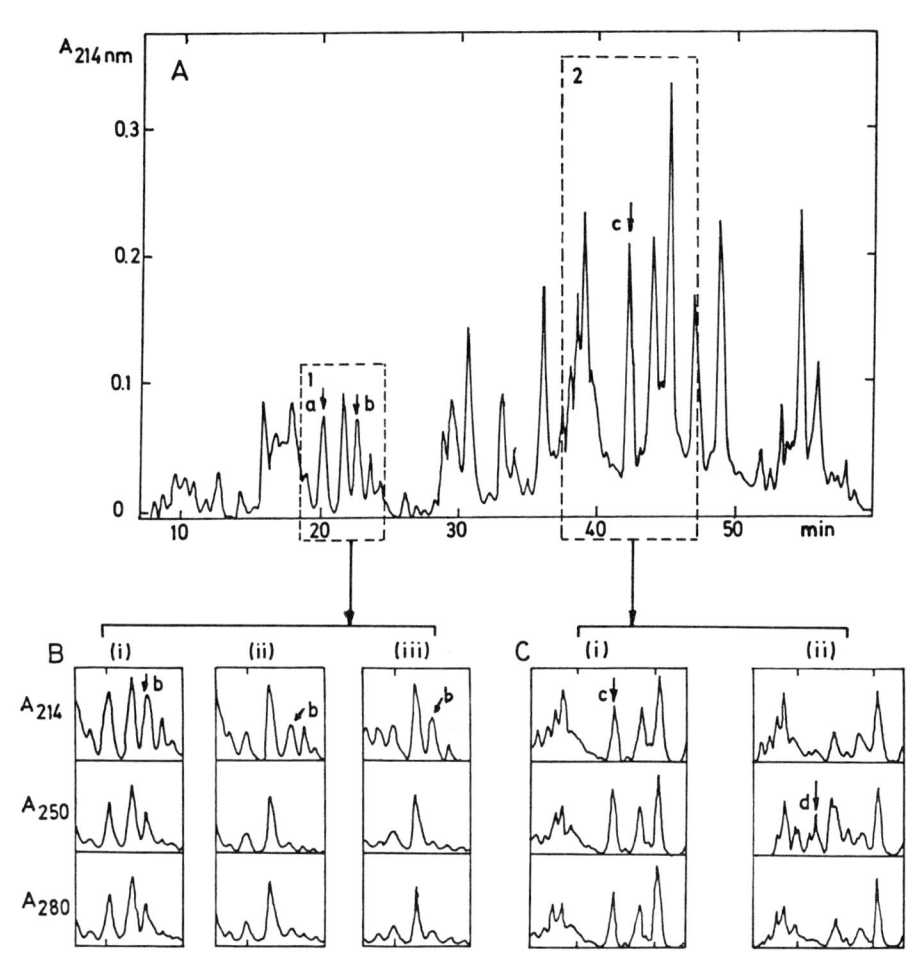

Figure 3. Tryptic peptide maps (HPLC): **A**. *Intact EG III (showing regions "1" and "2").* **B**. *Details of region "1" for tryptic proteolysates of (i) intact, (ii) oxydized EG III (2 Trp) and (iii) core protein (214, 250 and 280 nm tracings).* **C**. *Details of region "2" for (i) intact and (ii) oxydized EG III (2 Trp) (214, 250 and 280 nm tracings).*

mass and sequence data (Table II) give evidence for the presence of a peptide with a Trp containing sequence, intact in fraction c and oxydized in d. Chymotryptic cleavage at Phe[263] is deduced in both cases. Thus sufficient evidence is obtained to locate the second modifiable residue of EG III at Trp[255].

Discussion

Endoglucanase III from *Trichoderma reesei* contains 11 Trp residues. Its naturally occurring core-protein lacks the cellulose binding-domain and therefore counts two Trp less (*13,15*). These residues (Trp[5] and Trp[13]) are well conserved in all *Trichoderma* CBD's (*16*) and also in other fungal CBD's such as that of CBH I from *Phanerochaete chrysosporium* (*28*). Thus, function and location of these residues in the CBD's could be similar. The 3-D structure of the CBD from *T. reesei* CBH I has been determined (*17*). Tyr[466], equivalent to EG III Trp[5], is one of the three aromatic residues located on a flat hydrophilic face exposed to the solvent, while Tyr[474], equivalent to EG III Trp[13], is buried in the hydrophobic inner region of the CBH I CBD.

From our data Trp[5] can tentatively be considered to be the first target of NBS modification in EG III. Its rapid and selective oxidation leads to a approx. 50 % loss of binding capacity onto Avicel. The importance of the equivalent Tyr[466] residue in CBH I has been proved by chemical modification (Claeyssens and Tomme, unpublished) whereas the importance of another conserved residue, Tyr[462], located at the tip of the wedge-shaped domain, has been demonstrated by point-mutation experiments (*18*).

Modification of a second Trp residue (Trp[255]) does not further affect the binding behavior of the protein but leads to distinct changes in catalytic properties against soluble substrates. The same is valid when the first Trp in the EG III core protein is oxydized. Whereas the activity on cellodextrins is lowered (approx. 35 % residual), the hydrolysis pattern for these substrates is not affected. On the other hand, modification of Trp[255] has a different effect on the hydolysis of chromophoric substrates (Figure 2). Intact EG III cleaves MeUmb(Glc)$_3$ preferentially at the heterosidic bond while in CNP(Glc)$_3$ this is at the holosidic bond (Table I). Thus, oxidation of Trp[255] leads to an opposite effect in the hydrolysis patterns for these substrates. Although no evidence for gross overall changes in the cellulase could be found, this putative active site modification produces a dramatic change in specificity. It probably indicates specific interactions between the aromatic and/or glycosidic moieties of the substrates with the different subsites in the EG III active site.

Abbreviations

EG III, endoglucanase III from *Trichoderma reesei* ; CBD, cellulose-binding domain; NBS, N-bromosuccinimide; CNPGlc, 2'-chloro,4'-nitrophenyl ß-D-glucopyranoside; CNP(Glc)$_3$, 2'-chloro,4'-nitrophenyl ß-D-cellotrioside; MeUmbGlc, 4'-methylumbelliferyl ß-D-glucopyranoside, MeUmb(Glc)$_3$, 4'-methylumbelliferyl ß-D-cellotrioside; HPLC, high performance liquid chromatography; SDS-PAGE, sodium dodecyl sulfate-polyacrylamide gel electrophoresis; PDMS, plasma desorption mass spectrometry.

Acknowledgments

R. M. was recipient of a fellowship from the Spanish Ministerio de Educación y Ciencia. The authors wish to thank Dr. E. Messens for help with the peptide mapping. R. M. is indebted to Dr. C. Acebal and Dr. M. P. Castillón for encouragement. B. H.

thanks Organibio for a CM2AO grant and J. v. B. is indebted to the Concerted Research Action of the Flemish Government (contract 12052293).

Literature cited

1. Quiocho, F. A. *Ann. Rev. Biochem.* **1986**, *55*, 287.
2. Spande, T. F.; Witkop, B. *Methods Enzymol.* **1967**, *11*, 498.
3. Lundblad, R. L.; Noyes, C. M. *Chemical Reagents for Protein Modification*, CRC Press, West Palm Beach, FL, 1984, Vol. 2, pp. 47-71.
4. Keskar, S. S.; Srinivasan, M. C.; Deshpande, V. V. *Biochem. J.* **1989**, *261*, 49.
5. Mathew, C. D.; Balasubramaniam, K. *Phytochemistry* **1986**, *25*, 2439.
6. Clarke, A. J.; Svensson, B. *Carlsberg Res.Commun.* **1984**, *49*, 559.
7. Iwama, M.; Inokuchi, N.; Okazaki, Y.; Takahashi, T.; Yoshimoto, A.; Irie, M. *Chem. Pharm. Bull.* **1986**, *34*, 1355.
8. Privat, J.-P.; Lotan, R.; Bouchard, P.; Sharon, N.; Monsigny, M. *Eur. J. Biochem.* **1976**, *68*, 563.
9. Absar, N.; Yamasaki, N.; Funatsu, G. *Agr. Biol. Chem.* **1986**, *50*, 3071.
10. Hurst, P. L.; Sullivan, P. A.; Shepherd, M. G. *Biochem. J.* **1977**, *167*, 549
11. Clarke, A. J. *Biochim. Biophys. Acta* **1987**, *912*, 424.
12. Ozaki, K.; Ito, S. *J. Gen. Microbiol.* **1991**, *137*, 41.
13. Saloheimo, M.; Lehtovaara, P.; Penttilä, M.; Teeri, T. T.; Stahlberg, J.; Johansson, G.; Pettersson, G.; Claeyssens, M.; Tomme, P.; Knowles, J. K. C. *Gene* **1988**, *63*, 11.
14. Henrissat, B.; Claeyssens, M.; Tomme, P.; Lemesle, L.; Mornon, J.-P. *Gene* **1989**, *81*, 83.
15. Stahlberg, J.; Johansson, G.; Pettersson, G. *Eur. J. Biochem.* **1988**, *173*, 179.
16. Teeri, T. T.; Lehtovaara, P.; Kauppinen, S.; Salovuori, I.; Knowles, J. *Gene* **1987**, *51*, 43.
17. Kraulis, P. J.; Clore, G. M.; Nilges, M.; Jones, T. A.; Pettersson, G.; Knowles, J.; Gronenborg, A. M. *Biochemistry* **1989**, *28*, 7241.
18. Reinikainen, T.; Ruohonen, L.; Nevanen, T.; Laaksonen, L.; Kraulis, P.; Jones, T.A.; Knowles, J.K.C.; Teeri, T.T. *Proteins* **1992**, *14*, 475.
19. Macarrón, R.; Acebal, C.; Castillón, M. P.; Domínguez, J. M.; de la Mata, I.; Pettersson, G.; Tomme, P.; Claeyssens, M. *Biochem. J.* **1993**, *289*, 867.
20. Acebal, C.; Castillón, M. P.; Estrada, P.; Mata, I.; Costa, E.; Aguado, J.; Romero, D.; Jiménez, F. *Appl. Microbiol. Biotechnol.* **1986**, *24*, 218.
21. Claeyssens, M. *FEMS Symp.* **1989**, *43*, 393.
22. van Tilbeurgh, H.; Loontiens, F. G.; De Bruyne, C. K.; Claeyssens, M. *Methods Enzymol.* **1988**, *160*, 45.
23. Habeeb, A. F. S. A. *Methods Enzymol.* **1972**, *25*, 457.
24. Sokolovsky, M.; Riordan, J.F.; Vallee, B.L. *Biochemistry* **1966**, *5*, 3582.
25. Spande, T. F.; Green, N. M.; Witkop, B. *Biochemistry* **1966**, *5*, 1926.
26. Green, N. M.; Witkop, B. *Trans. N. Y. Acad. Sci.* **1964**, *26*, 659.
27. Ohnishi, M.; Kawagishi, T.; Abe, T.; Hiromi, K. *J. Biochem.* **1980**, *87*, 273.
28. Gilkes, N. R.; Henrissat, B.; Kilburn, D. G.; Miller Jr.; R. C.; Warren, R. A. J. *Microbiol. Rev.* **1991**, *55*, 303.

RECEIVED August 17, 1995

Chapter 12

Cellulomonas fimi Cellobiohydrolases

Hua Shen, Andreas Meinke, Peter Tomme, Howard G. Damude, Emily Kwan, Douglas G. Kilburn, Robert C. Miller, Jr., R. Antony J. Warren, and Neil R. Gilkes[1]

Department of Microbiology and Immunology, University of British Columbia, 300–6174 University Boulevard, Vancouver, British Columbia V6T 1Z3, Canada

The cellulolytic bacterium *Cellulomonas fimi* produces two cellobiohydrolases, CbhA and CbhB. These enzymes are major extracellular components during growth on cellulose. They correspond to previously identified cellulose-binding proteins, cbp95 and cbp120, respectively. Both comprise an N-terminal catalytic domain joined to three fibronectin type III modules and a C-terminal cellulose-binding domain. Amino acid sequence comparison shows that the CbhA catalytic domain is closely related to the catalytic domains of *Trichoderma reesei* CBH II and other fungal cellobiohydrolases in β-1,4-glucanase family B. CbhB is a member of a new β-1,4-glucanase family, designated family L (also called family 48). On the basis of data for CbhB, hydrolysis by family L enzymes proceeds with inversion of configuration at the anomeric carbon. CbhA and CbhB produce cellobiose from cellulose. Viscometric analysis of carboxymethylcellulose hydrolysis shows that both enzymes are predominantly exohydrolytic but Congo Red staining shows they also have weak endoglucanase activity. CbhB attacks cellohexaose from the reducing end. We suggest that *C. fimi* and *T. reesei* use similar strategies to hydrolyze cellulose because both appear to produce two types of cellobiohydrolase: one that attacks cellulose from the reducing end and another that attacks from the non-reducing end.

One of the most extensively studied cellulolytic microorganisms is *Trichoderma reesei*, a soft-rot fungus. The mechanism of crystalline cellulose hydrolysis by *T. reesei* is still not understood in detail but there is general agreement that it involves the concerted action of cellobiohydrolases (i.e., β-1,4-glucanases that remove cellobiosyl groups from the ends of cellulose

[1]Corresponding author

chains) and endo-β-1,4-glucanases. Two cellobiohydrolases are the major components of the *T. reesei* cellulase system. About 60% of the total extracellular protein is cellobiohydrolase I (CBH I). A second cellobiohydrolase, CBH II, accounts for a further 20% (1, 2).

Other soft-rot and white-rot fungi probably have similar cellulase systems. For example, *Penicillium pinophilum* and *Phanerochaete chrysosporium* both produce enzymes related to CBH I and CBH II (1-3). However, the relationship between these systems and cellulase systems from bacteria is less obvious. Do cellulolytic bacteria use a similar strategy, involving cellobiohydrolases related to CBH I and CBH II, or have they evolved different mechanisms to hydrolyze cellulose?

In this paper we review data on two cellobiohydrolases from the bacterium *Cellulomonas fimi*. CbhA is an enzyme whose structure and catalytic properties are closely related to *T. reesei* CBH II and is the first such enzyme to be identified in bacteria. The second cellobiohydrolase, CbhB, is not related to either CBH I or II: it is a member of a newly recognized family of β-1,4-glucanases designated family L. The family contains CelS, a major cellobiohydrolase from the bacterium *Clostridium thermocellum*, and several other β-1,4-glucanases from cellulolytic *Clostridium* spp..

Isolation of CbhA and CbhB

When *C. fimi* is grown on carboxymethylcellulose (Cm-cellulose) or other cellulosic substrates, it secretes five major proteins that bind to cellulose (Figure 1). The 53, 110 and 75 kDa cellulose-binding proteins (cbps) are endo-β-1,4-glucanases that we have already characterized (CenA, CenB and CenD, respectively) but N-terminal amino acid sequence analysis of the 95 and 120 kDa proteins (cbp95 and cbp120) showed that both were novel *C. fimi* proteins (4).

In order to characterize these proteins, their genes were isolated from a *C. fimi* genomic library and expressed in *Escherichia coli*. Expression in *E. coli* allowed the enzymes to be prepared free of contaminating β-1,4-glucanase activity. The analyses described below showed that cbp95 is a cellobiohydrolase; consequently, the protein was renamed CbhA. Initially, Cbp120 was thought to be an endo-β-1,4-glucanase and the enzyme was named CenE (5). Later, it was recognized that cbp120 is also a cellobiohydrolase and it was renamed CbhB.

General structural and functional organization of CbhA and CbhB

CbhA and CbhB are both multidomain enzymes and the arrangement of the domains in the two enzymes is similar. The same arrangement is seen in other *C. fimi* β-1,4-glucanases (Figure 2). Both enzymes have a C-terminal cellulose-binding domain (CBD) with amino acid sequences that resemble the CBDs of other *C. fimi* β-1,4-glucanases (6, 7). In both enzymes, the CBD is joined to the N-terminal catalytic domain by three fibronectin type III (Fn3) modules, each containing 95 to 98 amino acid residues. Fn3 modules are also found in *C. fimi*

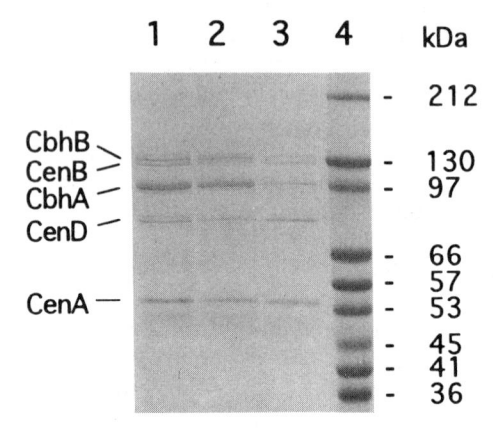

Figure 1. Cellulose-binding proteins in *C. fimi* culture supernatants. *C. fimi* was grown in basal medium containing 1% Cm-cellulose for 8 days. Cellulose-binding proteins (cbps) were adsorbed with Avicel (lane 1), bacterial microcrystalline cellulose (lane 2), or Sephadex (lane 3) and resolved by SDS-polyacrylamide gel electrophoresis. Molecular mass standards are shown in lane 4. CbhA and CbhB were originally identified as cbp95 and cbp120, respectively (4). (Reproduced with permission from Ref. 4. Copyright 1993 The American Society for Microbiology.)

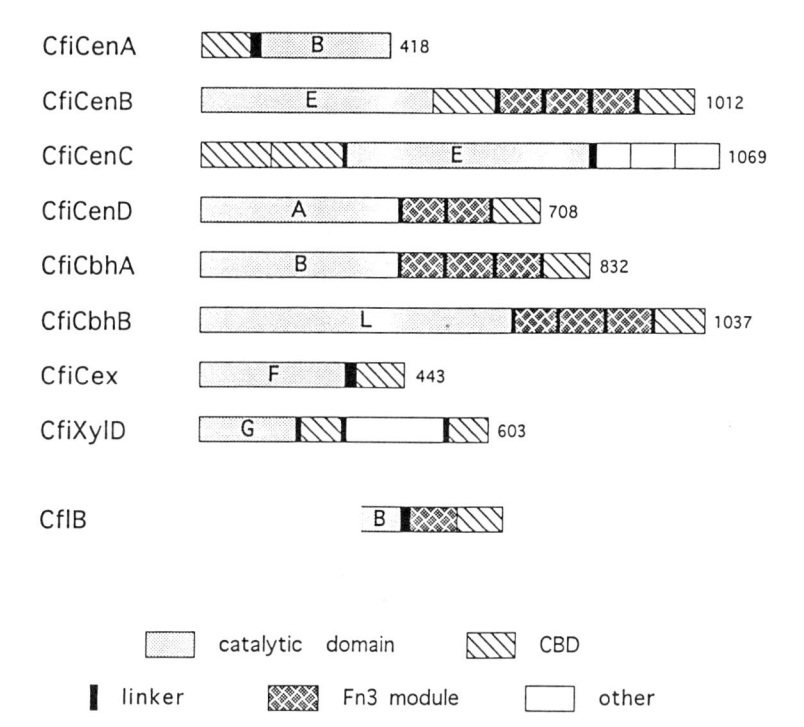

Figure 2. Structural and functional organization of C. *fimi* β-1,4-glycanases. The mature enzymes are represented as linear maps drawn approximately to scale. CenA (*56*), CenB (*57*), CenC (*58*) and CenD (*4*) are endoglucanases. CbhA (*16*) and CbhB (*5*) are cellobiohydrolases. Cex (*59*) is an exoglucanase and xylanase. XylD (*60*) is a xylanase. CflB, a translated open reading frame (ORFB) from C. *flavigena* (*11*), is also shown.

CenB and CenD (Figure 2), in several bacterial chitinases, and in other bacterial enzymes that catalyze hydrolysis of insoluble polymeric substrates (8-10). Bacterial hydrolases with Fn3 modules usually contain two or three modules in tandem although a single Fn3 module is encoded by an open reading frame from *Cellulomonas flavigena* (11) (Figure 2). The function of Fn3 modules in bacterial hydrolases is presently unknown.

Relationship of the CbhA and CbhB to other β-1,4-glucanases

Classification of β-1,4-glucanases and β-1,4-xylanases. The primary structures of more two hundred β-1,4-glucanases and β-1,4-xylanases from bacteria and fungi are now known and nearly all can be classified into one of a few families based on amino acid sequence comparison of their catalytic domains. Six families of β-1,4-glucanases and β-1,4-xylanases (families A - F) were originally recognized (6, 12). More recently, the classification was extended to include five more (G, H, I, J and K) (13, 14). The alternative family nomenclature, described in ref. 13, is as follows: A, 5; B, 6; C, 7; D, 8; E, 9; F, 10; G, 11; H, 12; I, 26; J, 44; K, 45. Some families (e.g., family A) contain only endoglucanases; others (e.g., family B) contain both endoglucanases and cellobiohydrolases.

Relationship of CbhA to *T. reesei* CBH II and other fungal family B cellobiohydrolases. The CbhA catalytic domain contains 431 amino acid residues. Amino acid sequence alignment reveals that CbhA belongs to β-1,4-glucanase family B (15, 16). The family contains five fungal cellobiohydrolases, including *T. reesei* CBH II, and six bacterial endo-β-1,4-glucanases (Figure 3). Three acidic residues in CBH II (D175, D221 and D401) implicated in catalysis and the cysteine residues that form two disulfide bonds in CBH II, CenA and *Thermomonospora fusca* E2 (17-19) are all conserved in CbhA (Figures 3 and 4).

The CBH II catalytic domain is a single domain α-β protein with a central β barrel of seven parallel strands (17). The active site is on the carboxyl side of the β barrel. Two large surface loops, one N-proximal, the other C-proximal, enclose the active site to give a tunnel-shaped structure (Figure 5). This structure is thought to restrict access to the ends of the β-1,4-glucan chains in cellulose and thus determine the exohydrolytic activity of the enzyme.

The family B amino acid sequence alignment reveals several discrete insertions in the catalytic domains of CBH II and the other fungal cellobiohydrolases that are absent from all the family B endoglucanases (15-17, 20). The locations of these insertions, in relation to the secondary structure of the CBH II catalytic domain, are shown in Figure 4. Most of them correspond to loops on the carboxyl side of the β barrel. Among them are two insertions that contribute to the formation the two large surface loops enclosing the CBH II active site tunnel (Figure 5). The insertion that forms the C-proximal loop over the CBH II active site, together with the corresponding regions from the family B endoglucanases, is shown in the alignment in Figure 3.

Bacterial cellobiohydrolases

```
CfiCbhA   ...AWCNPLGAGIGRFPEATPSGYAASHLDAFVWIKPGESDGASTDIPNDQGKRFDRMCDPTFVSPKLNNQLTGATPNAPLAGQWFEEQ...  418
CfIB         ...PGESDGASTEIENDQGKRFDRMCDPTFVSPKLSNNLTGATPNAPLAGQWFEDQ...  434
```

Fungal cellobiohydrolases

```
TreCBHII  ...DWCNVIGTGFGIRPSANTGD---SLLDSFVWVKPGGECDGTSDSSAP------RFDSHCALP-------------DALQPAPQAGAWFQAY...  434
AbiCel3A  ...DWCNVKGAGFGQRPTTNTGS---SLIDAIVWKPGGECDGTSDNSSP------RFDSHCSLS-------------DAHQPAPEAGTWFQAY...  426
AbiCel3B  ...DWCNVKGAGFGQRPTTNTGS---SLIDAIVWKPGGECDGTSDSSSP------RFDSHCSLS-------------DAHQPAPEAGTWFQAY...  426
FoxCel    ...DWCNAKGTGFGLRPSTNTGD---ALADAFVWVKPGGESDGTSDTSAA------RYDYHCGLD-------------DALKPAPEAGTWFQAY...  449
PchCbhII  ...DWCNIKGAGFGTRPTNTGS---QFIDSIVWVKPGGECDGTSNSSSP------RYDSTCSLP-------------DAAQPAPEAGTWFQAY...  447
```

Bacterial endoglucanases

```
CfiCenA   ...EWCNPRGRALGERPVAVNDG---SGLDALLWVKLPGESDGA----------------CNGG--------------PAAGQWWQEI..  408
MbiCelA   ...EWCDPPGRATGTWSTTDTGD---PAIDAFLWIKPPGEADG----------------CIAT--------------PGVFVPDR...  279
MceCenA   ...DWCADDNTDRRIGQYPTTNT-GDANIDAYLWVKPPGEADG----------------CATR                GSFQPDL...  292
ShaCelA   ...EWCNPSGRRIGTPTRTGGG-------AEMLLMIKTPGESDGN--------------CGVG--------------SGSTAGQFLPEV...  286
SspCasA   ...EWCDPPGRLVGNNPTVNGV---PGVDAFLWIKLPGELDG----------------CDGP-------------AGSFSPAK...  281
TfuEgE2   ...EWCDPSGRAIGTPSTTNTGD---PMIDAFLWIKLPGEADG----------------CIAG--------------AGQFVPQA...  278
                                                                                ↑
                   *                                                            *
```

Figure 3. Comparison of the amino acid sequences from C-proximal regions of the catalytic domains of family B β-1,4-glucanases. The underlined portion of the CBHII sequence corresponds to the C-proximal loop covering the active site (see Fig. 5). CfiCbhA (GenBank L25809) and CfiCenA (GenBank M15823) are C. fimi CbhA and CenA, respectively; CfIB (see Fig. 2) is a translation of open reading frame B from C. flavigena (11); TreCBH II (GenBank M16190) is T. reesi CBH II; AbiCel3A (GenBank L24519) and AbiCel3B (GenBank L24520) are Agaricus bisporus cellobiohydrolases Cel3A and Cel3B, respectively; FoxCel (GenBank L29377) is a Fusarium oxysporum β-1,4-glucanase; PchCBH II is P. chrysosporium cellobiohydrolase II (61); MbiCelA (SWISS-PROT P26414) is M. bispora CelA; MceCenA is Micromonospora cellulolyticum McenA (62); ShaCelA (SWISS-PROT P33682) is Streptomyces halstedii CelA1; SspCasA (GenBank L03218) is Streptomyces sp. CasA; TfuE2 (GenBank M73321) is T. fusca endoglucanase 2. Each sequence is numbered from the first residue of the mature polypeptide. Hyphens indicate gaps introduced to improve the alignment. Asterisks mark a pair of cysteine residues that form a disulfide bond (18). An arrow marks the putative base catalyst (17, 63). Problems with sequence alignment of CasA (17) were resolved by re-sequencing the gene (64).

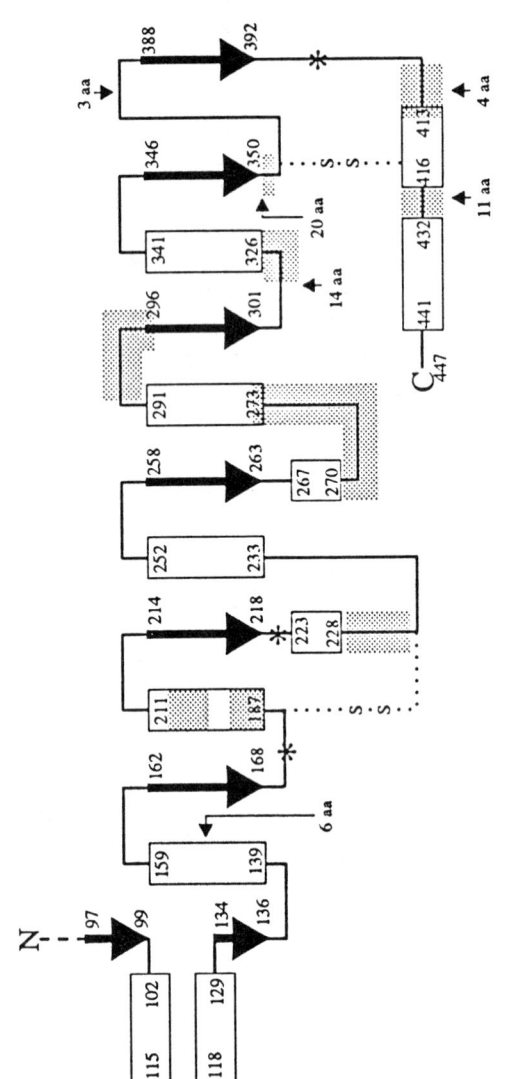

Figure 4. Schematic representation of the secondary structure of the *T. reesei* CBH II catalytic domain and its relationship to the catalytic domains of *C. fimi* CbhA and the family B endo-β-1,4-glucanases. Rectangles represent helices, bold arrows represent β-strands, solid interconnecting lines represent loops, disulfide bonds are indicated by dotted lines, and numbers refer to amino acid residues in CBH II. Asterisks show the positions of CBH II residues implicated in catalysis (D175, D221 and D401). Regions in CBH II and CbhA that are not found in the family B endo-β-1,4-glucanases are shaded. The positions and sizes (number of amino acid residues) of the extra insertions found in CbhA, relative to CBH II, are indicated by arrows. (Reproduced with permission from Ref. 16. Copyright 1994 Blackwell Science.)

CBHII   E2

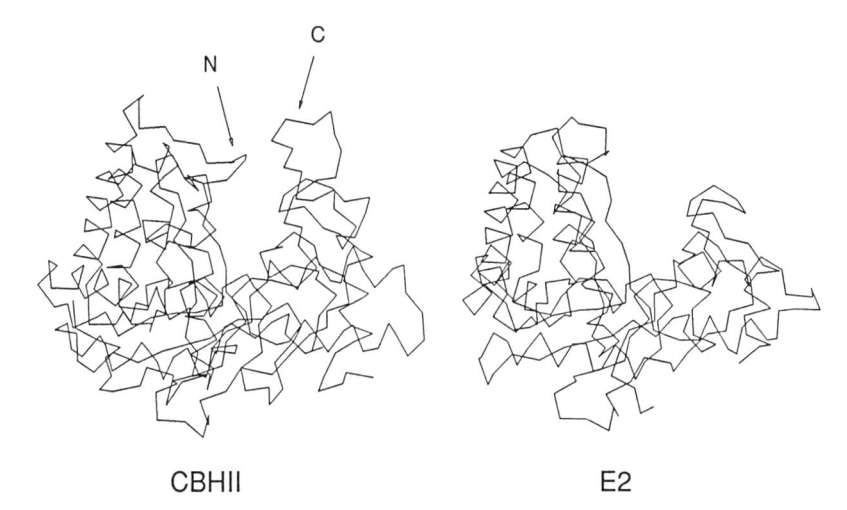

Figure 5. The α-carbon skeletons of *T. reesei* cellobiohydrolase II (CBH II) and *T. fusca* endoglucanase 2 (E2). The views are chosen to illustrate differences in the accessibilities of the two active sites. The labels "C" and "N", respectively, indicate the carboxyl- and amino-proximal loops that cover the active site of CBH II. (Reproduced with permission from Ref. 65. Copyright 1986 The American Society for Biochemistry and Molecular Biology.)

The absence of these insertions in the family B endoglucanases led to the prediction that the active sites of these enzymes are not enclosed but have an open conformation permitting access to internal glucosidic bonds (17). The prediction was subsequently confirmed by X-ray crystallographic analysis of *T. fusca* E2, a family B endoglucanase (20). A comparison of the structures of the E2 and CBH II active sites is shown in Figure 5.

The amino acid sequence of the *C. fimi* CbhA catalytic domain contains all the insertions that distinguish the fungal cellobiohydrolases from the endoglucanases in family B, including those that contribute to the active site loops (Figures 3-5). On this basis, it was predicted that CbhA is a cellobiohydrolase. Analysis of the catalytic properties of CbhA subsequently confirmed this prediction, as described below.

The 36 amino acid insertion that forms the C-proximal active site loop in CBH II is extended to 51 amino acids in CbhA (Figures 3 and 4). Several of the other insertions that distinguish the family B cellobiohydrolases are also longer in CbhA (Figure 4). It is also noted that an open reading frame from *C. flavigena* encodes a 66 amino acid region that appears to correspond to the C-terminus of another family B cellobiohydrolase (Figures 2 and 3) because it also contains the extended C-proximal insertion seen in *C. fimi* CbhA. It remains to be seen whether extended loops like these occur in all family B cellobiohydrolases from bacteria.

Other differences between CbhA and CBH II are evident when the whole enzymes are compared. The catalytic domain and the CBD of CbhA are separated by three Fn3 modules (Figure 2), but in CBH II they are joined directly by a short, glycosylated linker polypeptide (2). The CBD of CbhA is type II (7) and C-terminal while the CBD of CBH II is from a different family (type I) and N-terminal (2). Further analyses of fungal and bacterial family B cellobiohydrolases are needed determine how these differences influence enzyme activities.

Relationship of CbhB to other β-1,4-glucanases. DNA sequence analysis of the *cbhB* gene (5) showed that the N-terminal amino acid sequence of CbhB shared significant similarity with three β-1,4-glucanases from *Clostridium* spp. (Figure 6). On the basis of these and other reported sequence similarities, we proposed that *C. fimi* CbhB (then called CenE), *C. thermocellum* CelS (21), *C. thermocellum* S8 (22), *Clostridium stercorarium* Avicelase II (23), *Clostridium cellulolyticum* CelCCF (24), *Clostridium cellulovorans* P70 (25), and the C-terminal catalytic domain of *Caldocellum saccharolyticum* CelA (26), are members of a new family of β-1,4-glycanases designated family L (5). The alternative nomenclature for family L is family 48 (Henrissat, B., personal communication, 1994). An open reading frame (ORF1) from *Clostridium josui* encodes another related polypeptide (27). The sequences of the catalytic domains of CbhB, CelS and CelA are 47% identical (Shen, H. et al., Biochem. J., *in press*); only partial amino acid sequence data are currently available for the other family L enzymes.

CelS is the major enzymatic component of the cellulosome, the highly efficient extracellular cellulase complex produced by *C. thermocellum*. Like

```
CfiCbhB    *AVDGEYAORFLAOYDKIKDPANGYFSAQ-GIPYHAVETLMVEAPDYGHETTSEA...
CsaCelA    ...SGLGKYGQRFMWLWNKIHDPASGYFNQD GIPYHSVETLICEAPDYGHLTTSEA...
CstAviII   *XSDDPYKORFLDLWDDLHDPSNGYFSXH-GIPYHAV...
CthCelS   *GTPKAPTKDGTSYKDLFLELYGKIKDPKNGYFSPDEGIPYHSIETLIVEAPDYGHVTTSEA...
Cths8     *GTPKAPTKDGTSYKDLFXE...
```

Figure 6. Alignment of N-terminal amino acid sequences of *C. fimi* CbhB and related bacterial β-1,4-glucanases. Sequences were deduced from corresponding nucleotide sequences or (underlined) determined by Edman degradation. Unidentified residues are represented by the letter X. An asterisk indicates the N-terminus of the mature enzyme. CfiCbhB (GenBank L38827) is *C. fimi* CbhB; CsaCelA (GenBank L32742) is *C. saccharolyticum* CelA; CstAviII is *C. stercorarium* Avicelase II (23); CthCelS (GenBankL06942) is *C. thermocellum* ATCC 27405 CelS; Cths8 is *C. thermocellum* YS S8, a CelS analogue (22).

CbhB, CelS is a cellobiohydrolase (22, 28). It is possible that the family L enzymes from other *Clostridium* spp. are also cellobiohydrolases but it is not known whether family L contains only cellobiohydrolases or, like families B and C, both cellobiohydrolases and endoglucanases. So far, family L has no fungal members (Table I).

Catalytic activities of CbhA and CbhB

CbhA had weak activity on Cm-cellulose. Its molar specific activity was about 2000-fold lower than that of CenA (Table II). Comparable activity was shown by CbhB (Shen, H. et al., Biochem. J., in press). CbhA, also had low activity against barley β-glucan, a soluble, linear (1-3, 1-4)-β-D-glucan, but hydrolysis of this substrate by CbhB was not detectable (Shen. H. et al., Biochem. J., in press).

Exoglucanase activities. Viscometry can be used to determine whether a β-glucanase acts randomly or prefers terminal bonds. The analysis involves simultaneous determination of the increase in specific fluidity (φ, = η_{sp}^{-1}) and the appearance of reducing sugar groups during hydrolysis of Cm-cellulose. The hydrolysis of any β-1,4-glucosidic bond generates a new reducing group but the random activity of an endoglucanase gives a greater increase in specific fluidity than the action of an exoglucanase.

CbhA and CbhB gave only small increases in φ per reducing sugar group generated (17 and 20 $\varphi.ml.mmol^{-1}$, respectively), relative to the *C. fimi* endoglucanase CenA (96 $\varphi.ml.mmol^{-1}$) (16, Shen. H. et al., Biochem. J., in press). The activities of CbhA and CbhB were similar to the activity of Abg, a β-1,4-glucosidase from *Agrobacterium* sp. (13 $\varphi.ml.mmol^{-1}$) (Figure 7). Abg hydrolyzes glucosyl residues from the non-reducing ends of cellotriose and longer β-1,4-glucans by an exclusively exohydrolytic mode of attack (29). All these data are consistent with a mechanism in which CbhA and CbhB preferentially hydrolyze unsubstituted residues from the ends of Cm-cellulose molecules.

Data for CBH I are also included in Figure 7. CBH I, the major extracellular enzyme produced by *T. reesei*, is a family C cellobiohydrolase. The CBH I preparation used in this study was purified from a *T. reesei* culture supernatant (Novoclast) by ion-exchange chromatography on DEAE-Sepharose CL6B (30) and affinity chromatography using *p*-amino-1-thio-β-D-cellobioside linked to Sepharose 4B (31). The increase in φ per reducing sugar group generated for CBH I (60 $\varphi.ml.mmol^{-1}$) was lower than that for CenA but much higher than that for CbhA or CbhB. This showed that the CBH I preparation had significant endoglucanase activity.

Endoglucanase activities. Other investigators have also shown that *T. reesei* cellobiohydrolase preparations have endoglucanase activity (32, 33). Is this because cellobiohydrolases have a low level of intrinsic endoglucanase activity or are cellobiohydrolases derived from fungal culture supernatants still contaminated with endoglucanases, even after extensive purification

Table I. Family L β-1,4-glucanases

Enzyme	Organism and strain	Molecular mass (kDa)	Reference
CelA [a]	Caldocellum saccharolyticum	65	(26)
CbhB	Cellulomonas fimi ATCC 484	110	(5)
CelCCF [a]	Clostridium cellulolyticum ATCC 35319	78	(24)
P70	Clostridium cellulovorans ATCC 35296	70	(25)
ORF1 [b]	Clostridium josui FERM P-9684	-	(27)
CelS	Clostridium thermocellum ATCC 27405	81	(21)
S8	Clostridium thermocellum strain YS	81	(22)
Avicelase II	Clostridium stercorarium NCIB 11745	87	(23)

[a] Catalytic domain 2.
[b] Putative enzyme encoded by open reading frame.

Table II. Comparison of the activities of CbhA and CenA on various substrates

Enzyme	Bacterial cellulose [b]	Cm-cellulose [c]	Barley β-glucan [c]
CbhA	1.43	3.7	1.1
CenA	0.13	8129	1084

Substrate [a]

[a] Details of assay conditions are contained in ref. 16.
[b] Bacterial microcrystalline cellulose. Activity is expressed as μg soluble sugar released.min^{-1}.nmol enzyme^{-1}.
[c] Activity is expressed as mol reducing sugar group generated.min^{-1}.mol enzyme^{-1}.

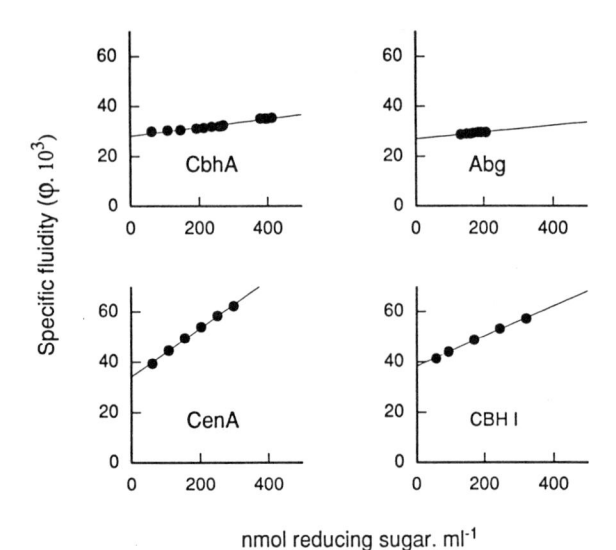

Figure 7. Specific fluidity *versus* reducing sugar production during hydrolysis of Cm-cellulose by *C. fimi* CbhA, *C. fimi* CenA, *Agrobacterium* sp. β-glucosidase (Abg). and *T. reesei* CBH I. The *C. fimi* and *Agrobacterium* enzymes were prepared from *E. coli* transformed with appropriate recombinant plasmids; CBH I was purified by affinity chromatography, as described in the text. Specific fluidity ($\phi, = \eta_{sp}^{-1}$) was determined with an Cannon-Fenske viscometer, as previously described (37). Reducing sugars were determined using hydroxybenzoic acid hydrazide reagent (4).

(34)? It was possible to assay the intrinsic endoglucanase activities of *C. fimi* CbhA and CbhB because the enzymes were available as recombinant gene products, free of contaminating β-1,4-glucanase. Recombinant Cex, CenA and Abg were also assayed for comparison. Cex is a *C. fimi* β-1,4-glucanase and β-1,4-xylanase (Figure 2) that behaves like an exoglucanase in the viscometric assay (15). CenA, a *C. fimi* endo-β-1,4-glucanase (Figure 2), was included as a positive control. Abg is a strictly exo-acting β-glucosidase, as described above.

Congo Red, a dye that binds to polymeric carbohydrates (35, 36), was used to assay for endoglucanase activity. Aliquots of enzyme solution were allowed to absorb into the surface of an agar plate containing 0.1% Cm-cellulose. After overnight incubation at 30°C, the plate was flooded with Congo Red and the unbound dye was removed by washing. Hydrolyzed Cm-cellulose appears as a zone of clearing against a dark red background in this assay. The assay is sensitive to endoglucanase activity but not exoglucanase activity (37, 38). Endoglucanases produce clearing because they can cleave Cm-cellulose randomly into oligomers that are too small to bind the dye. Exoglucanases do not give clearing because small oligomers are not produced by enzymes that are strictly exo-acting. Presumably, hydrolysis by exoglucanases is limited to the few unsubstituted residues at the ends of Cm-cellulose polymers and further hydrolysis is prevented when the enzyme meets carboxymethyl substituents.

No clearing was observed in the Congo Red assay with 0.02 nmol aliquots of CbhB, CbhA or Cex but a large clear zone was produced by 0.02 nmol of the CenA control. However, 2 nmol aliquots of CbhB, CbhA or Cex did produce clearing after overnight incubation, showing that all three enzymes had weak endoglucanase activity. As expected, no clearing was seen with 2 nmol of Abg (Shen. H. et al., Biochem. J., in press).

The catalytic domains of CbhA and *T. reesei* CBH II have closely related primary structures (see above) so it is possible that CBH II also has intrinsic endoglucanase activity. Such activity would require the loops enclosing the tunnel-shaped active site to move in order to allow entry of glucan chains and hydrolysis internal glucosidic bonds. The active site of *T. reesei* CBH I is also tunnel-shaped (39) but it remains to be seen whether the endoglucanase activity of our CBH I preparation is real or the result of contamination. The active site of Cex is an open cleft (40) so access to internal bonds seems relatively unrestricted. In this case, it is not yet clear why the enzyme normally shows a preference for terminal glucosidic bonds in Cm-cellulose. There is presently no information on the three-dimensional structure CbhB or other enzymes in family L.

Hydrolysis of cellulose and cellooligosaccharides. CbhA was about 10-fold more active on bacterial microcrystalline cellulose than CenA, the *C. fimi* family B endoglucanase (Table II). Approximately 98% of the total soluble sugar produced from bacterial cellulose following incubation with CbhA for 2 h was cellobiose. Only trace levels of glucose and cellotriose were seen (Figure 8). A similar product profile was obtained with phosphoric acid-swollen cellulose (so-called "amorphous cellulose"). Cellotetraose did not

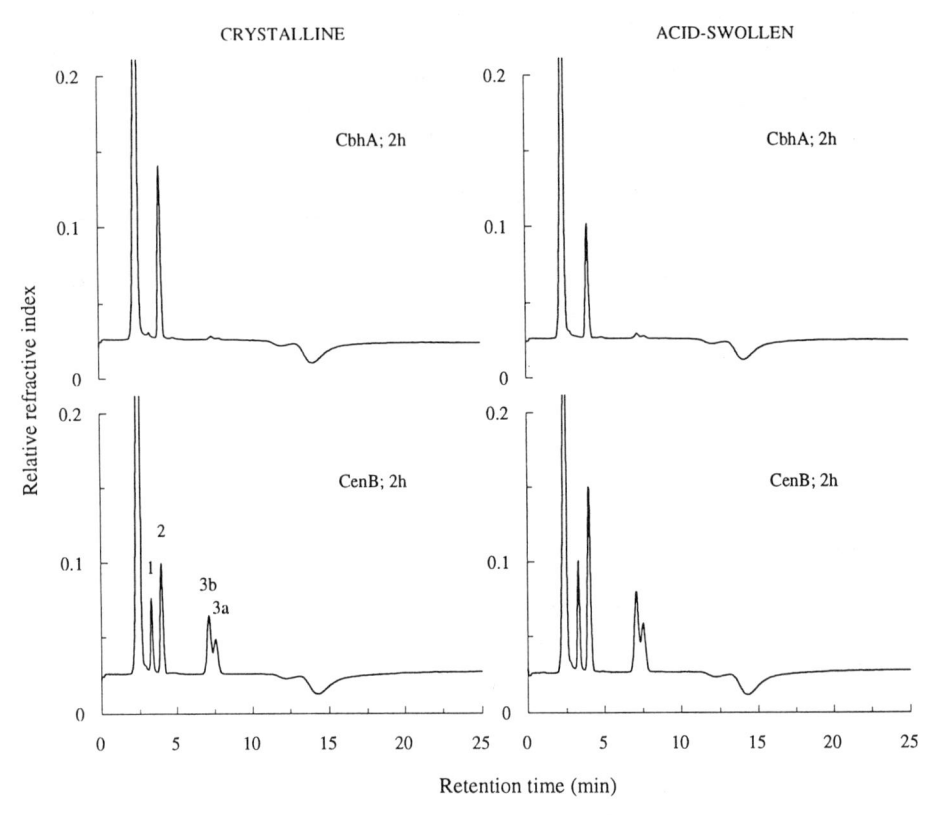

Figure 8. Analysis of soluble sugars released from cellulose by CbhA and the *C. fimi* endoglucanase CenB. Soluble sugars produced from bacterial microcrystalline cellulose or acid-swollen cellulose by CbhA or CenB after 2 h incubation at 37° C were analyzed high performance liquid chromatography (4). Reaction mixtures contained 1.0 mg. ml^{-1} cellulose, 5 mM K phosphate, pH 7 and 1.0 µM enzyme. Cellulose was removed by centrifugation and the samples were incubated at 100° C for 5 min to prevent further hydrolysis prior to analysis. The total amounts of soluble sugars were also determined using phenol sulfuric acid reagent (66). Glucose (peak 1) and cellobiose (peak 2) were resolved as single peaks; cellotriose was resolved into its α- and β-anomers (peak 3a and peak 3b, respectively). Standard α- and β-cellotetraose (not shown) had retention times of 23 and 21 min, respectively.

accumulate in reactions with either substrate. Similar product profiles were obtained after 24 h incubation (16). These profiles are consistent with cellobiohydrolase activity. In contrast, CenA, CenB and CenD produced significant amounts of glucose and cellotriose (Figure 8 and Meinke, A. et al., unpublished data).

The chromatographic column used to analyze soluble sugars in these experiments resolves the α- and β-anomers of cellotriose and cellotetraose (4, 41). This allowed the stereoselectivity of hydrolysis to be determined. Chromatographic resolution of the anomeric products from cellohexaose showed that hydrolysis by CbhA proceeds with inversion of anomeric carbon configuration, characteristic of a single-displacement mechanism (16). All the family B members that have been examined are "inverting" enzymes (14, 42). These data are consistent with the proposal that members of a given β-1,4-glucanase family have a similar general fold and active site topology and use the same catalytic mechanism (42).

The molar specific activities of CbhB on bacterial microcrystalline cellulose and acid-swollen cellulose were similar to those of CbhA. At least 95% of the soluble sugar produced from both substrates was cellobiose (Shen. H. et al., Biochem. J., in press). CbhB was originally described as an endoglucanase and named CenE because of its activity in the Congo Red assay (5). However, the more detailed analysis described here demonstrates that the enzyme has similar properties to CbhA. Therefore, CbhB is more accurately described as a cellobiohydrolase.

CbhB hydrolyzed cellohexaose to produce a mixture of cellotetraose, cellotriose and cellobiose; only a trace of glucose was produced (Figure 9). The amount of α-cellotriose seen after 5 min incubation at 37°C was greater than the amount of β-cellotriose (Figure 9A). After a further 40 min incubation to allow mutarotation of the products, α- and β-cellotriose were present in the equilibrium ratio of 1:1.8 (43) (Figure 9B). Therefore, hydrolysis by CbhB also proceeds with inversion of anomeric configuration. It is anticipated that all family L enzymes will have the same stereoselectivity.

Additional information about CbhB can be deduced from the data shown in Figure 9. Production of cellotriose showed that hydrolysis of cellohexaose occurs at site 1 (Figure 9B, inset). Production of cellotetraose showed that hydrolysis also occurs at site 2 or 3. Hydrolysis at site 2 was evidently preferred because the ratio of α- and β-cellotetraose seen after the 5 min incubation (approximately 1:0.9, Figure 9A) changed to the equilibrium ratio of approximately 1:2 after mutarotation (Figure 9B). These data are consistent with a mechanism in which CbhB attacks cellohexaose from its reducing end. It is noted that data for another family L enzyme, Avicelase II from *C. stercorarium*, shows hydrolysis of cellohexaitol at sites 2 and 3 (23). These data suggest that this enzyme also attacks from the reducing end.

Similarities between bacterial and fungal cellulase systems

Aerobic bacteria like *C. fimi*, *T. fusca* and *Streptomyces* spp. have cellulase systems that resemble systems in the soft-rot and white-rot fungi because the

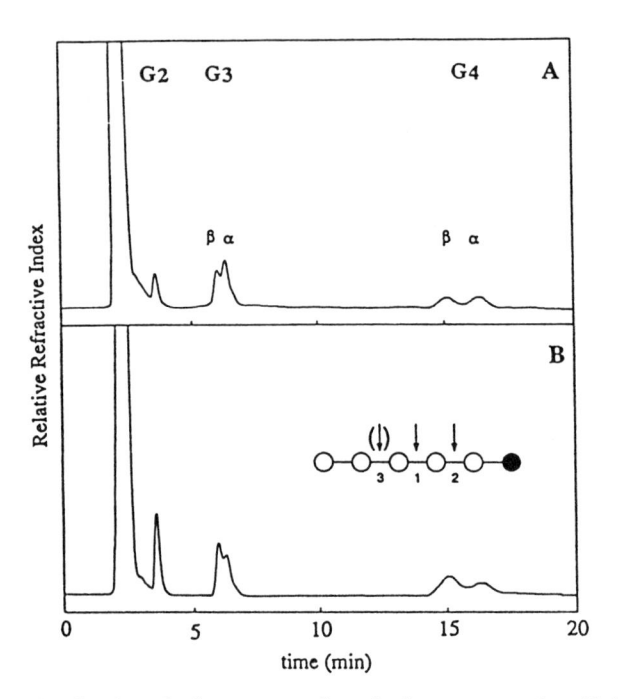

Figure 9. Analysis of the stereochemical course of cellohexaose hydrolysis by CbhB. Cellohexaose (4 mM) and CbhB (~1 μM) in 10 mM potassium phosphate buffer, pH 7, were incubated with enzyme at 37° C for 5 or 45 min and the products analyzed by high performance liquid chromatography (4). Panel A and panel B, respectively, show hydrolysis products before and after mutarotation. The α and β anomers of cellotriose (G3) and cellotetraose (G4) are resolved; those of cellobiose (G2) are not. The inset in panel B shows the deduced cleavage sites of cellohexaose (arrows); the oligosaccharide is represented as glucosyl units (O) linked by β-1,4-glucosidic bonds (—) and the terminal reducing sugar is shaded black. (Reproduced with permission from Ref. 5. Copyright 1994 Academic Press.)

various enzyme components do not appear to form stable physical associations. These types of systems are described as non-complexed (44). It is difficult to compare non-complexed systems from bacteria with those of fungi because we still have no detailed understanding of how either one operates. Nevertheless, characterization of the *C. fimi* cellulase system is now sufficiently advanced that a general comparison with fungal systems is possible.

The major components of the *T. reesei* cellulase system are the two cellobiohydrolases, CBH I and CBH II, belonging to β-1,4-glucanase families C and B, respectively. It appears that the cellulase systems of several aerobic fungi contain both family B and family C cellobiohydrolases and that both types are required for efficient hydrolysis of crystalline cellulose (1, 2, 44). Our studies show that a family B cellobiohydrolase, CbhA, is also a major component of the *C. fimi* cellulase system. Until recently, CbhA was the only known bacterial family B cellobiohydrolase but a similar enzyme has now been identified in *T. fusca* (45). The alignment shown in Figure 3 implies that *C. flavigena* also produces a family B cellobiohydrolase. These observations suggest that family B cellobiohydrolases are not uncommon in cellulolytic bacteria.

Do bacteria also produce family C cellobiohydrolases? Although there are at least fifteen CBH I homologues from fungi in the GenBank database, there are no examples of family C enzymes from bacteria (44). Cellobiohydrolases have been described in *C. stercorarium* (46), *Microbispora bispora* (47, 48). *Ruminococcus flavefaciens* (49) and *Streptomyces flavogriseus* (50) but there are no amino acid sequence data for any of these enzymes. All other known bacterial cellobiohydrolases belong to either family B or family L. Bacterial family C cellobiohydrolases may yet be discovered. Alternatively, family L cellobiohydrolases could be the functional equivalents of these enzymes in the bacteria.

The need for two distinct types of cellobiohydrolase is not obvious but a possible explanation is that one type of cellobiohydrolase (family B) attacks cellulose chains at their non-reducing end while the other type (family C in fungi, family L in bacteria) attacks at the reducing end. Synergistic interactions in cellulase systems are usually explained by a mechanism in which endoglucanases initiate attack at amorphous regions on cellulose chains, providing sites for cellobiohydrolase activity. In systems containing two types of cellobiohydrolase, "nicks" created by endoglucanase activity would spread simultaneously in both directions, quickly exposing underlying chains to further enzyme action (Figure 10).

Evidence for the attack of cellulose molecules from the reducing and non-reducing ends by *T. reesei* CBH I and CBH II, respectively, comes from analyses of cellooligosaccharide hydrolysis (51) and the tertiary structures of the CBH I and CBH II catalytic domains are consistent with the two different mechanisms (17, 39). It is reasonable to propose that *C. fimi* CbhA and other cellobiohydrolases in family B also attack cellulose from the non-reducing end. There is less evidence for attack from the reducing end by CbhB and

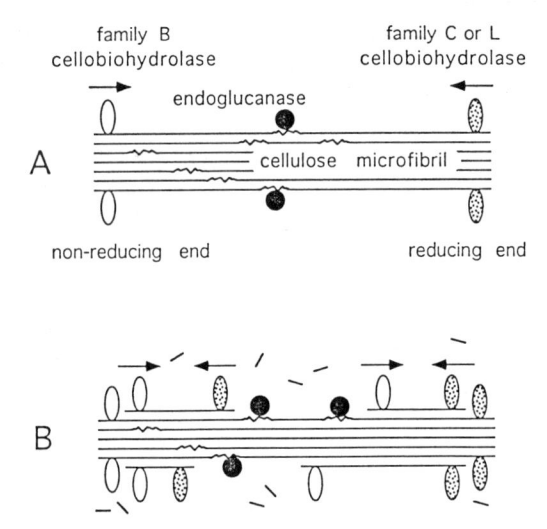

Figure 10. Hypothetical model for non-complexed cellulase systems of bacteria and fungi. Cellobiohydrolases that attack cellulose chains in opposite directions are shown as a common feature of bacterial and fungal systems. The family C (fungi) or family L (bacteria) cellobiohydrolase attacks at the reducing end of cellulose chains. The family B cellobiohydrolase (bacteria and fungi) attacks at the non-reducing end. As a result, "nicks" created by endoglucanase activity in disordered regions of cellulose chains are rapidly propagated in both directions. This activity reveals further sites for endoglucanase attack.

other family L cellobiohydrolases but all available data are consistent with this mechanism.

CbhA and CbhB were shown to have a low but significant level of endoglucanase activity on Cm-cellulose. If cellobiohydrolases also have endoglucanase activity on cellulose, there is no apparent requirement for additional endoglucanases in models like the one shown in Figure 10. The inability of genetic probes to detect endoglucanase-like sequences in *P. chrysosporium* led to suggestion that CBH I variants do indeed provide endoglucanase activity in this fungus (52).

Intrinsic endoglucanase activity would explain the observed synergistic interaction of *T. reesei* CBH I and CBH II (53). However, the synergistic interaction of two cellobiohydrolases may be simply the result of endoglucanase contamination in some cases. For example, the synergistic interaction of *P. pinophilum* CBH I and CBH II disappeared when a trace endoglucanase contaminant was removed (54). In this case at least, it appears that synergistic interaction requires the participation of three enzymes, as shown in Figure 10. Moreover, synergism between *P. pinophilum* CBH I and CBH II required a specific endoglucanase; others were not effective. Endoglucanase specificity has also been noted in other synergistic interactions (1). This aspect of synergy remains to be explained and is not addressed in our model.

Conclusion

Comparison of fungal and bacterial systems may help us to understand more about the mechanisms of cellulose hydrolysis by microorganisms. On the basis of data for *C. fimi*, we suggest that non-complexed cellulase systems from bacteria and fungi use a similar strategy involving two types of cellobiohydrolase that attack at different ends of cellulose chains.

It is not clear how non-complexed cellulase systems are related to the systems in anaerobic microorganisms, particularly those of the cellulolytic *Clostridium* spp. in which cellulases are assembled into multimolecular complexes called cellulosomes (55). Most of the known family L cellulases are from *Clostridium* spp. and at least some of these are cellobiohydrolases. However, cellobiohydrolases from other cellulase families have not yet been identified in these bacteria.

Acknowledgements

Our work on *C. fimi* is supported by grants from the Natural Sciences and Engineering Research Council of Canada. We thank Curtis Braun and Gary Lesnicki for expert technical assistance.

Literature Cited

1. Wood, T. M.; Garcia-Campayo, V. *Biodegradation*, 1990, 1, 147-161.

2. Teeri, T. T.; Pentillä, M.; Keränen, S.; Nevalainen, H.; Knowles, J. K. C. In *Biotechnology of Filamentous Fungi;* D. B. Finkelstein and C. Ball, Eds.; Butterworth-Heinemann: London, 1991, pp 417-444.
3. Broda, P. *Biodegradation,* **1992,** *3,* 219-238.
4. Meinke, A.; Gilkes, N. R.; Kilburn, D. G.; Miller, R. C., Jr.; Warren, R. A. J. *J. Bacteriol.* **1993,** *175,* 1910-1918.
5. Shen, H.; Tomme, P.; Meinke, A.; Gilkes, N. R.; Kilburn, D. G.; Warren, R. A. J.; Miller, R. C. Jr. *Biochem. Biophys. Res. Comm.* **1994,** *199,* 1223-1228.
6. Gilkes, N. R.; Henrissat, B.; Kilburn, D. G.; Miller, R. C., Jr.; Warren, R. A. J. *Microbiol. Rev.* **1991,** *55,* 303-315.
7. Coutinho, J. B.; Gilkes, N. R.; Warren, R. A. J.; Kilburn, D. G.; Miller, R. C., Jr. *Molec. Microbiol.* **1992,** *6,* 1243-1252.
8. Meinke, A.; Gilkes, N. R.; Kilburn, D. G.; Miller, R. C., Jr.; Warren, R. A. J. *J. Bacteriol.* **1991,** *173,* 7126-7135.
9. Hansen, C. K. *FEBS Lett.* **1992,** *305,* 91-96.
10. Blaak, H.; Schnellmann, J.; Walter, S.; Henrissat, B.; Schrempf, H. *Eur. J. Biochem.* **1993,** *214,* 659-669.
11. Al-Tawheed, A. R. *Molecular Characterization of Cellulase Genes from Cellulomonas flavigena;* M.Sc. thesis, University of Dublin, 1988.
12. Henrissat, B.; Claeyssens, M.; Tomme, P.; Lemesle, L.; Mornon, J.-P. *Gene,* **1989,** *81,* 83-95.
13. Henrissat, B.; Bairoch, A. *Biochem. J.* **1993,** *293,* 781-788.
14. Schou, C.; Rasmussen, G.; Kaltoft, M. B.; Henrissat, B.; Schülein, M. *Eur. J. Biochem.* **1993,** *217,* 947-953.
15. Meinke, A.; Gilkes, N. R.; Kilburn, D. G.; Warren, R. A. J.; Miller, R. C., Jr. In *Genetics, Biochemistry and Ecology of Lignocellulose Degradation;* K. Shimada, K. Ohmiya, Y. Kobayashi, S. Hoshino, K. Sakka and S. Karita, Eds.; Uni Publishers: Tokyo, 1993, pp 286-297.
16. Meinke, A.; Gilkes, N. R.; Kwan, E.; Kilburn, D. G.; Warren, R. A. J.; Miller, R. C., Jr. *Molec. Microbiol.* **1994,** *12,* 413-422.
17. Rouvinen, J.; Bergfors, T.; Teeri, T.; Knowles, J. K. C.; Jones, T. A. *Science,* **1990,** *249,* 380-386.
18. Gilkes, N. R.; Claeyssens, M.; Aebersold, R.; Henrissat, B.; Meinke, A.; Morrison, H. D.; Kilburn, D. G.; Warren, R. A. J.; Miller, R. C., Jr. *Eur. J. Biochem.* **1991,** *202,* 367-377.
19. McGinnis, K.; Wilson, D. B. *Biochemistry,* **1993,** *32,* 8151-8156.
20. Spezio, M.; Wilson, D. B.; Karplus, P. A. *Biochemistry,* **1993,** *32,* 9906-9916.
21. Wang, W. K.; Kruus, K.; Wu, J. H. D. *J. Bacteriol.* **1993,** *175,* 1293-1320.
22. Morag, E.; Bayer, E. A.; Hazlewood, G. P.; Gilbert, H. J.; Lamed, R. *Appl. Biochem. Biotechnol.* **1993,** *43,* 147-151.
23. Bronnenmeier, K.; Rücknagel, K. P.; Staudenbauer, W. L. *Eur. J. Biochem.* **1991,** *200,* 379-385.
24. Belaich, J.-P.; Gaudin, C.; Belaich, A.; Bagnara-Tardif, C.; Fierobe, H.-P.; Reverbel, C. In *Genetics, Biochemistry and Ecology of Lignocellulose Degradation;* K. Shimada, K. Ohmiya, Y. Kobayashi, S. Hoshino, K. Sakka and S. Karita, Eds.; Uni Publishers: Tokyo, 1993, pp 53-62.

25. Doi, R. H.; Goldstein, M. A.; Park, J.-S.; Lui, C.-C.; Matano, Y.; Takagi, M.; Hashida, S.; Foong, F. C.-F.; Hamamoto, T.; Segel, I.; Shoseyov, O. In *Genetics, Biochemistry and Ecology of Lignocellulose Degradation;* K. Shimada, K. Ohmiya, Y. Kobayashi, S. Hoshino, K. Sakka and S. Karita, Eds.; Uni Publishers: Tokyo, 1993, pp 43-52.
26. Bergquist, P. L.; Gibbs, M. D.; Saul, D. J.; Te'O, V. S. J.; Dwivedi, P. P.; Morris, D. In *Genetics, Biochemistry and Ecology of Lignocellulose Degradation;* K. Shimada, K. Ohmiya, Y. Kobayashi, S. Hoshino, K. Sakka and S. Karita, Eds.; Uni Publishers: Tokyo, 1993, pp 276-285.
27. Fujino, T.; Karita, S.; Ohmiya, K. In *Genetics, Biochemistry and Ecology of Lignocellulose Degradation;* K. Shimada, K. Ohmiya, Y. Kobayashi, S. Hoshino, K. Sakka and S. Karita, Eds.; Uni Publishers: Tokyo, 1993, pp 67-75.
28. Morag, E.; Halevy, I.; Bayer, E. A.; Lamed, R. *J. Bacteriol.* **1991,** *173,* 4155-4162.
29. Day, A. G.; Withers, S. W. *Biochem. Cell. Biol.* **1986,** *64,* 914-922.
30. Bhikhabhai, R.; Johansson, G.; Pettersson, G. *J. Appl. Biochem.* **1984,** *6,* 336-347.
31. Tomme, P.; McCrae, S.; Wood, T.; Claeyssens, M. *Meth. Enzymol.* **1988,** *160,* 187-193.
32. Okada, G.; Tanaaka, Y. *Agric. Biol. Chem.* **1988,** *52,* 2981-2984.
33. Ståhlberg, J.; Johansson, G.; Pettersson, G. *Biochim. Biophys. Acta,* **1993,** *1157,* 107-113.
34. Reinikainen, T. *The Cellulose-Binding Domain of Cellobiohydrolase I from Trichoderma reesei;* VTT Publications 206; Technical Research Centre of Finland: Espoo, 1994.
35. Wood, P. J. *Carbohydr. Res.* **1980,** *85,* 271-287.
36. Wood, P. J.; Erfle, J. D.; Teather, R. M. *Meth. Enzymol.* **1988,** *160,* 59-74.
37. Gilkes, N. R.; Langsford, M. L.; Kilburn, D. G.; Miller, R. C., Jr.; Warren, R. A. J. *J. Biol. Chem.* **1984,** *259,* 10455-10459.
38. Bartley, T. D.; Murphy-Holland, K.; Eveleigh, D. E. *Anal. Biochem.* **1984,** *140,* 157-161.
39. Divne, C.; Ståhlberg, J.; Reinikainen, T.; Ruohonen, L.; Petterson, G.; Knowles, J. K. C.; Teeri, T. T.; Jones, T. A. *Science,* **1994,** *265,* 524-528.
40. White, A.; Withers, S. G.; Gilkes, N. R.; Rose, D. R. *Biochemistry,* **1994,** *33,* 12546-12552.
41. Braun, C.; Meinke, A.; Ziser, L.; Withers, S. *Anal. Biochem.* **1993,** *212,* 259-262.
42. Gebler, J.; Gilkes, N. R.; Claeyssens, M.; Wilson, D. B.; Béguin, P.; Wakarchuk, W. W.; Kilburn, D. G.; Miller, R. C., Jr.; Warren, R. A. J.; Withers, S. G. *J. Biol. Chem.* **1992,** *267,* 12559-12561.
43. Stoddart, J. F. *Stereochemistry of Carbohydrates;* Wiley-Interscience: New York, NY: 1971.
44. Tomme, P.; Warren, R. A. J.; Gilkes, N. R. *Adv. Microbial Physiol.* **1995,** *37,* 1-81.
45. Zhang, S.; Lao, G.; Wilson, D. B. *Biochemistry,* **1995,** *34,* 3386-3395.

46. Creuzet, N.; Berenger, J.-F.; Frixon, C. *FEMS Microbiol. Lett.* **1983**, *20*, 347-350.
47. Yablonsky, M. D.; Bartley, T.; Elliston, K. O.; Kahrs, S. K.; Shalita, Z. P.; Eveleigh, D. E. In *Biochemistry and Genetics of Cellulose Degradation*; J.-P. Aubert, P. Béguin and J. Millet, Eds.; Academic Press: London, 1988, pp 249-266.
48. Hu, P.; Kahrs, S. K.; Chase, T.; Eveleigh, D. E. *J. Industr. Microbiol.* **1992**, *10*, 103-110.
49. Gardner, R. M.; Doerner, K. C.; White, B. A. *J. Bacteriol.* **1987**, *169*, 4581-4588.
50. MacKenzie, C. R.; Bilous, D.; Johnson, K. G. *Can. J. Microbiol.* **1984**, *30*, 1171-1178.
51. Biely, P.; Vranska, M.; Claeyssens, M. In *Trichoderma reesei Cellulases and Other Hydrolases*; P. Suominen and T. Reinikainen, Eds.; Foundation for Biotechnical and Industrial Fermentation Research, Vol. 8; Helsinki, 1993, pp 99-108.
52. Sims, P. F. G.; Soares-Felipe, M. S.; Wang, Q.; Gent, M. E.; Tempelaars, C.; Broda, P. *Mol. Microbiol.* **1994**, *12*, 209-216.
53. Fägerstam, L. G.; Pettersson, L. G. *FEBS Lett.* **1980**, *119*, 97-101.
54. Wood, T. M.; McCrae, S. I.; Bhat, K. M. *Biochem. J.* **1989**, *260*, 37-43.
55. Béguin, P.; Aubert, J.-P. *FEMS Microbiol. Rev.* **1994**, *13*, 25-58.
56. Wong, W. K. R.; Gerhard, B.; Guo, Z. M.; Kilburn, D. G.; Warren, R. A. J.; Miller, R. C., Jr. *Gene*, **1986**, *44*, 315-324.
57. Meinke, A.; Braun, C.; Gilkes, N. R.; Kilburn, D. G.; Miller, R. C., Jr.; Warren, R. A. J. *J. Bacteriol.* **1991**, *173*, 308-314.
58. Coutinho, J. B.; Moser, B.; Kilburn, D. G.; Warren, R. A. J.; Miller, R. C., Jr. *Molec. Microbiol.* **1991**, *5*, 1221-1233.
59. O'Neill, G., Goh, S.H., Warren, R.A.J., Kilburn, D.G. and Miller, R.C., Jr. *Gene*, **1986**, *44*, 325-330.
60. Millward-Sadler, S. J.; Poole, D. M.; Henrissat, B.; Hazlewood, G. P.; Clarke, J. H.; Gilbert, H. J. *Mol. Microbiol.* **1994**, *11*, 375-382.
61. Tempelaars, C.; Birch, P. R. J.; Sims, P. F. G.; Broda, P. *Appl. Environ. Microbiol.* **1994**, *60*, 4387-4393.
62. Lin, F.; Marchenko, G.; Cheng, Y.-R. *J. Industr. Microbiol.* **1994**, *13*, 344-350.
63. Damude, H. G.; Withers, S. G.; Kilburn, D. G.; Miller, R. C., Jr.; Warren, R. A. J. *Biochemistry*, **1995**, *34*, 2220-2224.
64. Damude, H. G.; Gilkes, N. R.; Kilburn, D. G.; Miller, R. C., Jr.; Warren, R. A. J. *Gene*, **1993**, *123*, 105-107.
65. Meinke, A.; Damude, H.G.; Tomme, P.; Kwan, E.; Kilburn, D.G.; Miller, R.C., Jr.; Warren, R.A.J.; Gilkes, N.R. *J. Biol. Chem.* **1995**, *270*, 4383-4386.
66. Chaplin, M. F. In *Carbohydrate Analysis: a Practical Approach*; M. F. Chaplin and J. F. Kennedy, Eds.; IRL Press: Oxford, 1986, pp 1-36.

RECEIVED September 12, 1995

Chapter 13

Thermostable β-Glucosidases

Badal C. Saha[1], Shelby N. Freer, and Rodney J. Bothast

Fermentation Biochemistry Research Unit, National Center for Agricultural Utilization Research, Agricultural Research Service, U.S. Department of Agriculture, Peoria, IL 61604

>Interest in ß-glucosidase has increased in recent years because of its application in the conversion of cellulose to glucose for the subsequent production of fuel alcohol. Cellulolytic enzymes in conjunction with ß-glucosidase act sequentially and cooperatively to degrade cellulose to glucose. Product inhibition, thermal inactivation, low product yield and high cost of the enzyme constitute some problems to develop enzymatic hydrolysis of cellulose as a commercial process. A thermostable ß-glucosidase from *Aureobasidium pullulans*, a yeast-like fungus, was optimally active at 75°C and pH 4.5 against p-nitrophenyl-ß-D-glucoside, cellobiose and a series of higher cellooligosaccharides. Recent developments in thermostable ß-glucosidase research particularly the biochemical and kinetic properties of the enzyme, mode of action and its use in the conversion of cellulose to glucose are described.

The search for lower cost raw materials for the production of biofuels has led to increasing interest in the enzymatic hydrolysis of cellulose. Currently, over one billion gallons of ethanol are produced annually in the United States, with approximately 95% derived from the fermentation of corn starch. With increased attention to clean air and oxygenates for fuels, opportunities exist for expansion of the fuel ethanol industry. Lignocellulosic biomass, particularly corn fiber, represents a renewable resource that is available in sufficient quantities from the

[1]Permanent address: Department of Biochemistry, Michigan State University, East Lansing, MI 48824

corn wet milling industry to serve as a low cost feedstock. Advances in enzyme technology are necessary if conversion of cellulosic biomass to ethanol is to be a commercial reality.

Role of ß-Glucosidase in Enzymatic Hydrolysis of Cellulose

Cellulose is a linear polymer of 8,000-12,000 D-glucose units linked by 1,4-ß-D-glucosidic bonds. The enzyme system for the conversion of cellulose to glucose comprises endo-1,4-ß-glucanase (EC 3.2.1.4), exo-1,4-ß-glucanase (EC 3.2.1.91) and ß-glucosidase (ß-D-glucosidic glucohydrolase, EC 3.2.1.21). Cellulolytic enzymes in conjunction with ß-glucosidase act sequentially and cooperatively to degrade crystalline cellulose to glucose. Endoglucanase acts in a random fashion on the regions of low crystallinity of the cellulosic fiber whereas exoglucanase removes cellobiose (ß-1,4 glucose dimer) units from the non-reducing ends of cellulose chains. ß-Glucosidase hydrolyzes cellobiose and in some cases cellooligosaccharides to glucose. The enzyme is generally responsible for the regulation of the whole cellulolytic process and is a rate limiting factor during enzymatic hydrolysis of cellulose as both endoglucanase and cellobiohydrolase activities are often inhibited by cellobiose (*1,2,3*). Thus, the ß-glucosidase not only produces glucose from cellobiose but also reduces cellobiose inhibition, allowing the cellulolytic enzymes to function more efficiently. However, like ß-glucanases, almost all ß-glucosidases are subject to end-product (glucose) inhibition.

Problems of Current ß-Glucosidase

Product inhibition, thermal inactivation, substrate inhibition, low product yield and high cost of ß-glucosidase constitute some barriers to commercial development of the enzymatic hydrolysis of cellulose. There is an increasing demand for the development of a thermostable, environmentally compatible, product and substrate tolerant ß-glucosidase with increased specificity and activity for application in the conversion of cellulose to glucose in the fuel ethanol industry. A thermostable ß-glucosidase offers certain advantages such as higher reaction rate, increased product formation, less microbial contamination, longer shelf-life, easier purification and better yield.

Production of Thermostable ß-Glucosidases

ß-Glucosidases having temperature optima of at least 60°C will be considered thermostable. Thermostable ß-glucosidases reported to date include those purified from thermophiles such as *Clostridium thermocellum* (*4*), *Thermoascus aurantiacus* (*5,6*), *Talaromyces emersonii* (*7,8*), *Thermotoga* sp. (*9*), *Microbispora bispora* (*10*), *Mucor miehei* (*11*), *Thermoanaerobacter ethanolicus* (*12*), *Sporotrichum thermophile* (*13*), a thermophilic cellulolytic anaerobe Tp8 (*14*), an extremely thermophilic anaerobic strain Wai21W.2 (*15*) and the hyperthermophilic archaeon *Pyrococcus furiosus* (*16*). Some thermophilic organisms such as *Thermomonospora* sp. produce less thermostable ß-glucosidase (17,18). On the other hand, some mesophilic fungi such as *Aspergillus* sp. and *Sclerotium* sp. produce ß-glucosidase having an optimum temperature at 60-70°C (19,20). *Aureobasidium pullulans*, a yeast-like fungus, produces a thermostable ß-glucosidase (231 mU/ml culture broth) when grown on corn bran (21). Lactose was also a good carbon source for the production of ß-glucosidase. A time course study of ß-glucosidase production by *A. pullulans* grown on corn bran at 28°C showed that the ß-glucosidase production increased gradually up to 4 days, after which it remained constant. *P. furiosus* cells grown on cellobiose contained very high levels of ß-glucosidase (18 U/mg at 80°C) (16).

Physico-chemical Characteristics

The specific activity of purified extracellular ß-glucosidase from *A. pullulans* was 315 U/mg protein at pH 4.5 and 75°C using p-nitrophenyl ß-D-glucoside (pNPßG) as assay substrate (21). The specific activity of purified ß-glucosidase from various microorganisms varies from 9 to 468 U/mg protein with pNPßG as substrate. (9, 22-24). The native ß-glucosidase from *P. furiosus* had a molecular weight of 230,000 and was composed of four subunits each with a molecular weight of 58,000 (16). The native ß-glucosidase from *Pisolithus tinctorius* had a molecular weight of 450,000 with three subunits each of molecular weight 150,000 (25). The ß-glucosidase from *A. pullulans* was a glycoprotein having a molecular weight of 340,000 with two similar subunits of molecular weight of 165,000 (21). The ß-glucosidase from *S. thermophile* had a molecular weight of 240,000 with two similar subunits (13). The ß-glucosidase from an extremely thermophilic anaerobic bacterium had a molecular weight of 43,000 (15). A ß-glucosidase from *Clostridium stercorium* was a monomer with a molecular weight of 85,000 (26). The cloned ß-glucosidase (BglB) from *M. bispora* had a molecular weight of 52,000 (10). Thus, there is considerable diversity in enzyme structure for different thermostable ß-glucosidases.

Thermostability and Thermoactivity

The half-lives of some thermostable ß-glucosidases are given in Table I. The purified ß-glucosidase from *P. furiosus* showed optimum activity at pH 5.0 and 102-105°C, and was remarkably thermostable with a half life of 85 h at 100°C and 13 h at 110°C (16). The thermostability of a ß-glucosidase from the thermophilic bacterium Tp8 cloned in *Escherichia coli* was greatest at pH 6.0-6.5, and the enzyme had a half-life value of 11 min at 90°C, 105 min at 85° and 900 min at 80°C (14). The ß-glucosidase from an extremely thermophilic anaerobic bacterium strain Wai21W.2 was inactivated with a half-life of 45 h at 65°C, 47 min at 75°C and 1.4 min at pH 6.2 and 85°C (15). At pH 7.0, which was the optimum pH for thermostability, half-life of the enzyme was 130 min at 75°C. The thermostability of this enzyme was enhanced 8 fold by 10% glycerol, 6-fold by 0.2 M cellobiose and 3 fold by 5 mM dithiothreitol and 5 mM 2-mercaptoethanol at pH 6.2 and 75°C. A partially purified ß-glucosidase from *Thermotoga* sp. had a half-life of

Table I. Half-lives of some thermophilic ß-glucosidases

Source	ß-Glucosidase	Half-life
Thermotoga sp. (*9*)	Partially pure	8 h at 90°C (pH 7.0)
	Pure	2.5 h at 98°C (pH 7.0)
Thermophilic bacterium Tp8 enzyme cloned in *Escherichia coli* (*14*)	Pure	11 min at 90°C (pH 6.0-6.5)
Thermophilic anaerobic bacterium Wai21W.2 (*15*)	Pure	45 h at 65°C (pH 6.2)
		47 min at 75°C (pH 6.2)
		130 min at 75°C (pH 7.0)
Pyrococcus furiosus (*16*)	Pure	85 h at 100°C (pH 5.0)
		13 h at 110°C (pH 5.0)
Aureobasidium pullulans (*21*)	Crude	72 h at 75°C (pH 4.5)
		24 h at 80°C (pH 4.5)

8 h at 90°C (*9*). The pure enzyme had a half-life of 2.5 h at 98°C in the presence of bovine serum albumin (40 µg/ml). The thermostability of the enzyme was increased further by addition of either trehalose or betaine. Immobilization of ß-glucosidase from *A. phoenicis* increased the half-life at 65°C from 0.5 to 252 h (*27*). One ß-glucosidase (BglB) from *M. bispora* expressed in *E. coli* showed an optimum activity at 60°C and pH 6.2 (*10*). The cloned enzyme was thermostable retaining about 70% activity after 48 h at 60°C. In our work (*21*), the optimum temperature of the crude ß-glucosidase from *A. pullulans* was 80°C (Figure 1). The crude enzyme had a half-life of 72 h at 75°C and 24 h at 80°C. However, the optimum temperature of the purified ß-glucosidase was 75°C. Similar findings were reported for the ß-glucosidase from *Neocallimastix frontalis,* in which case the pure enzyme had an optimum temperature at 45°C, whereas the crude enzyme showed optimum activity at 55-60°C (*28*). While the crude ß-glucosidase from *C. stercorarium* exhibited a half-life of 3 h at 60°C, the purified enzyme was rapidly inactivated at this temperature (*26*). However, the thermostability of the purified enzyme could be increased by Mg^{+2}, Ca^{+2} or DTT. By adding $MgCl_2$ and DTT, the half-life of the purified enzyme at 60°C was increased to more than 5 h. The enzyme showed optimal activity at pH 5.5 and 65°C. A ß-glucosidase from *S. thermophile* was optimally active at pH 5.4 and 65°C (*13*). The ß-glucosidase from *Trichoderma reesei* QM 9414 exhibited optimal activity towards cellobiose at pH 4.5 and 70°C (*29*). Characteristics of some recently described thermostable ß-glucosidases are presented in Table II.

Catalytic Properties

ß-Glucosidases constitute a very diverse family of enzymes capable of hydrolyzing

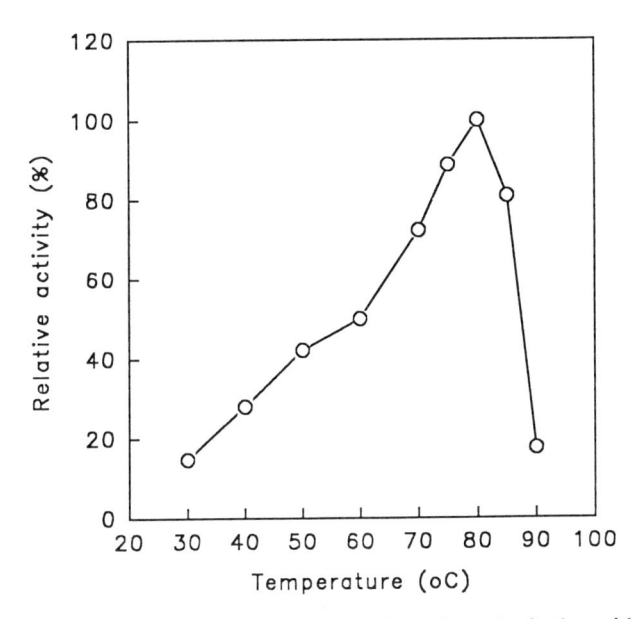

Figure 1. Effect of temperature on activity of crude ß-glucosidase from *Aureobasidium pullulans*. The enzyme activity was assayed at various temperatures at pH 4.5 using p-nitrophenyl ß-D-glucoside as substrate (30 min reaction).

Table II. Characteristics of some recently described thermostable β-glucosidases

Organism	Specific activity (U/mg protein)[a]	Optimum temp. (°C)	Optimum pH	Glucose inhibition (K_i, mM))
Thermotoga sp. (9)	195	na	7.0	0.42
Sporotrichum thermophile (13)	89	65	5.4	0.5
Pyrococcus furiosus (16)	389	102-105	5.0	300
Candida cacaoi (22)	9	60	4.0-5.5	8
Clostridium thermocellum (23)	125	65	6.5	na
Aspergillus nidulans (24)	468	60	5.5	5.48
Pisolithus tinctorius (25)	128	65	4.0	na
Aspergillus niger (30)	-	65	4.6	3.22

[a] Specific activity was determined using p-nitrophenyl β-D-glucoside (pNPβG) as substrate. One unit (U) of β-glucosidase was defined as the amount of enzyme required to liberate 1 μmol p-nitrophenol per min from pNPβG under certain assay conditions.
na, not available.

a broad range of β-glucosides. The hydrolytic mechanism of a β-glucosidase is considered to be by general acid catalysis (31). β-Glucosidases may be divided into three groups on the basis of substrate specificity: 1) aryl-β-glucosidases hydrolyze exclusively aryl-β-glucosides, 2) cellobiases hydrolyze cellobiose and cello-oligosaccharides only, and 3) broad specificity β-glucosidases hydrolyze both substrate types and are most common in cellulolytic microbes (1,2,32). The β-glucosidase from *P. furiosus* had higher affinity for pNPβG than cellobiose with K_m values of 0.15 and 20 mM, respectively (16). It also exhibited β-galactosidase, β-xylosidase and some β-mannosidase activity. The β-glucosidase from *C. stercorarium* hydrolyzed a variety of substrates including pNPβG, cellobiose and disordered cellulose (26). K_m values were determined to be 0.8 mM for pNPβG and 33 mM for cellobiose. The K_m values of β-glucosidase from thermophilic anaerobic strain Wai21W.2 were 0.15 and 0.73 mM for hydrolysis of pNPβG and cellobiose, respectively (15). Purified β-glucosidase from *A. pullulans* hydrolyzed cellobiose and pNPβG effectively (21). The purified enzyme had very little (< 5%) or no activity on lactose, maltose, sucrose and trehalose. It also had no or very little activity on pNP-α-D-glucoside, pNP-β-D-xyloside, pNP-β-D-cellobioside, pNP-α-L-arabinofuranoside and pNP-β-D-glucuronide (< 5%). K_m values of 1.17 and 1.00 mM and V_{max} values of 897 and 800 U mg^{-1} protein of β-glucosidase from *A. pullulans* were obtained for the hydrolysis of pNPβG and cellobiose, respectively, at pH 4.5 and 75°C. The purified β-glucosidase hydrolyzed cello-oligosaccharides. K_m values for hydrolysis of cellotriose, cellotetraose, cellopentaose, cellohexaose and celloheptaose by this enzyme were 0.34, 0.36, 0.64, 0.68 and 1.65 mM, respectively. The enzyme preparation did not hydrolyze lactose, although the organism produced very high level of ß-glucosidase when grown on lactose. It, however, hydrolyzed pNP-ß-D-galactoside at 7.6% of that of pNPßG. The ß-glucosidase from *P. furiosus* hydrolyzed lactose very well (*16*). The intracellular ß-glucosidase from *Evernia prunastri* was considered to be a true cellobiase because of its great affinity toward cellobiose (*33*). The K_m values for the hydrolysis of cellobiose and pNPßG were 0.244 and 0.635 mM, respectively. K_m values for the hydrolysis of cellobiose and pNPßG by ß-glucosidase from *T. reesei* strain QM 9414 were 0.5 mM and 0.3 mM, respectively (*29*). This enzyme hydrolyzed cellodextrins by sequentially splitting off glucose units from the non-reducing end of the oligomers. It is interesting that the major role of a thermolabile ß-glucosidase from *Butyrivibrio fibrisolvens* cloned in *E. coli* in the degradation of cellulose was the cleavage of cellodextrins rather than cellobiose (*34*). The K_m values for the hydrolysis of pNPßG and cellobiose by some thermophilic ß-glucosidases are presented in Table III. The ß-glucosidase from *Bacillus polymyxa* expressed in *E. coli* produced glucose and cellotriose from cellobiose (*35*). The cellotriose was formed by transglycosylation. Metal ions such as Ca^{2+}, Mg^{2+}, Mn^{2+} or Co^{2+} (5 mM) did not stimulate or inhibit the ß-glucosidase activity of *A. pullulans*, and thiol was not essential for activity (*21*). The ß-glucosidase from *P. furiosus* was also unaffected by thiol-specific inhibitors (*16*). Galactose, mannose, arabinose, fructose, xylose

Table III. Comparison of K_m values of some thermostable ß-glucosidases for the hydrolysis of p-nitrophenyl ß-D-glucoside (pNPßG) and cellobiose

Source	K_m value (mM)	
	pNPßG	Cellobiose
Thermotoga sp. (*9*)	0.10	19
Sporotrichum thermophile (*13*)	0.29	0.83
Pyrococcus furiosus (*16*)	0.15	20
Aureobasidium pullulans (*21*)	1.17	1
Candida cacaoi (*22*)	0.44	87
Clostridium thermocellum (*23*)	2.20	77
Aspergillus nidulans (*24*) P-1	0.84	1
P-II	0.47	0.8
Pisolithus tinctorius (*25*)	0.87	-
Trichoderma reesei QM9414 (*29*)	0.30	0.5
Aspergillus niger (*30*) USBD 0827	0.75	0.89
USBD 0828	1.23	1.64

and lactose at 1% (w/v) concentration did not inhibit ß-glucosidase activity from *A. pullulans*. Some ß-glucosidases preferentially utilize alcohols rather than water as acceptors for the glycosyl moiety during catalysis, yielding ethyl ß-D-glucoside in the reaction (*9,35-37*). The ß-glucosidase of *Dekkera intermedia* is activated by 2 M ethanol using pNPßG as substrate, suggesting that ethanol increases the hydrolysis rate of pNPßG by acting as an acceptor molecule for the intermediary glucosyl[1] cation (*37*). The effect of ethanol can be attributed to ß-glucosyl transferase activity - the ethanol acting as a suitable acceptor for this reaction (*9*). The initial activity of the ß-glucosidase from *A. pullulans* was slightly stimulated by ethanol. HPLC analysis of the reaction products in the presence of 6% ethanol after 24 h indicated the formation of an additional unidentified peak (*21*).

Substrate Inhibition

Substrate inhibition by cellobiose is a common property of ß-glucosidase from *Trichoderma* sp. and other microorganisms (*29,38,39*). Cellobiose strongly inhibited its own hydrolysis by ß-glucosidase from *Pyromyces* sp. at concentrations above 0.2 mM (*38*). The inhibition constant (K_i) for cellobiose was 0.62 mM. The ß-glucosidase from *A. pullulans* was not inhibited by up to 20 mM pNPßG or 6% (w/v) cellobiose (*21*). It was shown that the ß-glucosidase from *A. niger* normally inhibited by cellobiose concentration greater than 10 mM, was not apparently subject to inhibition by cellobiose concentration as high as 100 mM if it was immobilized and entrapped with calcium alginate gel spheres (*40*).

Glucose Inhibition

Inhibition by glucose, a common characteristic of ß-glucosidases (*1,9,41,42*) although there are exceptions (*10,43,44*), is an important constraint to overcome if this enzyme is to have industrial applications. Most of the ß-glucosidases studied were competitively inhibited by glucose. Glucose inhibited the ß-glucosidase catalyzed reaction of cellulase of *T. viride* in a mixed inhibition pattern with a competitive character (*45*). The ß-glucosidase from *A. pullulans* was competitively inhibited by glucose with an inhibition constant (K_i) of 5.65 mM (*21*). The intracellular ß-glucosidase from *E. prunastri* was competitively inhibited by glucose with a K_i value of 1.26 mM (*33*). ß-Glucosidase from *S. thermophile* was competitively inhibited by glucose with a K_i of 0.5 mM (*13*). A ß-glucosidase from *Streptomyces* sp. was not only resistant to glucose inhibition but it was stimulated two-fold by 0.1 M glucose (*44*). A cloned ß-glucosidase (BglB) from *M. bispora* was also activated two to three fold in the presence of 2-5% (0.1 - 0.3M) glucose and did not become inhibited until the glucose concentration reached about 40% (*10*). The ß-glucosidase from thermophilic anaerobic bacterium strain Wai21W.2 was insensitive to glucose inhibition (*15*). The inhibition of ß-glucosidase activity from *P. furiosus* by glucose was almost negligible with a K_i of 300 mM (*16*).

Synergism with Cellulases

Supplementation of ß-glucosidase from *S. thermophile* stimulated cellulose hydrolysis by cellulases where there was no accumulation of cellobiose in the reaction mixture (*13*). The ß-glucosidase from *A. pullulans* showed synergistic interaction with cellulase to increase the efficiency of glucose production from cellulose by converting cellobiose to glucose (Table IV). This ß-glucosidase may have utility in the enzymatic hydrolysis of cellulose from corn fiber and other cellulosic biomasses for the subsequent production of fuel ethanol. The addition of cloned ß-glucosidase from *C. thermocellum* increased the degradation of crystalline cellulose by *C. thermocellum* cellulase complex (*3*). Addition of ß-glucosidase from *A. niger* to a simultaneous saccharification/fermentation resulted in a 20% increase in percent conversion of cellulose to ethanol while addition to saccharification resulted in a 53% increase in percent conversion (*36*). The presence of immobilized ß-glucosidase from *A. phoenicis* during enzymatic hydrolysis of cellulosic materials significantly increased the concentration of glucose by converting cellobiose effectively to glucose (*45*).

Table IV. Cellulose hydrolysis by a commercial cellulase preparation supplemented with purified ß-glucosidase from *Aureobasidium pullulans*

Enzyme	Hydrolysis (%)[a]
Cellulase	58.5
Cellulase + ß-Glucosidase	66.4
ß-Glucosidase	0.0

[a] At pH 4.5 and 50°C. Reaction time, 48 h. Substrate used: Sigmacell 50 2% (w/v). Enzyme used: cellulase, 5 U/ml; ß-glucosidase, 0.1 U/ml. The reaction products were quantified by analyzing in HPLC (*21*).

Concluding Remarks

In recent years, a lot of research effort has been directed toward finding a suitable ß-glucosidase for application in the enzymatic conversion of cellulose to glucose. A number of ß-glucosidase genes have been cloned from different organisms and more than fifteen nucleotide sequences are now available, allowing for the identification of enzyme families, domain, putative catalytic sites, evolutionary traits, and other features (*10,47,48*). Our approach has been to develop an improved ß-glucosidase that is temperature and pH compatible with process conditions and to screen for a glucose tolerant ß-glucosidase. The high activity of the *A. pullulans* ß-glucosidase on cellobiose, its ability to hydrolyze a variety of cellooligosaccharides, high substrate tolerance, its non-dependence on metal ions or thiol compounds, and the high thermoactivity make the enzyme a suitable candidate for application in the enzymatic hydrolysis of cellulose to glucose. The glucose inhibition of ß-glucosidase could possibly be overcome by employing a combined saccharification and fermentatation process using a glucose fermenting organism.

Literature cited

1. Woodward, J.; Wiseman. A. *Enzyme Microb. Technol.* **1982**, 2, 73-79.
2. Coughlan, M. P. *Biotechnol. Genet. Eng. Rev.* **1985**, 3, 39-109.
3. Kadam, S. K.; Demain, A. L. *Biochem. Biophys. Res. Commun.* **1989**, 161, 706-711.
4. Ait, N.; Creuzet, N.; Cattaneo, J. *J. Gen. Microbiol.* **1982**, 128, 569-577.
5. Tong, C. C.; Cole, A. L.; Shepherd, M. G. *Biochem. J.* **1980**, 191, 83-94.
6. Shepherd, M. G.; Tong, C. C.; Cole, A. L.. *Biochem. J.* **1981**, 193, 67-74.
7. McHale, A.; Coughlan, M. P. *Biochim. Biophys. Acta* **1981**, 662, 152-159.
8. McHale, A.; Coughlan, M. P. *J. Gen. Microbiol.* **1982**, 128, 2327-2331.
9. Rutthersmith, L. D.; Daniel, R. M. *Biochim. Biophys. Acta* **1993**, 1156, 167-172.
10. Wright, W. M.; Yablonsky, M. D.; Ahalita, Z. P.; Goyal, A. K.; Eveleigh, D. E. *Appl. Environ. Microbiol.* **1992**, 58, 3455-3465.
11. Yoshioka, H.; Hayashida, S. *Agric. Biol. Chem.* **1980**, 44, 1729-1735.
12. Mitchell, R. W.; Hahn-Hagerdal, B.; Ferchak, J. D.; Kendall-Pye, E. *Biotechnol. Bioeng. Symp.* **1982**, 12, 461-467.
13. Bhat, K. M.; Gaikward, J. S.; Maheshwari, R. *J. Gen. Microbiol.* **1993**, 139, 2825-2832.
14. Plant, A. R.; Oliver, J.; Platchett, M.; Daniel, R.; Morgan, H. *Arch. Biochem. Biophys.* **1988**, 262, 181-188.
15. Patchett, M. L.; Daniel, R. M.; Morgan, H. W. *Biochem. J.* **1987**, 243, 779-787.
16. Kengen, S. W. M.; Luesink, E. J., Stams, A. J. M.; Zehnder, A. J. B. *Eur. J. Biochem.* **1993**, 213, 305-312.
17. Hagerdal, B.; Harris, H.; Pye, E. K. *Biotechnol. Bioeng.* **1980**, 22, 1515-1526.
18. Bernier, R; Stutzenberger, F. *MIRCEN J. Appl. Microbiol. Biotechnol.* **1989**, 5, 15-25.
19. Sternberg, D.; Vijaykumar, P.; Reese, E. T. *Enzyme Microb. Technol.* **1977**, 6, 508-512.
20. Rapp, P. *J. Gen. Microbiol.* **1989**, 135, 2847-2858.
21. Saha, B. C.; Freer, S. N.; Bothast, R. J. *Appl. Environ. Microbiol.* **1994**, 60, 3774-3780.
22. Drider, D.; Pommares, P; Chemardin, P.; Arnaud, A.; Galzy, P. *J. Appl. Bacteriol.* **1993**, 74, 473-479.
23. Katayeva, A.; Golovchenko, N. P.; Chuvilskaya, N. A.; Akimenko, V. K. *Enzyme Microb. Technol.* **1992**, 14, 407-412.
24. Kwon, K.-S.; Kang, H. G.; Hah, Y, C. *FEMS Microbiol. Letts.* **1992**, 97, 149-154.

25. Cao, W.; Crawford, D. L. *Can. J. Microbiol.* **1993**, 39, 125-129.
26. Bronnenmeier, K.; Staudenbauer, W. L. *Appl. Microbiol. Biotechnol.* **1988**, 28, 380-386.
27. Bissett, F.; Sternberg, D. *Appl. Environ. Microbiol.* **1978**, 35, 750-755.
28. Li, X.; Calza, R. E. *Enzyme Microb. Technol.* **1991**, 13, 622-628.
29. Schmid, G.; Wandrey, C. *Biotechnol. Bioeng.* **1987**, 30, 571-585.
30. Hoh, Y. K.; Yeoh, H.-H.; Tan, T. K. *Appl. Microbiol. Biotechnol.* **1992**, 37, 590-593.
31. Sinnott, M. L. *Chem. Rev.* **1990**, 90, 1171-1202.
32. Paavilainen, S.; Hellman, J.; Korpela, T. *Appl. Environ. Microbiol.* **1993**, 59, 927-932.
33. Yague, E.; Estevez, M. P. *Eur. J. Biochem.* **1988**, 175, 627-632.
34. Lin, L.-L.; Rumbak, E.; Zappe, H.; Thomson, J. A.; Woods, D. R. *J. Gen. Microbiol.* **1990**, 136, 1567-1576.
35. Painbeni, E.; Valles, S.; Polaina, J.; Flors, A. *J. Bacteriol.* **1992**, 174, 3087-3091.
36. Pemberton, M. S.; Brown, R. D. JR.; Emert, G. H. *Can. J. Chem. Eng.* **1980**, 58, 723-734.
37. Blondin, B.; Ratomahenina, R.; Arnaud, A.; Galzy, P. *Eur. J. Appl. Microbiol. Biotechnol.* **1983**, 17, 1-6.
38. Teunissen, M. J.; Lahaye, D. H. T. P.; Huis in't Veld, J. H. J.; Vogels, G. D. *Arch. Microbiol.* **1992**, 158, 276-281.
39. Witte, K.; Wartenberg, A. *Acta Biotechnol.* **1989**, 9, 179-182.
40. Lee, J. M.; Woodward, J. *Biotechnol. Bioeng.* **1983**, 25, 2441-2451.
41. Kohchi, C.; Hayashi, M.; Nagai, S. *Agric. Biol. Chem.* **1985**, 49, 779-784.
42. Lo, A. C.; Barbier, J.-R.; Willick, G. E. *Eur. J. Biochem.* **1990**, 192, 175-181.
43. Freer, S. N. *Arch. Biochem. Biophys.* **1985**, 243, 515-522.
44. Ozaki, H.; Yamada, K. *Agric. Biol. Chem.* **1992**, 55, 979-987.
45. Montero, M.; Romeu, A. *Appl. Microbiol. Biotechnol.* **1992**, 38, 350-353.
46. Sundstorm, D. W.; Klel, H. E., Coughlin, R. W., Biederman, G. J., Brouwer, C. A. *Biotechnol. Bioeng.* **1981**, 23, 473-485.
47. Perez-pons, J. A.; Cayetano, A.; Rebordosa, X.; LLoberas, J.; Guasch, J.; Querol, E. *Eur. J. Microbiol.* **1994**, 223, 557-565.
48. Henrissat, H.; Bairoch, A. *Biochem. J.* **1993**, 293, 781-788.

RECEIVED September 11, 1995

Chapter 14

Initial Approaches to Artificial Cellulase Systems for Conversion of Biomass to Ethanol

Steven R. Thomas, Robert A. Laymon, Yat-Chen Chou,
Melvin P. Tucker, Todd B. Vinzant, William S. Adney, John O. Baker,
Rafael A. Nieves, J. R. Mielenz, and Michael E. Himmel[1]

Alternative Fuels Division, Applied Biological Sciences Branch,
National Renewable Energy Laboratory, 1617 Cole Boulevard,
Golden, CO 80401-3393

The process of converting low-value biomass to ethanol via fermentation depends keenly on the development of economic biocatalysts to achieve effective depolymerization of the cellulosic content of biomass. This process has few features in common with contemporary large-scale uses of cellulases, such as food processing and detergent augmentation. Cellulase preparations used in an ethanol-from-biomass process must hydrolyze crystalline cellulose completely, operate effectively at mildly acidic pH, withstand process stress, and they need not be derived from microorganisms that are generally recognized as safe (GRAS). Furthermore, the ideal cellulase complex used in biomass processing should be highly active on the intended feedstock and obtainable at the lowest possible cost. Clearly, the ability to "engineer" cellulase systems in anticipation of each application is key to successful optimization and commercialization. To achieve this goal, we are collaborating with other researchers to purify individual cellulase enzymes and their genes from a variety of bacteria and fungi. We are also comparing purified cellulases, both individually and in mixtures, using key criteria, to design and assemble optimized, use-specific, artificial cellulase systems that can be produced by recombinant host microorganisms on an industrial scale. This paper describes both strategies and progress to date.

The U.S. Department of Energy (DOE) supports a research, development, and deployment (R&D&D) program directed at supporting industrial processes for conversion of lignocellulosic biomass to ethanol transportation fuel. The project has been divided into three generations of technology, which differ mainly in the feedstock expected to be used and the date targeted for piloting the scaled-up technology (i.e., operating a process development unit). First-generation technology

[1]Corresponding author

is directed at the conversion of three feedstocks considered by DOE to be of near-term significance: waste paper, hardwood saw mill waste, and corn residue. Second-generation technology will employ the first "energy crop," a selected agricultural residue. Third-generation technology is directed at conversion of short-rotation hardwood trees, such as hybrid poplar.

Cellulose and hemicellulose are the principle sources of fermentable sugars in lignocellulosic feedstocks; however, nature has designed woody tissue for effective resistance to microbial attack. To be effective, cellulose-degrading microorganisms typically produce cellulase enzyme systems characterized by multiple enzymatic activities that work synergistically to reduce cellulose to cellobiose, and cellobiose to glucose. At least three different enzymatic activities are required to accomplish this task. β-1,4-endoglucanases (EC 3.2.1.4; also called endocellulases) cleave β-1,4-glycosidic linkages randomly along the cellulose chain. β-1,4-exoglucanases (EC 3.2.1.91; also called cellobiohydrolases, CBH) cleave cellobiose from either the reducing or the nonreducing end of a cellulose chain. 1,4-β-D-glucosidases (EC 3.2.1.21) hydrolyze aryl- and alkyl-β-D-glucosides and various disaccharides. Those enzymes that preferentially cleave cellobiose are called cellobiases. Interestingly, Kong et al. (1) have recently shown that another activity, xylan acetyl esterase, enhances cellulase action. Furthermore, two nonenzymatic factors are implicated in augmenting cellulose hydrolysis: the "yellow affinity substance" associated with *Clostridium thermocellum* cellulases (2) and the cellulose binding domains (CBD) isolated from *Cellulomonas fimi* cellulases (3). Thus, the efficient degradation of cellulose to glucose in nature may require as many as 5 or 6 different enzyme-types, peptides, or factors.

Filamentous fungi are well known as a cost-effective resource for industrial cellulases. For example, the fed-batch *Trichoderma reesei* cellulase fermentations reported by Watson et al. (4) have reached the spectacular level of 427 FPU/L·h. However, Philippidis (5) has recently shown that most cellulase productivities reported for similar systems fall between 100 and 200 FPU/L·h. In a recent monograph, Eveleigh (6) proposed that the maximum theoretical value obtainable from fungi cannot exceed 600 FPU/L·h. In contrast, the highest rate for cellulase production from a native cellulolytic bacterium is only 40 FPU/L·h; i.e., *Thermomonospora fusca* YX (7). However, early work with the *T. fusca* YX endoglucanase system, for example, showed that very high specific activities are obtainable; i.e., 2200 carboxymethylcellulose units/mg purified protein (8). These values are approximately 100-fold greater than specific activities from *T. reesei* endoglucanases. Furthermore, several bacterial cellulases have the desirable property of thermal stability; i.e., endoglucanases from *Acidothermus cellulolyticus* (9) and *Thermotoga neapolitana* (10). Therefore, the benefits of developing heterologous cellulase expression systems in rapidly growing bacteria include substantial enhancement of enzyme stability and specific activity, the potential for greater cell densities using fed-batch cultures, a dramatic reduction in cell-growth time, and the potential for protein overproduction. Some bacteria also appear to produce cellulase component enzymes with important properties such as resistance to end-product inhibition (6,11).

Many naturally occurring, cellulose-degrading fungi, bacteria, and protozoa have been described to date (11,12). In addition, carboxymethylcellulases (CMCases) have been reported from several higher plants (13). The most studied, and thus well-characterized, cellulase system is undoubtedly that from the filamentous fungus T. reesei, but several bacterial systems are rapidly becoming very well understood (12). T. reesei is a prolific natural producer of secreted mesophilic cellulases; however, it may not be the most effective cellulase system for use in biomass conversion processes that essentially require complete hydrolysis of the feedstock for economic viability.

Review of the Cellulase Literature

As a result of intense study for nearly 40 years, more than 60 cellulolytic fungi have been reported (11,14), representing the soft-rot, brown-rot, and white-rot fungi. Members of the latter group are able to modify and degrade lignin and cellulose in woody substrates. Bacterial cellulase systems are currently the focus of considerable study. A major review of bacterial cellulases has identified 46 unique bacterial producers of cellulases (11). One-third of this group belongs to the anaerobic bacteria; the remainder are aerobic. Most bacterial cellulase enzymes are closely associated with the cell wall or are found in multi-enzyme aggregates called "cellulosomes." These high-molecular-weight (MW) cellulolytic organelles are characteristic of most "ruminant" type bacterial cellulase systems. A few aerobic bacterial cellulase systems have been identified that secrete freely soluble cellulase component enzymes. Examples of these systems belong to the genera *Thermomonospora*, *Microbispora*, *Acidothermus*, *Pseudomonas*, *Thermotoga*, *Erwinia*, and *Acetivibrio*.

An exhaustive examination of the cellulase research literature has shown that very little information is suitable for comparison of the activity characteristics of endoglucanases or exoglucanases. This has been a persistent and aggravating deficit in the field since study began in the early 1950s. The problem is caused by the lack of universally accepted assay procedures and unwillingness or inability on the part of researchers to employ the few existing procedures considered rigorous, such as Ghose's International Union of Pure and Applied Chemists (IUPAC) method (15) and the more recent method of Sattler et al. (16). Data shown in Table I represent a list of endoglucanase-specific activities that are comparable because the authors have used purified enzymes and equivalent assay methods.

Review Parameters. For this review, we have chosen to compare published specific activities because the ultimate production system for endoglucanases in the biomass-to-ethanol process will likely use a recombinant host for heterologous expression of an endoglucanase gene. Specific activity is defined as the catalytic activity produced by a known quantity of purified enzyme. Many researchers express one endoglucanase unit of activity as the amount of enzyme that releases one micromole of glucose-reducing sugar equivalent from carboxymethycellulose (CMC) in one minute. This type of initial rate information, usually gathered at arbitrary protein concentration and at less than 1% total hydrolysis of an artificial substrate (CMC), is clearly less than ideal. In order to be universally comparable, endoglucanase activity must be assessed with a substrate of standardized solution concentration, degree of substitution, and degree of polymerization. Reducing sugars released must also be measured consistently and reflect standardized extents of hydrolysis.

Existing literature reviews ranking specific activities of more than a few native endoglucanases were not identified. Reviews reporting values for activities of commercial cellulase preparations were found during the course of this study (*17,18*), but were not considered useful for this evaluation because they are mixtures of several cellulolytic activities and are often adulterated or modified before sales. The purpose of this comparison is to assess purified individual enzymes.

Conclusions from Tables I and II. Tables IA, IB, and II list the characteristics of 61 native and recombinant, fungal and bacterial cellulases. Pure enzyme preparations were required to directly compare specific activities. With the exception of the cloned bacterial cellulases, these enzymes were purified with a minimum of two steps and a maximum of eight steps.

Tables IA and IB support the general observation that native fungal endoglucanases are less active than their native bacterial counterparts (*6,11*). The specific activities listed in Table I clearly show that one of the most active bacterial endoglucanases, *Pseudomonas fluorescens* var. *cellulosa* CELC (1860 units/mg), is nearly 5 times more active than the most active fungal enzyme, *Talaromyces emersonii* EG I (425 units/mg), and nearly 50 times more active than the primary endoglucanase from *T. reesei* EG I (40 units/mg). Table I also shows that the endoglucanases from *Thermotoga neapolitana* and *T. fusca* are highly active against CMC.

It is noteworthy that the *Erwinia* and *Microbispora* species cloned in *Escherichia coli* and *Zymomonas mobilis* tended to be produced with low specific activities (see Table II). The authors of these papers attributed their results to nonglycosylation of the recombinant proteins. Encouragingly, both *C. fimi* and *C. thermocellum* endoglucanases were expressed in moderately active forms from *E. coli*.

Enzymes Not Included in Table I. Only endoglucanase enzymes were considered in Table I, because measurements of whole-system cellulase activity (i.e., filter paper activity) were reported in the literature by at least five different and nonconvertible methods. Because most bacterial cellulose degraders require the presence of the cell (or at least the cellulosome) to hydrolyze crystalline cellulose, the data base for specific activities of microcrystalline cellulose hydrolysis is limited to only a few well-studied fungi. Furthermore, the concept of specific activity, when applied to filter paper-type assays, becomes nonstandard due to the multiplicity of enzymes required for such hydrolysis (e.g., EGs and CBHs).

While endoglucanases from many different fungal and bacterial systems have been fairly well characterized, the dearth of information on the corresponding exoglucanases assumed to be associated with these systems is stunning. This may be explained by the observation that exoglucanases are often much more difficult to monitor during the course of a purification. Commonly used, soluble, chromogenic substrates, which are hydrolyzed by some exoglucanases, are not cleaved by all

Table IA.
Characteristics of Purified Native Fungal Endoglucanases

Source	Reference	Given Name	Specific Activity (CMCU*)	M_r (kDa)	Purification Steps	Optimal pH	Optimal T°C
Aspergillus aculeatus	Murao et al. (19)	FII	40	66	6	5.0	70
Aspergillus niger	Okada (20)	cellulase	117	31	6	5-8	55
Eupenicillium javanicum	Tanaka et al. (21)	B-5-E	70.8	41	4	5.0	55
Fusarium solani	Wood (22)	endo C_x	25.5	37	3	5.5	55
Humicola insolens YH-8	Hayashida et al. (23)	endoglucanase	217	57	7	5.0	50
Irpex lacteus	Kanda et al. (24)	S-1	1.1	56	8	4.5	50
Irpex lacteus	Kubo & Nisizawa (25)	CMCase E2-B	76.4	35	6	4.0	50
Talaromyces emersonii	Moloney et al. (26)	EG I	414	68	3	5.6	60
		EG II	425				
		EG III	385				
		EG IV	325				
Trichoderma koningii	Halliwell & Vincent (27)	endoglucanase	5	ND	2	5.5	60
	Wood (28)	E1	29	13	5	5.0	60
	Wood (28)	E3a	33	48	2	5.0	60
Trichoderma viride							
QM9414	Voragen et al. (29)	EG II	20.1	45	4	4.0	50
	Shoemaker (30)	EG II	29.1	37	4	4.2	50
Trichoderma reesei	Schulein (31)	EG I	60	52	3	4.0	50
		EG II	40	48	3	4.0	50

*carboxymethylcellulose units (CMCU) = µmol released from CMC/min · mg.

Table IB.
Characteristics of Purified Native Bacterial Endoglucanases

Source	Reference	Given Name	Specific Activity (CMCU*)	M_r (kDa)	Purification Steps	Optimal pH	Optimal T°C
anaerobic bacterium BW	Creuzet & Frixon (32)	endoglucanase	45.1	90	2	6.4	60
Bacillus subtilis DLG	Robson & Chambliss (11)	endoglucanase	550	35	2	5-6	60
Bacillus succinogenes	Groleau & Forsberg (33)	endoglucanase	42.5	45	4	6.0	45
Cellulomonas uda	Nakamura & Kitamura (34)	cellulase IV	62	48	3	6.0	47
Clostridium papyrosolvens	Garcia et al. (35)	CMCase	6.7	ND	2(partial)	4.8	45
Clostridium stercorarium	Creuzet et al. (36)	endoglucanase	92	91	partial	6.4	ND
Clostridium thermocellum	Ng & Zeikus (37)	endoglucanase I	65	83	5	5.2	62
Pseudomonas fluorescens var. cellulosa	Yamane & Suzuki (38)	cellulase A	1150	40	4	8.0	30
		cellulase C	1860	ND	4	7.0	30
Thermomonospora fusca YX	Calza et al. (8)	E1	770	96	4	6.0	58
		E2	77	46	4	6.0	58
Thermomonospora fusca YX	Wilson (39)	E1	1100	108	4	6.0	58
		E2	100	42	4	ND	ND
		E4	6	106	4	ND	ND
		E5	150	45	4	ND	ND
Thermotoga neapolitana	Bok et al. (10)	endoglucanase A	1200	30	3	7.0	110
		endoglucanase B	1500	24	2	7.0	110

*carboxymethylcellulose units (CMCU) = μmol glucose released from CMC/min · mg.

Table II
Characteristics of rCellulases Expressed from Heterologous Systems

Source (genes)	Reference	CMC Specific Activity (μmol RS/min·mg)	Given name	M_r (kDa)	Purification Steps	Optimal pH	Optimal T°C
Bacteroides succinogenes cloned in *E. coli*	Taylor et al. (40)	0.025	*cel* gene endo.	43	none	5.9	47
Caldocellum saccharolyticum cloned in *E. coli*	Love & Streiff (41)	1.31[1]	β-D-glucosidase	52	none	6.25 (periplasmic)	85
Bacillus subtilis cloned in *Zymomonas mobilis*	Yoon et al. (42)	0.025	endoglucanase	~39	none	ND (cytosol)	ND
Pseudomonas fluorescens var. *cellulosa* cloned in *Z. mobilis*	Lejeune et al. (43)	0.35	endoglucanase	ND	none	ND	ND
Clostridium thermocellum cloned in *E. coli*	Beguin et al. (44)	140	endoglucanase A	56	3(cytosol)	6	60+
		60	endoglucanase B	66	3.	6	ND
		27	endoglucanase C	40	3	6	65+
		428	endoglucanase D	74	3	6	ND
B. subtilis cloned in *Bacillus megaterium*	Kim & Pack (45)	452	endoglucanase	33	4	5.5	60

thermophilic anaerobe cloned in E. coli	Honda et al. (46)	0.31	endoglucanase	ND	none	6.0	80
Erwinia chrysanthemi cloned in E. coli	Brestic-Goachet et al. (47)	0.8	endoglucanase Z	45	none	7.0	52
E. chrysanthemi cloned in Z. mobilis	Brestic-Goachet et al. (47)	2.5	endoglucanase Z	45	none	7.0	52
Bacteroides ruminicola subsp. brevis cloned in E. coli	Woods et al. (48)	0.02-0.04	endoglucanase	ND	none	5.5-6	37-42
Microbispora bispora cloned in E. coli	Eveleigh et al. (49)	6.6	endoglucanase A	44	none	6.5	60
Bacillus polymyxa cloned in E. coli	Baird et al. (50)	ND	endoglucanase	37	none	ND	ND
Bacillus circulans cloned in E. coli	Baird et al. (50)	ND	endoglucanase	52	none	ND	ND

Continued on next page

Table II (continued)
Characteristics of rCellulases Expressed from Heterologous Systems

Source (genes)	Reference	CMC Specific Activity (μmol RS/min · mg)	Given name	M_r (kDa)	Purification Steps	Optimal pH	Optimal T°C
Bacteroides ruminicola cloned in *E. coli*	Matsushita et al. *(51)*	340-840[2]	endoglucanase	88	1	ND	ND
M. bispora cloned in *E. coli*	Wright et al. *(52)*	10 (130)[3]	BglB β-D-glucosidase	52	none	6.2	60
Trichoderma reesei cloned in *S. cerevisiae*	Bailey et al. *(53)*	25[4] 1560[4]	CBH II cellobiohydrol. EG I endoglucanase	>native >native	2 2	ND ND	ND ND
Clostridium thermocellum cloned with *E. coli*	Lemaire and Béguin *(54)*	50	CelG endoglucanase	36-42	4	ND	ND
M. bispora cloned in *Streptomyces lividans*	Hu et al. *(55)*	0.25-0.56[5]	CBHII cellobiohydrol.	93	none	6.5	60
Thermomonospora fusca cloned in *S. lividans*	Irwin et al. *(56)*	5410[2] 369 122 2840	E1 endoglucanase E2 endoglucanase E4 endoglucanase E5 endoglucanase	101 43 90 46	3 4 3 3	ND ND ND ND	ND ND ND ND

[1] Units/mg protein using p-nitrophenyl-β-D-glucoside (pNPG) [2] μmol cellobiose/min per μmol protein
[3] Units/mL culture supernatant using p-nitrophenyl-β-D-glucoside (units/mL using cellobiose)
[4] nKatal/mL culture supernatant using 4-methylumbelliferyl-β-D-cellobiose (MUC)
[5] mUnits/mL using hydroxyethylcellulose (HEC)

similar enzymes, an example being *T. fusca* E$_3$ (*56*). This problem forces the use of more cumbersome assays (i.e., synergy or immunochemical) for the purification and characterization of (in particular) bacterial exoglucanases. Exoglucanases are, perhaps, often mis-identified as "weak" endoglucanases when they cleave pNP-cellobioside or methylumbelliferyl-cellobioside substrates at low rates. This problem is further exacerbated by the general paucity of CMC fluidity data and information regarding reaction product anomeric carbon stereochemistry. These kinds of results have caused some cellulase researchers to conclude that exoglucanases are simply not present in some cellulolytic systems, despite the fact that no well characterized crystalline cellulose-hydrolyzing system is known that does not exhibit one or more distinct exoglucanase activities, i.e., *T. reesei, Humicola insolens, Phanerochaete chrysosporium, T. fusca, C. thermocellum,* and *Microbispora bispora* (*6,12*). Recently, an exoglucanase from *C. fimi* was also reported (*57*).

Microbial β-D-glucosidases differ from both cases described above in that it is possible to find quite extensive kinetic characterizations of these enzymes, and associated assays are not considered problematic (*58*).

Strategy for Cellulase System at NREL

The initial focus of NREL's biomass-to-ethanol technology will probably employ proprietary commercial preparations of cellulase derived from mutant *Trichoderma* strains such as Rut-C30, RL-P37, or L27 (*59*), in a conversion process known as simultaneous saccharification and fermentation (SSF). Later process designs will be based on engineered components and mixes of cellulases. Figure 1 shows a logic diagram describing NREL's cellulase development strategy. The following four-part strategy was developed to accomplish this goal.

1. **Acquire Purified Cellulase Enzymes.** Cellulase producers will be chosen for detailed testing based on a study of published information summarized in Table I. Cellulases purified from these bacteria and fungi will be procured from DOE subcontractors or produced at NREL. Standard chromatographic protocols for cellulase purification (*60*) will be followed where possible, and all preparations will be subjected to tests of homogeneity (i.e., SDS-polyacrylamide gel electrophoresis and Fast Protein Liquid Chromatography).

2. **Verify Activities and Biochemical Characteristics of Cellulases.** To date, no concerted, national effort has been funded to assess the efficacy of specific bacterial and fungal cellulases for biomass hydrolysis via a "head-to-head" comparison. Candidate cellulase genes will be identified for incorporation into recombinant cellulase expression systems based on this performance testing. Although conceptually possible in any one laboratory, the identification and purification of many selected cellulases in a limited time frame is best accomplished by the coordinated effort of multiple, highly focused research laboratories. Thus, DOE has subcontracted much of the enzyme purification work to academic and corporate laboratories.

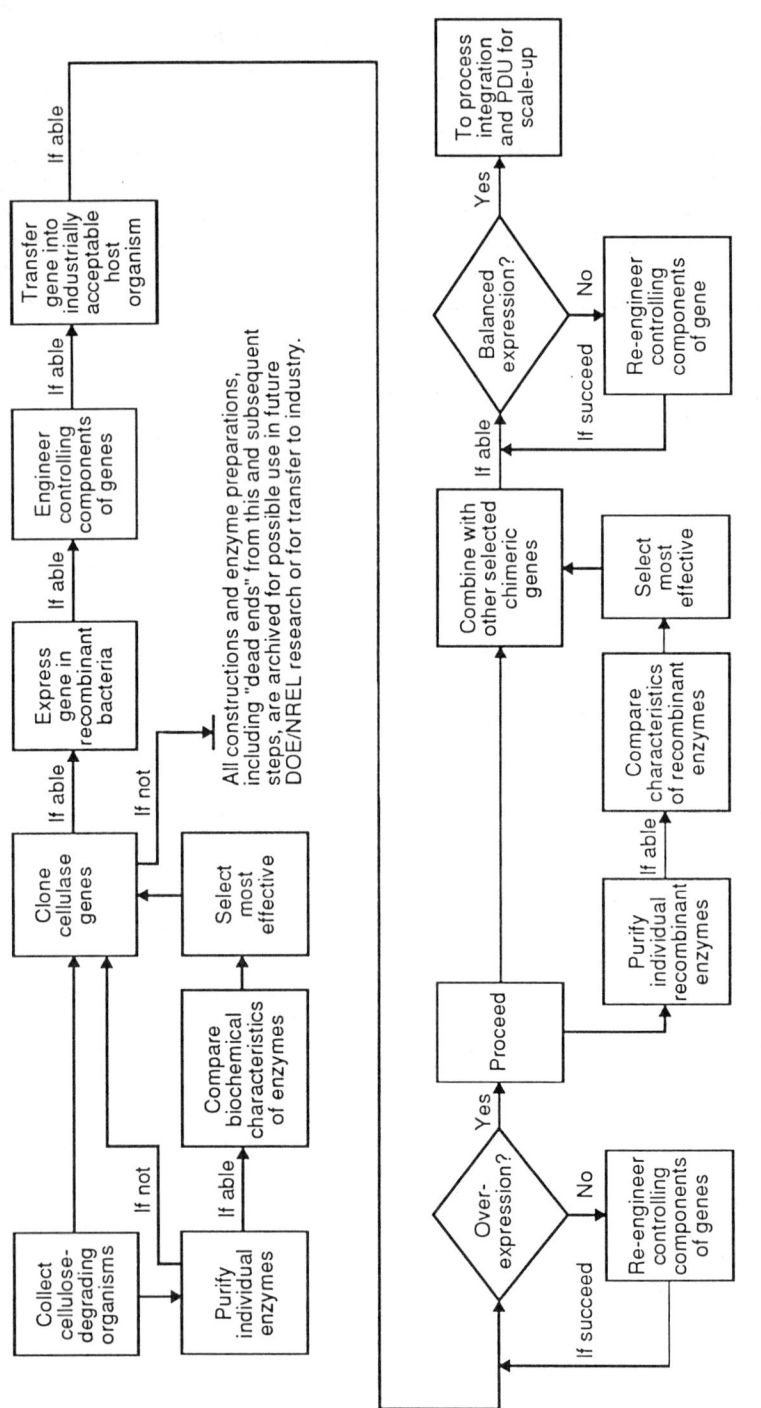

Figure 1. NREL's strategy for development of engineered cellulase systems.

These efforts will be limited to selected endoglucanases and exoglucanases, specifically those chosen from the comprehensive literature study discussed above. Candidate enzymes will be subjected to studies of kinetics, temperature and pH tolerance, resistance to end product inhibition, and the synergistic effect.

3. **Maximize Expression of Cellulases in Heterologous Host Systems.** Based on the efforts described in Step 2 to determine which combinations of cellulases and β-D-glucosidases work effectively together under process conditions, genes encoding these enzymes will be used to construct recombinant cellulase-producing organisms.

Since the first heterologous expression of a *C. fimi* endoglucanase gene was reported from *E. coli* in 1978, a strong potential has existed for the successful expression of many cellulase genes in a number of rapidly growing bacteria (*61*). Indeed, a review by Lejeune et al. (*62*) tabulates examples of heterologous expression of endoglucanases, exoglucanases, and β-D-glucosidases in *E. coli*, *Bacillus subtilis*, and *Streptomyces lividans* hosts before 1987. Table II provides examples of successful expression of recombinant cellulases reported since 1987. Table II also shows that, apart from a wide variety of gene source and host combinations producing enzymes of marginal specific activity, a few examples exist for the expression of recombinant cellulases displaying very high specific activities (e.g., *C. thermocellum* genes in *E. coli* and *T. fusca* genes in *S. lividans*).

Opportunities to clone β-D-glucosidases are complicated because these enzymes are often of high MW and are not secreted by the native organism (*63*). There are, nonetheless, several examples of heterologous expression of β-D-glucosidases in the literature [i.e., *C. thermocellum* enzyme expressed in *E. coli* (*64*), *C. fimi* enzyme expressed in *E. coli* (*65*), *Caldocellum saccharolyticum* enzyme expressed in *E. coli* and *B. subtilis* (*41*), and *M. bispora* enzyme expressed in *E. coli* (*52*)].

4. **Construct Multi-Gene Expression Systems.** While in the past, researchers have endeavored to use cellulase cloning to answer a variety of diverse scientific questions, our immediate goal is the overexpression of highly active cellulases that act synergistically to degrade crystalline cellulose. The ultimate goal of this concerted research effort is to develop a multiple-gene expression system in a suitable bacterial or fungal host that produces high levels of endoglucanase, exoglucanase, and β-D-glucosidase activities in optimal proportions for the rapid and efficient degradation of cellulosic biomass.

Possible problems anticipated from a recombinant bacterial cellulase system include limited productivity, low specific activities of recombinant enzymes, ineffective multi-gene expression, identification of exo/endoglucanases that act synergistically on crystalline cellulose, and differences in enzyme processing that affect enzyme function. These factors and others render this effort a formidable task. Yet, clear evidence exists from the biotechnology industry that the manipulation of multi-gene systems for the improvement of cellular activities is achievable in a reasonable time frame; an example is ethanol production by transformed *E. coli* (*66*).

The recent successes of Yoo and Pack (*65*) in demonstrating the expression of a *B. subtilis* endoglucanase and a *C. fimi* β-D-glucosidase in *E. coli* using a bi-cistronic plasmid under *tac* promotor control are encouraging. They showed that 15% to 20% of the total protein produced by these transformed cells was cellulase. Earlier examples of successful constructions of multiple cellulase genes begins with the 1987 work of Penttilä et al. (*67*) to clone two *T. reesei* endoglucanases in *Saccharomyces cerevisiae*. In 1988, Wong et al. (*68*) reported the cloning of CenA and Cex from *C. fimi* in *S. cerevisiae* using a tandem cartridge with two identical yeast promotors on one plasmid. That same year, Penttilä et al. (*69*) cloned *T. reesei* CBH I and CBH II on different plasmids in *S. cerevisiae* using yeast PGK promotors. In 1989, Huang et al. (*70*) reported cloning three *Ruminococcus albus* cellulases into *E. coli* using one lambda vector. In 1993, a prokaryotic endoglucanase (*C. thermocellum*, endo E) was expressed in active form in mammalian cells by Soole et al. (*71*), further demonstrating the range of heterologous expression possible.

Progress Toward Cellulase System Development

Research and development of an engineered cellulase system for the DOE Biofuels Program was initiated in 1991 and has followed the logic diagram outlined in Figure 1. At the close of 1994, the primary effort regarding collection and characterization of cellulase enzymes will near completion. The outline below describes current progress toward the four key technical objectives.

1. **Identify and Acquire Purified Cellulase Enzymes.** Forty years of intense study have clearly established cellulase producers as a truly diverse group of organisms. Unlikely ecosystems recently shown to harbor cellulolytic microorganisms include extreme environments, such as oceanic thermal vents (i.e., *T. neapolitana*), and the anaerobic rumen of cattle (i.e., the fungus *Neocallimastix frontalis*) (*72*). Figures 2 and 3 were developed to illustrate the impact of such diveristy. These figures show scatter plots of pH and temperature activity optima published for selected fungal and bacterial endoglucanases, respectively. These plots clearly demonstrate the relatively narrow range of characteristics exhibited by fungal enzymes, as opposed to the wide-ranging characteristics of bacterial enzymes. The most widely used industrial cellulase system, *T. reesei*, possesses a major endoglucanase, EG I, with temperature activity characteristics that partially overlap the "SSF window" (see the 50% activity box shown in Figure 1). The thermotolerant endoglucanase, *A. cellulolyticus* E1, demonstrates temperature activity characteristics that are useful for high-temperature saccharification, although clearly not optimal for direct use in SSF (see Figure 2). In general, pH and temperature effects on hydrolytic activity are used as a first test for cellulase suitability, and neither EG I nor E1 is ideal for this purpose.

    Tables IA and IB show the specific activities and native MWs of purified fungal and bacterial endoglucanases, respectively. Bacterial cellulases appear to

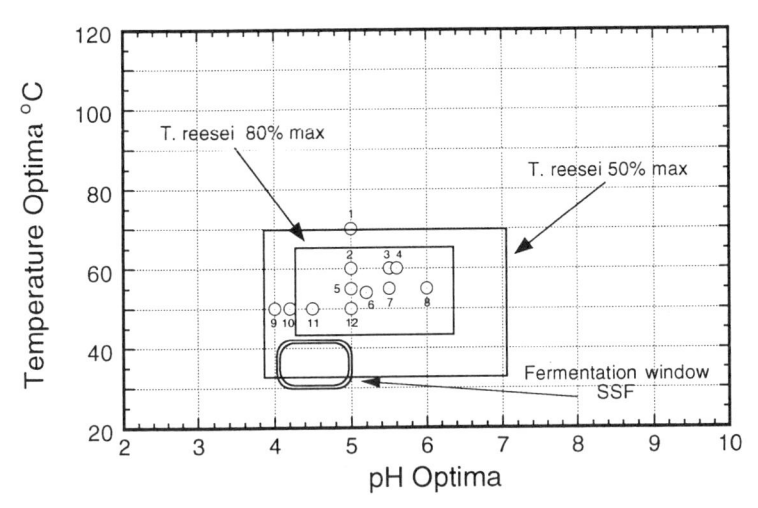

Figure 2. Scatter plot of pH and temperature activity optima for selected, purified fungal endoglucanases reported in the literature. The 50% and 80% activity windows for *Trichoderma reesei* EG I are shown. Plot key: **(1)** 66kD *Aspergillus aculeatus* (*19*); **(2)** *Trichoderma koningii* (*28*); **(3)** 48kD *T. koningii* (*27*); **(4)** 68kD *Talaromyces emersonii* (*26*); **(5)** 41kD *Eupenicillium javanicum* (*21*); **(6)** 12kD *Aspergillus fumigatus* (*73*); **(7)** 37kD *Fusarium solani* (*22*); **(8)** 31kD *Aspergillus niger* (*20*); **(9)** 52kD EG I *T. reesei* (*31*) and 78kD *Sclerotium rolfsii* (*74*); **(10)** 37kD EG II *Trichoderma viride* (*30*); **(11)** 56kD *Irpex lacteus* (*24*); **(12)** 57kD *Humicola insolens* YH8 (*23*).

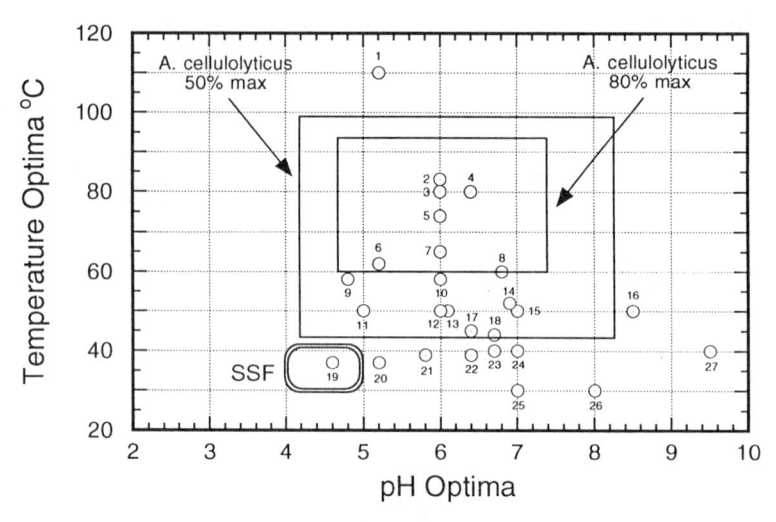

Figure 3. Scatter plot of pH and temperature activity optima for selected, purified bacterial endoglucanases reported in the literature. The 50% and 80% activity windows for *Acidothermus cellulolyticus* E1 are shown. Plot key: **(1)** 24kD *Thermotoga neapolitana* (*10*); **(2)** 69kD E1 *A. cellulolyticus* (*75*); **(3)** 56kD *Clostridium thermocellum* NCIB 10682 (*76*); **(4)** 91kD *Clostridium stercorarium* (*32*); **(5)** 94kD *Thermomonospora fusca* YX (*8*); **(6)** 83kD *C. thermocellum* LQRI (*77*); **(7)** 22kD *Thermomonospora curvata* (*78*); **(8)** 45kD *Clostridium josui* FERMP-9684 (*79*); **(9)** 35kD *Bacillus subtilis* DLG (*80*); **(10)** 46kD *T. fusca* YX (*8*); **(11)** 33kD *Acetivibrio cellulolyticus* (*81*); **(12)** 32kD *Streptomyces* strain KSM-9 (*82*); **(13)** 45kD *Fibrobacter succinogenes* S85 (*33*); **(14)** 45kD *Erwinia chrysanthemi* 3665 (*83*); **(15)** 77 kD *Cellulomonas* sp. (*84*); **(16)** 32kD *Streptomyces* strain KSM-9 (*82*); **(17)** 89kD *Ruminococcus flavefaciens* str. 67 (*85*); **(18)** 50kD *Ruminococcus albus* F-40 (*86*); **(19)** *Clostridium acetobutylicum* NRRL B527 (*87*); **(20)** *C. acetobutylicum* P270 (*87*); **(21)** 118kD *F. succinogenes* S85 (*88*); **(22)** 65kD *F. succinogenes* S85 (*88*); **(23)** *Pseudomonas* sp. (*89*); **(24)** 44kD *Cellulomonas* sp. (*84*); **(25)** 40kD cell A *Pseudomonas fluorescens* var. *cellulosa* (*38*); **(26)** cel B *P. fluorescens* var. *cellulosa* (*38*); **(27)** *Bacillus* sp. KSM-635 (*90*).

possess the important property of high specific activity. These parameters represent important criteria for initial study and, although chosen from literature reports, all require further experimental verification before final ranking.

Based on a combination of encouraging literature reports and reasonable availability, a group of 9 bacterial and fungal cellulase systems were selected for further examination by NREL and DOE subcontractors. These are *C. fimi, T. fusca, M. bispora, A. cellulolyticus, P. fluorescens* var. *cellulosa, C. thermocellum, T. reesei, Aspergillus niger,* and *T. neapolitana*.

2. **Verify Activities and Biochemical Characteristics of Cellulases.** In 1993 and 1994, these efforts were limited to the 21 selected endoglucanases, exoglucanases, and β-D-glucosidases available at NREL (see Table III). Candidate enzymes were subjected to kinetic and synergistic effect analysis, as described below.

Synergism. Different combinations of endoglucanases and exoglucanases isolated from heterologous systems show differing degrees of synergy. Our goal is to determine which combinations produce the greatest synergistic effect and the highest amount of glucose production per unit of time at equivalent enzyme loadings. To date, all experiments have been carried out using only the model cellulose substrate, Sigmacell-20. An ordered list of the best four binary mixtures of available cellulases in synergy assays has been developed using combinations of individual purified cellulases. Binary synergy reactions were carried out under standard conditions, with total molar cellulase loading (endoglucanase plus exoglucanase) held constant in all digestion mixtures. *A. niger* β-D-glucosidase was purified according to the method of Himmel et al. *(93)* and included in every reaction to prevent feedback inhibition by accumulated cellobiose. Bovine serum albumin (BSA) was added to each reaction mixture to attenuate loss of enzyme by surface adsorption. The synergistic effect is quantified by calculating the ratio of glucose released when both enzymes are present in a reaction versus the sum of equivalent amounts of enzymes in separate, individual reactions:

$$[(glucose_{endo+exo}) / (glucose_{endo} + glucose_{exo})].$$

Synergy values significantly greater than 1 indicate a synergistic effect between the two enzymes under examination. Synergy values around 1 indicate no synergistic effect. Synergy values significantly less than 1 indicate some degree of interference between the two enzymes. Tests were conducted with 16 endo/exo binary pairs using enzymes shown in Table III. Table IV lists the four best synergistic pairs under the conditions employed for these reactions.

Of the top five synergy pairs listed in Table IV, all include *T. reesei* CBH I paired with different bacterial endoglucanases. Interestingly, the *T. reesei* pair CBH I and EG I is missing from this set and ranks ninth out of the 16 reactions analyzed. This result highlights the potential value of using *T. reesei* CBH I in future recombinant systems. A surprising observation, however, was that the best synergy pair included the E1 endoglucanase from *A. cellulolyticus*.

Table III

Purified Cellulases and β-D-Glucosidases Archived at NREL for Testing

Microbial Source	Enzyme(s)	Institutional Source
Acidothermus cellulolyticus	E1, rE1 endoglucanase[1,*]	NREL
	E2 endoglucanase[1]	NREL
	β-D-glucosidase[1,*]	NREL
Aspergillus niger	β-D-glucosidase	NREL
	Endo A endoglucanase	NREL
Cellulomonas fimi	110 kDa endoglucanase	Cognis, Inc.[3]
Clostridium thermocellum	rCel D endoglucanase[2]	Univ. of Rochester[4]
	rCel S exoglucanase	Univ. of Rochester
Microbispora bispora	rEndo A endoglucanase[*]	Rutgers University[5]
Pseudomonas fluorescens var. *cellulosa*	rEndo X endoglucanase	Cognis, Inc./NREL
Thermomonospora fusca	rE_1 endoglucanase	Cornell University[6]
	rE_2 endoglucanase[*]	Cornell University
	E_3 exoglucanase[*]	Cornell University
	rE_4 endoglucanase	Cornell University
	rE_5 endoglucanase[*]	Cornell University
Thermotoga neapolitana	rEndo B endoglucanase[*]	Rutgers University
	rβ-D-glucosidase A	Rutgers Univeristy
Trichoderma reesei	CBH I cellobiohydrolase[*]	NREL
	CBH II cellobiohydrolase[*]	NREL
	EG I endoglucanase[*]	NREL
	EG II endoglucanase[*]	NREL

[1] DOE/NREL U.S. patents (*91,92*) or patents pending
[2] r denotes recombinant product
[3] Supplied to NREL by C. Paech
[4] Supplied to NREL by D. Wu
[5] Supplied to NREL by D. Eveleigh
[6] Supplied to NREL by D. Wilson
* Used for tests of synergy in binary mixtures.

Kinetic Analysis. A summary of the kinetic properties for the available cellulases is given in Tables V and VI. The conditions for assaying the endoglucanases, exoglucanases, and β-D-glucosidases were chosen for compatibility with biomass-to-ethanol process options, including traditional yeast SSF (i.e., pH 5 and 40°C) and separate hydrolysis and fermentation (SHF) using saccharification at elevated temperatures (i.e., pH 5 and 80°C). Data in Tables V and VI can thus be used to rank attractive enzyme candidates for the DOE/NREL process. At 40°C, *T. fusca* E_5 and *T. reesei* EG II endoglucanases are the most active enzymes tested by the CMC fluidity assay. At 75°C and 80°C, however, *A. cellulolyticus* E1 is the most active endoglucanase. The Michaelis-Menten kinetics for β-D-glucosidases shown in Table VI indicates that at 40°C the *A. cellulolyticus* and *T. neapolitana* enzymes have similar (and quite low) K_m values for MUG hydrolysis, although the *A. cellulolyticus* enzyme is more active on cellobiose at this temperature. If a process option requires high-temperature enzyme use, data in Table VI illustrate that of the group tested, only the β-D-glucosidase from *T. neapolitana* can be considered viable.

Table IV
Best Synergistic Endo/Exo Cellulase Pairs Tested[1]

Endo/Exo pair	Synergism Ratio	Composition for Maximum Synergy (% endo)
A. cellulolyticus E1 & *T. reesei* CBH I	2.74	20% E1
T. fusca E_5 & *T. reesei* CBH I	2.16	20% E_5
M. bispora rEndo A & *T. reesei* CBH I	2.04	20% Endo A
T. neapolitana rEndo B & *T. reesei* CBH I	1.90	60% Endo B

[1] Adapted from Baker et al. (*94*). Reactions were performed in 50 mM acetate pH 5.0 buffer with 0.36 µM total cellulase, 4.1 µg/mL β-glucosidase, 20 µg/mL BSA, and 50 mg/mL Sigmacell-20. Mixtures were held with agitation at 50°C for 120 h, which produced 9% or greater theoretical yields of glucose. Glucose was determined using the bicinchoninic acid method of Doner and Irwin (*95*).

3. **Maximize Expression of Cellulases in Heterologous Host Systems.** The genes shown in Table VII are those currently archived at NREL to construct the initial generation of recombinant cellulase-producing organisms. The immediate preferred host for these constructions, based on existing industrial example and a high potential for enhanced heterologous expression systems, is *S. lividans* TK24 (*96*).

STATUS. Two plasmids have been used during the course of this work, pIJ702 (*97*) and pFD666 (*98*), generously provided by Dr. Ryszard Brzezinski, Université de Sherbrooke, Quebec, Canada. Plasmid pIJ702 confers resistance

Table V
Characteristics of Purified Endoglucanases Archived at NREL

Microbial Source		Mr (kDa) from gene[2]	CMC Fluidity $(\Delta\phi)$[1] 40°C	75°C	80°C
A. cellulolyticus	rE1[3]	69	4284	48,544	39,927
A. niger	Endo A	29[4]	12575	inactive[7]	
C. fimi	110 kDa	110[4]	396	inactive	
C. thermocellum	rCel D	68[5]	2820	inactive	
M. bispora	rEndo A	44	16638	12,376	inactive
P. fluorescens var. cellulosa	rEndo X	38[4]	134[6]	inactive	
T. fusca	rE₁	94	5319	inactive	
	rE₂	42	2058	inactive	
	rE₄	90	443	inactive	
	rE₅	46	18752	inactive	
T. neapolitana	rEndo B	24[4]	2003	28,902	25,750
T. reesei	EG I	46	15758	inactive	
	EG II	42	37791	inactive	

Data from this study.
[1] Based on a 0.5-unit increase in the fluidity of a 0.11% w/w carboxymethylcellulose (CMC; Hercules 7HF) buffered solution. Units are: per mg enzyme per 30 m digestion time. Assays were performed with a Schott Automatic Capillary Viscometer (3 mL capacity) in 50 mM acetate buffer pH 5.0 with 100 µg/mL bovine serum albumin (BSA; Sigma Chemical).
[2] Molecular weight derived from the published gene sequence (Gen Bank), unless otherwise noted.
[3] r = Recombinant enzyme.
[4] From SDS-polyacrylamide gel electrophoresis.
[5] Native gene is 68 kDa, protein expressed is 68 kDa + (11 kDa non-native leader peptide) = 79 kDa.
[6] Performed in 50 mM phosphate buffer pH 7.0 with 100 µg/mL BSA.
[7] Enzymes labeled "inactive" were shown to undergo complete activity loss either immediately, or within 20 min, after introduction to the viscometer.

Table VI
Characteristics of Purified β-D-Glucosidases
Archived at NREL

Microbial Source (substrate)	Mr (kDa)	40°C K_m (mM)	40°C V_{max} (μmol/s)	75°C K_m (mM)	75°C V_{max} (μmol/s)
(MUG[1])					
A. cellulolyticus	52[2]	0.0221[3]	1.1x10^{-4}	inactive	$(T_{½}$~95s)[4]
A. niger	117	0.321	1.6x10^{-4}	inactive	$(T_{½}$~55s)
T. neapolitana, rβG-A	93	0.0103	1.2x10^{-4}	0.041	3.9x10^{-4}
(cellobiose)					
A. cellulolyticus		1.96	3.17x10^{-4}	inactive	$(T_{½}$~95s)
A. niger		8.63	9.18x10^{-3}	inactive	$(T_{½}$~55s)
T. neapolitana, rβG-A		29.13	1.77x10^{-3}	9.63	2.75x10^{-3}

Data from this study.
[1] MUG = 4-methylumbelliferyl-β-D-glucopyranoside.
[2] From SDS-polyacrylamide gel electrophoresis.
[3] K_m and V_{max} values were determined at 40°C and 75°C, respectively, using MUG and cellobiose as substrates over concentration ranges of 0.01-0.5 mM and 0.2-10 mM, respectively. The protein concentration in each reaction mixture was approximately 1 μg/mL. Assays were performed at pH 5.0 in 50 mM acetate buffer.
[4] Enzyme lability at this temperature renders kinetics unreliable ($T_{½}$=half life).

Table VII

Genes Encoding Cellulases and β-D-Glucosidases Undergoing Development at NREL

Microbial Source	Gene(s)(Enzyme Coded)	Institutional Source
Acidothermus cellulolyticus	*celE1* (endoglucanase)[1]	NREL
Caldocellum saccharolyticum	*bglA* (β-D-glucosidase)[2]	Univ. of Auckland New Zealand
Clostridium thermocellum	*celD* (endoglucanase)	Univ. of Rochester
Microbispora bispora	*celA* (endoglucanase)	Rutgers University
	bglA (β-D-glucosidase)	Rutgers University
	bglB (β-D-glucosidase)	Rutgers University
Pseudomonas fluorscens var. *cellulosa*	*eglX* (endoglucanase)	Rutgers Univeristy
Thermomonospora fusca	E_3 (exoglucanase)	Cornell University
Thermotoga neapolitana	*bglA* (β-D-glucosidase)	Rutgers University
	bglB (β-D-glucosidase)	Rutgers University

[1] DOE/NREL U.S. patent pending
[2] Property rights sold to DOE/NREL by University of Auckland, New Zealand.

to thiostrepton, whereas pFD666 confers resistance to kanamycin. Plasmid pSZ7, a shuttle plasmid containing a 3 kb fragment of *T. fusca* genomic DNA, encodes the E$_3$ exoglucanase and was supplied by Dr. David Wilson. The *A. cellulolyticus* E1 endoglucanase gene was cloned as a 3.7 kb genomic DNA fragment into pIJ702 at the unique BglII site, after ligation of BglII linkers, to produce plasmid E1-pIJ702. The same *A. cellulolyticus* DNA fragment containing the E1 endoglucanase was likewise cloned into the unique HindIII site of both pSZ7 and pFD666, producing plasmids E1-pSZ7 and E1-pFD666, respectively. Plasmid E1-pFD666 carries only the E1 endoglucanase genes. The single-gene plasmid constructs in *S. lividans* strain TK24, thus developed, were labeled TK24/E1-pIJ702, and TK24/pSZ7 (Cornell University).

Streptomyces strains were grown in tryptic soy broth (TSB) with or without antibiotics (i.e., thiostrepton: 5 µg/mL in TSB, 50 µg/mL in TSB agar plates; kanamycin: 50 µg/mL in liquid and solid media). Five-mL seed cultures of *S. lividans* were grown for 48 h at 30°C before transferring into 40 mL of fresh TSB. Five-mL aliquots were collected at various times during growth of this culture to estimate enzyme activity, cell mass, and antigen concentration with an ELISA assay developed for this purpose (99).

Preliminary small shake flask (40 mL) experiments with the *S. lividans* clones described above showed that the TK24/E1-pIJ702 culture produced 4.2 mg rE1/L in 40 h and the TK24/pSZ7 culture produced 40 mg rE$_3$/L in 29 h. In both cases, the enzymes were measured in culture broths using optimized ELISA assays (99). More recently, TK24/E1-pIJ702 cultures were grown at the 150-L scale in a New Brunswick fermenter. Cell densities of approximately 16 g/L of recombinant *S. lividans* were obtained from this controlled growth. This experiment yielded 18 mg rE1/L of growth supernatant. Although clearly in an early stage of molecular and fermentation optimization, these results show that cellulase component enzymes as diverse as an endoglucanase from *A. cellulolyticus* and an exoglucanase from *T. fusca* can be successfully expressed in *S. lividans* using unimproved native genes, albeit at modest levels.

4. **Construction of Multi-Gene Expression Systems**.

Strategies. Several possible strategies for developing multiple gene expression systems are depicted in Figure 4. In the first case (see Figure 4A), each strain produces only one recombinant product. The amount of product produced by each strain could be determined during and after each fermentation. Recombinant fermentation products could be custom blended according to previously determined optimal enzyme ratios to maximize cellulose degradation and minimize the use of recombinant enzymes. This approach offers the most freedom to adjust the relative amount of recombinant product that the host strains produce. Host strains could be of the same or different species. Plasmids could be equivalent or different replicons. Promotors P1, P2, and P3 also could be equivalent or different. Each gene is totally independent from the others.

The second example is based on three independent plasmids in one strain (see Figure 4B). In this case, all constructs co-exist in the same cell, requiring the use of compatible plasmids (i.e., 3 replicons). All plasmids should also have equivalent stabilities so that gene dosage is maintained at constant (desired)

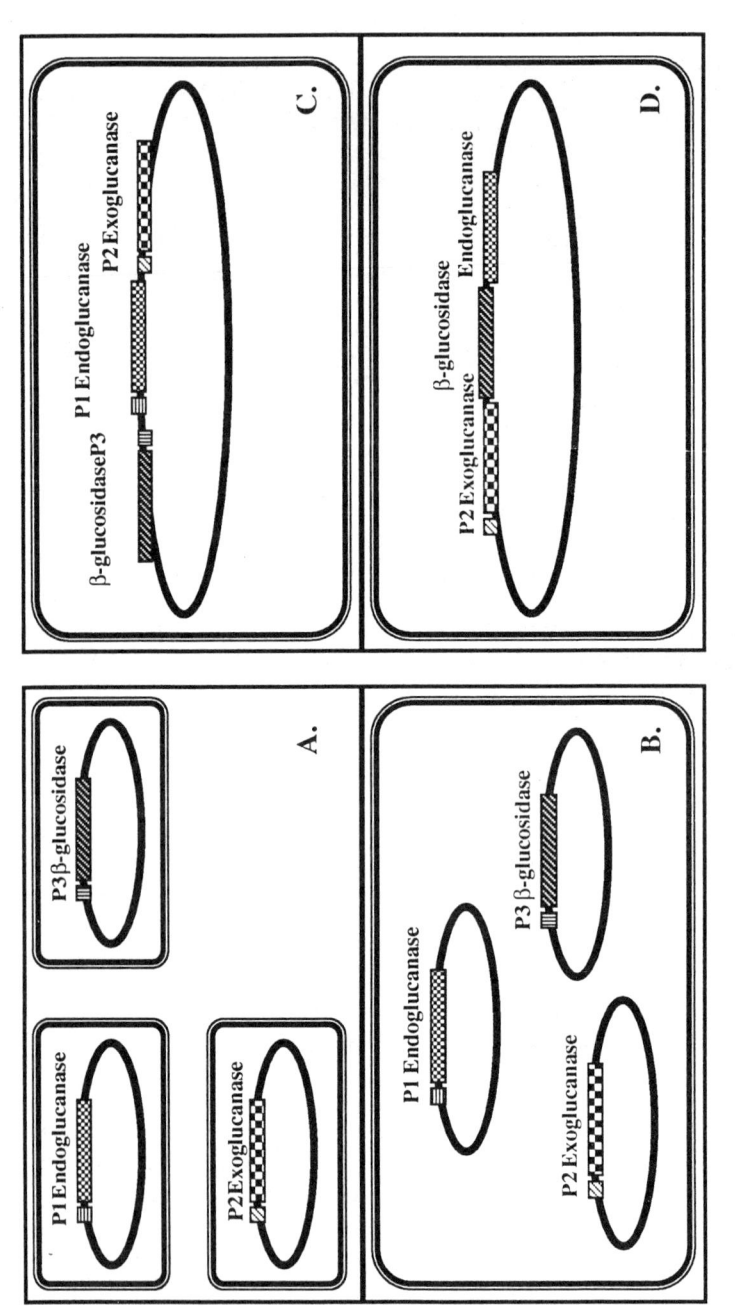

Figure 4. Depiction of the four possible strategies for cloning multiple cellulase genes in an industrial host. (A) Different genes cloned in separate host strains, (B) each gene cloned on separate, compatible plasmid in the same host cell, (C) each independently regulated gene cloned into a single plasmid, (D) polycistronic expression of all coding sequences regulated as a single transcription unit.

levels. Gene dosage may then be adjusted through plasmid copy number (a process not precisely defined or predictable *a priori*). Expression levels of different genes can be manipulated through use of different promotors for each gene to produce appropriate ratios of recombinant activities.

The third approach is to construct three independent genes on one plasmid (see Figure 4C). Here, use of the single plasmid removes plasmid compatibility considerations of concern in the previous option. Furthermore, relative gene dosage is more easily managed. Promotors 1, 2, and 3 could be the same or different, depending on required activity levels. Relative polarity of the genes may not be important (i.e., parallel or anti-parallel). However, increasing the number of genes on a single plasmid will result in a larger, perhaps less stable, plasmid.

The fourth option is a polycistronic gene harboring all three genes (see Figure 4D). In this example, a single promotor controls the expression of all three polypeptides, and gene dosage is maintained at controlled stoichiometric levels. Relative expression levels could vary considerably because downstream cistrons may be expressed at lower levels than upstream cistrons. In this scenario, the polarity of the genes must be parallel. Although the final construct is still rather large, this option should generate a somewhat smaller construct than that described in option 3.

STATUS. The two-gene constructs developed at NREL were labelled JT46/E1-pSZ7 (one plasmid) and TK24/E1-pFD666,pSZ7 (two compatible plasmids). Transformed *S. lividans* cultures were grown and monitored as described above. The recombinant culture containing two plasmids, TK24/E1-pFD666,pSZ7, was observed to produce 26 mg/L rE$_3$ in 40 h and nearly undetectable levels (< 1 mg/L) of rE1.

In a second case, plasmid E1-pSZ7 carried both the rE1 endoglucanase and rE$_3$ exoglucanase genes. JT46, a recombination deficient strain of *S. lividans*, was used for expression of this two-gene plasmid. The transformant JT46/E1-pSZ7 was found to produce 0.4 mg rE$_3$/L in 40 h and 0.2 mg rE1/L in 25 h from inoculation. The productivity for either gene in both strains is much lower than the single-gene-containing transformants described above. These results show that the present two-gene construct is obviously far from optimal, in that enzyme and biomass productivity are suppressed.

Conclusion

A substantial record of successful recombinant expression of numerous bacterial and fungal cellulases in bacterial and fungal hosts exists in the literature. A major obstacle to developing engineered cellulase systems that are efficient and economically viable has been the lack of a broad-scope, concerted effort to identify the best enzymes for biomass conversion. This goal is being actively pursued at NREL and is still somewhat hampered by the plethora of basic science issues embodied in the cellulase story.

In reviewing the cellulase literature reported over the past two decades, the lack of uniformity in assay methodology and enzyme nomenclature stand out as obvious sources of confusion for comparative studies. Fortunately, work pioneered by Ghose (*15*) and continued by Sattler et al. (*16*) to extend and improve cellulase assays has clarified our picture of the procedures necessary to correctly compare the activities of these enzymes. Henrissat and Mornon's hydrophobic cluster analysis method (*100*) is also key to the problem of resolving cellulase classification. As enzyme function becomes better understood, functional artificial cellulase systems may soon be assembled with a high degree of success.

A review of the cellulase cloning literature also illustrates inherent biases in the field, some of which have hindered translation of this technology into the commercial sector. Most reported studies of heterologous expression of cellulases, for example, deal with the required details of plasmid construction and refinement, whereas little effort has been expended on maximizing the expression of recombinant products or assessing the biochemical characteristics of those enzymes. Although there was doubt for some time, the work of Beguin et al. (*44*) and Irwin et al. (*56*) has clearly shown that bacterial cellulases can be expressed from bacterial hosts at specific activities equivalent to those enzymes produced by the native strains. Most researchers, however, do not focus on this level of the problem. Examples can even be found for the successful expression of bacterial endoglucanases in *S. cerevisae* (*101*).

The cellulase production effort at NREL will attempt to bring all aspects of this complex problem to successful resolution. So far, a series of cellulases has been evaluated for biochemical characteristics important to the DOE Ethanol Project. In most cases, the genes coding these cellulases also have been acquired. Specifically, a successful expression system must synthesize and secrete large amounts (1-10 g/L) of functional enzyme. This goal will require the judicious use of genetic control elements, such as signal peptides, compatible codons, promoters, and ribosome binding sites. Compatible types and levels of enzyme glycosylation and benign proteolytic environments may also be desirable (*102*). The current challenge is to design expression systems in industrially compatible hosts for the key enzymes listed in Table III, while keeping our eyes open for new candidate enzymes discovered during this process.

This compelling R&D project incorporates elements important to U.S. industrial competitiveness and to basic biotechnology. Indeed, the country possessing key technology that can deliver cost-effective hydrolytic enzymes to industry may possess an important economic advantage.

Acknowledgments

We would like to thank Dr. David Wilson for kindly supplying purified enzymes and plasmids, and Drs. David Wu, Douglas Eveleigh, and Christian Paech for supplying purified enzymes. We would also like to thank Christopher Reeves for *S. lividans* JT46. This work was funded by the Biochemical Conversion Element of the DOE Biofuels Energy Systems Program.

Literature Cited

1. Kong, F.; Engler, C.R.; Soltes, E.J. *Appl. Biochem. Biotech.* **1992**, *34/35*, 23-35.
2. Ljungdahl, L.G.; Pettersson, B.; Eriksson, K.-E.; Wiegel, J. *Current Microbiol.* **1983**, *9*, 195-200.
3. Greenwood, J.M.; Ong, E.; Gilkes, N.R.; Kilburn, D.G.; Miller, Jr., R.C.; Warren, R.A.J., In *Cellulose Sources and Exploitation*; Kennedy, J.F.; Phillips, G.O.; Williams, P.A., Eds.; Ellis Horwood: New York, NY, 1990; pp. 423-428.
4. Watson, T.G.; Nelligan, I.; Lessing, L. *Biotechnol. Lett.* **1984**, *6*, 667-672.
5. Philippidis, G.P. In *Enzymatic Conversion of Biomass for Fuels Production*; Himmel, M.E.; Baker, J.O.; Overend, R.P., Eds.; ACS Series 566, American Chemical Society: Washington, DC, 1994; pp. 188-217.
6. Eveleigh, D.E. *Phil. Trans. Royal Soc. Lond.* **1987**, *A321*, 435-447.
7. Meyer, H.P.; Humphrey, A.E. *Chem. Eng. Commun.* **1982**, *19*, 149-156.
8. Calza, R.E.; Irwin, D.C.; Wilson, D.B. *Biochemistry* **1985**, *24*, 7797-7804.
9. Tucker, M.P.; Mohagheghi, A.; Grohmann, K.; Himmel, M.E. *Bio/Technology* **1989**, *7*, 817-820.
10. Bok, J.D.; Goers, S.K.; Eveleigh, D.E. In *Enzymatic Conversion of Biomass for Fuels Production*; Himmel, M.E.; Baker, J.O.; Overend, R.P., Eds.; ACS Series 566, American Chemical Society: Washington, DC, 1994; pp. 54-65.
11. Robson, L.M.; Chambliss, G.H. *Enzyme Microb. Technol.* **1989**, *11*, 626-644.
12. Rapp, P.; Beermann, A. In *Biosynthesis and Biodegradation of Cellulose*; Haigler, C.H.; Weimer, P.J., Eds.; Marcel Dekker: New York, NY, 1990; pp. 535-597.
13. Brummell, D.A.; Lashbrook, C.C.; Bennett, A.B. In *Enzymatic Conversion of Biomass for Fuels Production*; Himmel, M.E.; Baker, J.O.; Overend, R.P., Eds.; ACS Series 566, American Chemical Society: Washington, DC, 1994; pp. 100-129.
14. Ljungdahl, L.G.; Eriksson, K.-E. In *Advances in Microbial Ecology*; Marshall, K.C., Ed.; Vol. 8; Plenum Press: New York, NY, 1985; pp. 237-299.
15. Ghose, T.K. *Pure & Appl. Chem.* **1987**, *59*, 257-268.
16. Sattler, W.; Esterbauer, H.; Glatter, O.; Steiner, W. *Biotech. Bioeng.* **1989**, *33*, 1221-1234.
17. Klesov, A.A.; Rabinovich, M.L.; Nutsubidze, N.N.; Todorov, P.T.; Ermolova, O.V.; Chernoglazov, V.M.; Mel'nik, M.S.; Kude, E.; Dzhafarova, A.N.; Kornilova, I.G.; Kvesitadze, E.G. *Biotekhnologiay* **1987**, *2*, 152-168.
18. Nutsubidze, N.N.; Todorov, P.T.; Kvesitadze, E.G.; Mazur, N.S. *Prikladnaya Biokhimiya i Mikrobiologiya* **1989**, *25*, 41-47.
19. Murao, S.; Sakamoto, R.; Arai, M. *Methods Enzymol.* **1988**, *160*, 274-299.
20. Okada, G. *Methods Enzymol.* **1988**, *160*, 259-264.
21. Tanaka, M.; Taniguchi, M.; Matsuno, R.; Kamikubo, T. *Methods Enzymol.* **1988**, *160*, 251-259.
22. Wood, T.M. *Biochem. J.* **1971**, *121*, 353-362.
23. Hayashida, S.; Ohta, K.; Mo, K. *Methods Enzymol.* **1988**, *160*, 323-332.
24. Kanda, T.; Wakabayashi, K.; Nisizawa, K. *J. Biochem.* **1976**, *79*, 977-988.
25. Kubo, K.; Nisizawa, K. *J. Ferment. Technol.* **1983**, *61*, 383-389.

26. Moloney, A.P.; McCrae, S.I.; Wood, T.M.; Coughlan, M.P. *Biochem. J.* **1985**, *225*, 365-374.
27. Halliwell, G.; Vincent, R. *Biochem. J.* **1981**, *199*, 409-417.
28. Wood, T.M. *Methods Enzymol.* **1988**, *160*, 221-234.
29. Voragen, A.G.J.; Beldman, G.; Rombouts, F.M. *Methods Enzymol.* **1988**, *160*, 243-251.
30. Shoemaker, S.P. *Characterization of structure and enzymic activity of endo-1,4-beta-D-glucanases purified from Trichoderma viride*, PhD Thesis, Virginia Polytechnic Institute and State University, Blacksburg, VA, 1976.
31. Schulein, M. *Methods Enzymol.* **1988**, *160*, 234-242.
32. Creuzet, N.; Frixon, C. *Biochimie* **1983**, *65*, 149-156.
33. Groleau, D.; Forsberg, C.W. *Can. J. Microbiol.* **1983**, *29*, 504-517.
34. Nakamura, K.; Kitamura, K. *Methods Enzymol.* **1988**, *160*, 211-216.
35. Garcia, V.; Madarro, A.; Pena, J.L.; Pinaga, F.; Valles, S.; Flors, A. *J. Chem. Tech. Biotechnol.* **1989**, *46*, 49-60.
36. Creuzet, N.; Berenger, J.-F.; Frixon, C. *FEMS Microbiol. Lett.* **1983**, *20*, 347-350.
37. Ng, T.K.; Zeikus, J.G. *Methods Enzymol.* **1988**, *160*, 351-355.
38. Yamane, K.; Suzuki, H. *Methods Enzymol.* **1988**, *160*, 200-210.
39. Wilson, D.B. *Methods Enzymol.* **1988**, *160*, 314-323.
40. Taylor, K.A.; Crosby, B.; McGavin, M.; Forsberg, C.W.; Thomas, D.Y. *Appl. Environ. Microbiol.* **1987**, *53*, 41-46.
41. Love, D.R.; Streiff, M.B. *Bio/Technology* **1987**, *5*, 384-387.
42. Yoon, K.H.; Park, S.H.; Pack, M.Y. *Biotechnol. Lett.* **1988**, *10*, 213-216.
43. Lejeune, A.; Eveleigh, D.E.; Colson, C. *FEMS Microbiol. Lett.* **1988**, *49*, 363-366.
44. Beguin, P.; Millet, J.; Grepinet, O.; Navarro, A.; Juy, M.; Amit, A.; Poljak, R.; Aubert, J.-P. In *Biochemistry and Genetics of Cellulose Degradation*; Aubert, J.-P.; Beguin, P.; Millet, J., Eds.; Academic Press: London, UK, 1988; pp. 267-282.
45. Kim, H.; Pack, M.Y. *Enzyme Microbiol. Technol.* **1988**, *10*, 347-351.
46. Honda, H.; Saito, T.; Iijima, S.; Kobayashi, T. *Appl. Microbiol. Biotechnol.* **1988**, *29*, 264-268.
47. Brestic-Goachet, N.; Gunasekaran, P.; Cami, B.; Baratti, J.C. *J. Gen. Microbiol.* **1989**, *135*, 893-902.
48. Woods, J.R.; Hudman, J.F.; Gregg, K. *J. Gen. Microbiol.* **1989**, *135*, 2543-2549.
49. Eveleigh, D.E., Rutgers University, personal communication, 1990.
50. Baird, S.D.; Johnson, D.A.; Seligy, V.L. *J. Bacteriol.* **1990**, *172*, 1576-1586.
51. Matsushita, O.; Russell, J.B.; Wilson, D.B. *J. Bacteriol.* **1990**, *172*, 3620-3630.
52. Wright, R.M.; Yablonsky, M.D.; Shalita, Z.P.; Goyal, A.K.; Eveleigh, D.E. *Appl. Environ. Microbiol.* **1992**, *58*, 3455-3465.
53. Bailey, M.J.; Siika-aho, M.; Valkeajärvi, A.; Penttilä, M.E. *Biotechnol. Appl. Biochem.* **1993**, *17*, 65-76.
54. Lemaire, M.; Béguin, P. *J. Bacteriol.* **1993**, *175*, 3353-3360.
55. Hu, P.; Chase, Jr., T.; Eveleigh, D.E. *Appl. Microbiol. Biotechnol.* **1993**, *38*, 631-637.
56. Irwin, D.C.; Spezio, M.; Walker, L.P.; Wilson, D.B. *Biotech. Bioeng.* **1993**, *42*, 1002-1013.

57. Meinke, A.; Gilkes, N.R.; Kwan, E.; Kilburn, D.G.; Warren, R.A.J.; Miller, R.C. *Mol. Microbiol.* **1994**, *12*, 413-422.
58. β-*Glucosidases, Biochemistry and Molecular Biology*; Esen, A., Ed.; ACS Symposium Series 533; American Chemical Society: Washington, DC, 1993.
59. Gogary, S.E.; Leite, A.; Crivellaro, O.; Dorry, H.E.; Eveleigh, D.E. In *Trichoderma reesei Cellulases*; Kubicek, C.P.; Eveleigh, D.E.; Esterbauer, H.; Steiner, W.; Kubicek-Pranz, E.M., Eds.; The Royal Society of Chemistry: Canbridge, UK, 1990; pp. 200-211.
60. Paech, C. In *Enzymatic Conversion of Biomass for Fuels Production*; Himmel, M.E.; Baker, J.O.; Overend, R.P., Eds.; ACS Series 566, American Chemical Society: Washington, DC, 1994; pp. 130-178.
61. Chakrabarty, A.M.; Brown, J.F. In *Genetic Engineering*; Chakrabarty, A.M., Ed.; CRC Press: Boca Raton, FL, 1978; p. 185.
62. Lejeune, A.; Colson, C.; Eveleigh, D.E. In *Biosynthesis and Biodegradation of Cellulose*; Haigler, C.H.; Weimer, P.J., Eds.; Marcel Dekker: New York, NY, 1990; pp. 623-671.
63. Woodward, J.; Wiseman, A. *Enzyme Microb. Technol.* **1982**, *4*, 73-79.
64. Kadam, S.; Demain, A.; Millet, J.; Beguin, P.; Aubert, J.-P. *Enzyme Microb. Technol.* **1988**, *10*, 9-13.
65. Yoo, Y.D.; Pack, M.Y. *Biotechnol. Lett.* **1992**, *14*, 77-82.
66. Ingram, L.O.; Conway, T.; Alterthum, F. *Ethanol Production by Escherichia coli Strains Co-Expressing Zymomonas PDC and ADH Genes*; United States Patent No. 5,000,000, 1991.
67. Penttilä, M.E.; André, L.; Saloheimo, M.; Lehtovaara, P.; Knowles, J.K.C. *Yeast* **1987**, *3*, 75-85.
68. Wong, W.K.R.; Curry, C.; Parekh, R.S.; Parekh, S.R.; Wayman, M.; Favies, R.W.; Kilburn, D.G.; Skipper, N. *Bio/Technology* **1988**, *6*, 713-719.
69. Penttilä, M.E.; André, L.; Lehtovaara, P.; Bailey, M.; Teeri, T.T.; Knowles, J.K.C. *Gene* **1988**, *63*, 103-112.
70. Huang, C.M.; Kelly, W.J.; Asmundson, R.V.; Yu, P.L; *Appl. Microbiol. Biotechnol.* **1989**, *31*, 265-271.
71. Soole, K.L.; Hirst, B.H.; Hazlewood, G.P.; Gilbert, H.J.; Laurie, J.L.; Hall, J. *Gene* **1993**, *125*, 85-89.
72. Wood, T.M.; Wilson, C.A.; McCrae, S.I.; Joblin, K.N. *FEMS Microbiol. Lett.* **1986**, *34*, 37-40.
73. Parry, J.B.; Stewart, J.C.; Hepinstall, J. *Biochem. J.* **1983**, *213*, 437.
74. Sadana, J.C.; Lachke, S.I.; Patil, R.B. *Carboh. Res.* **1984**, *133*, 297.
75. Thomas, S.R.; Adney, W.S.; Baker, J.O.; Chou, Y.-C.; Laymon, R.; Nieves, R.A.; Tucker, M.P.; Vinzant, T.B.; Himmel, M.E. "*A. cellulolyticus* E1 Endoglucanase: Characteristic of Native and Recombinant Forms," Presented at *The Sixteenth Symposium on Biotechnology for Fuels and Chemicals*; Gatlinburg, TN, 1994.
76. Petre, J.; Longin, R.; Millet, J. *Biochimie* **1981**, *63*, 629.
77. Ng, T.K.; Zeikus, J.G. *Biochem. J.* **1981**, *199*, 341.
78. Stutzenberger, F. *Appl. Microbiol. Biotechnol.* **1988**, *28*, 387.
79. Fujino, T.; Sukhumavasi, J.; Sasaki, T.; Ohmiya, K.; Shimizu, S. *J. Bacteriol.* **1989**, *171*, 4076.
80. Robson, L.M.; Chambliss, G.H. *Appl. Environ. Microbiol.* **1984**, *47*, 1039.

81. Saddler, J.N.; Khan, A.W. *Can. J. Microbiol.* **1981**, *27*, 288.
82. Nakai, R.; Horinouchi, S.; Uozumi, T.; Beppu, T. *Agr. Biol. Chem.* **1987**, *51*, 3061.
83. Boyer, M.H.; Chambost, J.P.; Magnan, M.; Cattaneo, J. *J. Biotechnol.* **1984**, *1*, 241.
84. Prasertsan, P.; Doelle, H.W. *Appl. Microbiol. Biotechnol.* **1986**, *24*, 326.
85. Pettipher, G.L.; Latham, M.J. *J. Gen. Microbiol.* **1979**, *110*, 21.
86. Ohmiya, K.; Maeda, K.; Shimizu, S. *Carb. Res.* **1987**, *166*, 145.
87. Lee, S.F.; Forsberg, C.W.; Gibbins, L.N. *Appl. Environ. Micro.* **1985**, *50*, 220.
88. McGavin, M.; Forsberg, C.W. *J. Bacteriol.* **1988**, *170*, 2914.
89. Tewari, H.K.; Chahal, D.S. *Ind. J. Microbiol.* **1977**, *11*, 88.
90. Ito, S.; Shikata, S.; Ozaki, K.; Kawai, S.; Okamoto, K.; Inoue, S.; Takei, A.; Ohta, Y.; Satoh, T. *Agr. Biol. Chem.* **1989**, *53*, 1275.
91. Himmel, M.E.; Adney, W.S.; Grohmann, K.; Tucker, M.P. *Thermostable Purified Endoglucanase from Acidothermus cellulolyticus*; United States Patent No. 5,275,944, 1994.
92. Adney, W.S.; Thomas, S.R.; Nieves, R.A.; Himmel, M.E. *Thermostable Purified Endoglucanase II from Acidothermus cellulolyticus*; United States Patent No. 5,366,884, 1994.
93. Himmel, M.E.; Adney, W.S.; Mitchell, D.J.; Baker, J.O. *Appl. Biochem. Biotechnol.* **1993**, *39/40*, 213-225.
94. Baker, J.O.; Thomas, S.R.; Adney, W.S.; Nieves, R.A.; Himmel, M.E. "The Cellulase Synergistic Effect: Binary and Ternary Systems," Presented at *The symposium on Enzymic Degradation of Insoluble Polysaccharides*, The Annual American Chemical Society Meeting, San Diego, CA, March, 1994.
95. Doner, L.W.; Irwin, P.L. *Anal. Biochem.* **1992**, *202*, 50-53.
96. Brawner, M.; Poste, G.; Rosenberg, M.; Westpheling, J. *Current Opinion in Biotechnol.* **1991**, *2*, 674-681.
97. Katz, E.; Thompson, C.J.; Hopwood, D.A. *J. Gen. Microbiol.* **1983**, *129*, 2703-2714.
98. Denis, F.; Brzezinski, R. *FEMS Microbiol Lett.* **1991**, *81*, 261-264.
99. Nieves, R.A.; Chou, Y.-C.; Thomas, S.R.; Himmel, M.E. *Appl. Biochem. Biotechnol.* **1995**, *51/52*, 211-223.
100. Henrissat, B,; Mornon, J.-P. In *Trichoderma reesei Cellulases*; Kubicek, C.P.; Eveleigh, D.E.; Esterbauer, H.; Steiner, W.; Kubicek-Pranz, E.M., Eds.; The Royal Society of Chemistry: Canbridge, UK, 1990; pp. 12-29.
101. Honda, H.; Iijima, S.; Kobayashi, T. *Appl. Microbiol. Biotechnol.* **1988**, *28*, 57-58.
102. Goeddel, D.V. *Methods Enzymol.* **1990**, *185*, 3-7.

RECEIVED September 11, 1995

Chapter 15

Enhanced Enzyme Activities on Hydrated Lignocellulosic Substrates

K. L. Kohlmann[1], A. Sarikaya[1,2], P. J. Westgate[3], J. Weil[1,4], A. Velayudhan[5], R. Hendrickson[1], and M. R. Ladisch[1,4,6]

[1]Laboratory of Renewable Resources Engineering and [2]Department of Food Science, Purdue University, West Lafayette, IN 47907–1295
[3]W. R. Grace and Company, Columbia, MD 21033–3098
[4]Agricultural Engineering Department, Purdue University, West Lafayette, IN 47907–1295
[5]Department of Bioresource Engineering, Oregon State University, Corvallis, OR 97331–3906

> Enzyme and substrate factors which limit hydrolysis include cellulose crystallinity and lignocellulose morphology, as well as enzyme activity, stability and inhibition. *Brassica napus* (rapeseed) is a biomass having large amounts of inedible material proposed for use in a controlled ecological life support system (CELSS) for human space flight. Mechanistic descriptions between morphological, chemical, and surface properties of this lignocellulose and enzyme hydrolysis are being developed. The goal is to define conditions for a cost effective pretreatment based on biological lignin removal followed by pressure cooking of the remaining cellulose in water at 180 to 220°C. Liquid water treatment of plant stems has resulted in a 6-fold improvement in cellulose hydrolysis during a 24 h incubation with commercial cellulases. When the water treatment is preceded by mycelial growth of the mushroom, *Pleurotus ostreatus*, further enhancement of enzymatic hydrolysis is achieved. Enzyme hydrolysis of plant material will be analyzed for its ability to sustain a CELSS.

Controlled ecological life support systems (CELSS) on space stations or on Moon or Mars colonies will likely produce exclusively vegetarian diets, and it has been estimated that inedible plant residues could constitute from 62-88% of the total waste generated in the CELSS (*1*). Enzymatic conversion of lignocellulosic materials to sugars is therefore of great interest in space research. Safe and efficient enzymatic hydrolysis of the inedible material lessens the disposal load, furnishes sugars which could be incorporated into human diets, and decreases the area needed to produce plant matter by increasing the edible fraction of the biomass (*2*). Enzymatic conversion of inedible plant material to edible sugars effectively increases the harvest index of plant candidates.

Brassica napus (rapeseed) is potentially attractive for a CELSS with seeds yielding 40%-45% canola oil and 20%-25% of a high quality protein meal lacking the beany flavor present in soybeans. Dwarf rapid-cycling *B. napus* plants are being

[6]Corresponding author

studied at Purdue. The harvest index of these plants is 20%-25%, with the remainder being the inedible stems, siliques (seed pods), leaves, and roots. The major component of the inedible material is carbohydrate (cellulose, hemicellulose, and lignin), approximately 60%-70%. Rapeseed leaves have a higher protein content and comparatively lower carbohydrate content than stems or pods and perhaps may be consumed directly.

When cellulases are added to lignocellulosic substrates, there is a dramatic decrease in the hydrolysis rate during the first few hours of the reaction. This rapid decline in the hydrolysis rate of cellulose has been widely reported and is more pronounced for pretreated substrates than native materials (3, 4). Reasons for this decline, as reported by Nidetzky and Steiner (5), include thermal instability of the enzyme, product inhibition by glucose or cellobiose, inactivation of the adsorbed enzyme (due to diffusion into the cellulose fibrils), transformation of the more susceptible portions of the substrate to sugars leaving a less digestible form, and the general heterogeneous structure of the substrate.

The heterogeneous structure of lignocellulosic materials is an extremely important factor in the development of pretreatment systems designed to enhance the enzymatic conversion of cellulose to glucose. In each cellulose microfibril the linear molecules of cellulose are bound laterally by hydrogen bonding to form crystalline cellulose in a highly ordered manner and amorphous cellulose in a less ordered manner. Hemicellulose and lignin surround the microfibril to form a matrix, i.e. a lignin seal (6). In these lignocellulosic complexes (LCC), cellulose is associated both chemically and physically with lignin and physically with hemicellulose. Covalent lignin-carbohydrate linkages can be ester bonds through the free carboxy of uronic and aromatic acids or ether linkages through sugar hydroxyls (7). Both the lignin and the hemicellulose protect the cellulose from hydrolysis (8, 6).

Pretreatments chemically and/or physically help to overcome resistance to enzymatic hydrolysis and are used to enhance cellulase action. Physical pretreatments for plant lignocellulosics include size reduction, steam explosion, irradiation, cryomilling, and freeze explosion. Chemical pretreatments include dilute acid hydrolysis, buffered solvent pulping, alkali or alkali/H_2O_2 delignification, organic solvents, ammonia, and microbial or enzymatic methods (9).

Pretreatments useful for a CELSS must avoid any potentially toxic components and complicated or dangerous equipment, while meeting weight requirements, and constraints imposed by limited manpower in a CELSS (10). These limitations prompted the investigation of a liquid water pretreatment of plant residues at elevated temperatures. Pulping processes use water to cook wood materials in order to remove lignin and obtain pulping grade celluloses suitable for making paper (11). Reagents added to assist pulping processes in acid sulfite pulping (140°C, pH 1-2) result in effective delignification. Lignin would remain with the wood using a neutral solution.

Water should be an ideal pretreatment agent for CELSS applications which demand strict safety requirements. Bobleter et al. (12) first used water for pretreatment to enhance susceptibility of lignocellulosic material to enzymatic hydrolysis. High temperature steam was previously used as a pretreatment agent in the well documented hydrothermolysis and steam explosion pretreatments. Hydrothermolysis studies, such as those by Haw et al. (13), Hormeyer et al. (14), and Walch et al. (15), have shown that the primary effects of hot water pretreatment are the removal and solubilization of hemicellulose, which is catalyzed by small quantities of acid and solubilization of some of the lignin at high temperatures (>180°C). Brownell and Saddler (16) reported that steam pretreatment of lignocellulosic material was at least as effective for pretreatment of aspen chips and that neither the explosion or temperatures above 190°C were necessary. Mok and Antal (17) found that at 230°C, amorphous cellulose was also solubilized. It has generally been accepted that crystalline cellulose is unaffected by hot liquid water pretreatment.

In steam explosion, steam penetrates the lignin, hemicellulose, and cellulose. The mixture is explosively decompressed, and the resulting expansion increases cellulose accessibility (*18*). The operating conditions promote acid formation and result in degradation of cellulose by autohydrolysis. This phenomenon is considered to be an important and necessary aspect of pretreatment (*18, 19*). Water pretreatment, by comparison, has the goal of maintaining a liquid phase (under pressure) during pretreatment, while keeping the pH above 5.0.

Many process schemes for biotechnological use of lignocellulosic materials ignore possible uses of lignin, other than suggesting that it be burned for energy recovery. Leisola and Fiechter (*20*) point out that an efficient mechanism for lignin degradation exists in nature, because no accumulation occurs (i.e., lignin either degrades to CO_2 and H_2O or is converted to humus). Lignin is degraded by a narrower array of microbes than any other major biopolymer (*21*). White-rot fungi (*Basidiomycotina*) are of special interest because lignin is attacked simultaneously with cellulose and hemicellulose; the wood becomes pale as the pigmented lignin is removed (*22*).

Because lignin comprises a considerable portion of inedible plant material which is not affected by cellulase action, alternative methods for lignin removal are needed for a CELSS. *Pleurotus ostreatus*, the oyster mushroom, and *Lentinus edodes*, the shiitake mushroom, are white rot fungi which produce edible fruiting bodies or mushrooms. A system using the growth of these fungi to break the lignin "seal" around cellulose and hemicellulose could be a biological pretreatment beneficial to a CELSS. Fungal growth requires minimal outside control and added moisture. During mycelial growth the amount of lignin present is reduced which may be a cellulose pretreatment. Additionally, the fruiting bodies or oyster mushrooms, high both in protein and micronutrient contents, as well as being flavorful, could potentially improve the quality of the diet in a CELSS.

This work focused on the development of a pretreatment scheme for inedible plant material. A biological pretreatment, mycelial growth of *P. ostreatus*, was followed by a hot water pretreatment in order to improve the enzymatic conversion of cellulose. Rapeseed, cowpea, and rice plants currently being evaluated for inclusion in a CELSS were grown by the Horticulture Department at Purdue University (*23*). Inedible plant portions were collected and subjected to proximate analysis, as well as analysis for carbohydrate composition. Portions of the plant material were ground and initial hydrolysis profiles determined using commercial cellulases. The plant materials were then treated in an effort to improve cellulose hydrolysis.

Materials And Methods

Plant Growth. Rapeseed and cowpea plants were grown in the Horticulture Department at Purdue University (*24, 25*). Plants were grown until maturity, and then harvested. Upon harvesting, the inedible portions of the plant material were air dried for several days prior to being oven dried for 2 days between 70-75°C. The plant materials, divided into stem, leaf, and seed pod (or hull portions), were then ground to between 20-40 mesh (0.84-0.43 mm). The ground material was stored in sealed glass containers until use.

Carbohydrate Analysis. Proximate analysis was completed in the Food Science Department at Purdue. Fat determination was by Soxhlet extraction (AOAC method 920.39B). Protein was by the microKjeldahl method (AOAC 960.52), and ash determination was per AOAC 923.03. Moisture contents were found by oven drying (104°C, 24 h). Since assays for protein and ash were conducted on defatted samples, the values were normalized to reflect the protein and ash contents of the original sample. The conversion factor 6.25 x N was used to obtain the protein content. Total carbohydrate (CHO) concentration was calculated from the moisture, protein and fat values: Total CHO = 100% - (% protein + % fat + % ash + % moisture).

Further investigations into the amounts of specific carbohydrate components were conducted with the help of the Animal Science Department at Purdue and were completed in the Laboratory of Renewable Resources Engineering (LORRE). These data were obtained on a dry weight basis using both the acid detergent fiber (ADF) procedure and the neutral detergent fiber (NDF) procedure of Goering and Van Soest (26), and Van Soest and Wine (27). ADF is comprised of lignin, cellulose, and insoluble minerals, while the NDF also includes hemicellulose. Therefore, subtracting the ADF value from the NDF value gives the weight of the hemicellulose in the sample. The amount of lignin in the samples was found using the permanganate lignin assay of Van Soest and Wine (28).

Enzyme Activity. Cellulase enzyme systems evaluated included Cytolase CL, Rhozyme HP-150, Multifect XL, and Cytolase M103S, all from Genencor (Genencor International, Schaumburg, IL). Conditions for the measurement of cellulase activities were as reported by Ghose (29) and included filter paper activity (FP), cellobiase activity, and CMCase activity. A glucose analyzer (Glucose Analyzer II, Beckman Instruments, Fullerton, CA) was also used to measure the amount of glucose present following hydrolysis. Alternatively, the amount of glucose, as well as other oligosaccharides, was calculated by liquid chromatography (LC) using appropriate standard curves, Lin et al. (30). The percentage hydrolysis of the plant cellulosic material was calculated by:

$(g\ glucose\ produced) \left[\dfrac{162}{180}\right]$ (100%) dry weight of the cellulose.

The cellulose hydrolysis assay developed for the plant material was performed as follows. The enzyme solution was diluted to between 1 to 90 filter paper units (FPU)/mL with 0.05 M citrate buffer, pH 4.8, and prewarmed to 50°C. One mL of the enzyme solution was added to 100 mg of the plant material. The contents of the tubes were mixed, covered, and incubated at 50°C. At time intervals, 30 µL portions were removed and microfuged for 1.5 minutes. The supernatant portion was then injected into the glucose analyzer for glucose concentration determination. In some cases, samples were frozen for LC investigation. Duplicate or triplicate samples were analyzed along with appropriate enzyme and substrate blanks.

Water Pretreatment. Both the microcrystalline cellulose and the plant materials were pretreated in a 300 mL pressure reactor (Autoclave Engineers, Erie, PA) as described by Kohlmann et al. (31). The microcrystalline cellulose (Avicel, FMC, Newark, DE) was sieved dry to give a material with an initial particle size of greater than 53 microns measured by a Malvern particle size analyzer (Malvern Instruments, Malvern, England). The reactor was loaded with 135 mL of deionized distilled water containing 1.5-5 weight % fiber particles. For the plant material, the reactor was heated to 180 or 200°C, the time required to reach these temperatures being 30-40 minutes. Microcrystalline cellulose was heated to 220°C over approximately 2 hours. The pretreated materials were collected through a sample port which emptied into a cooled coil, having sufficient back pressure to avoid flashing. The liquid portion of the treated material was removed by filtering. The plant materials were then exposed to air for 3-4 days (moisture content 6-8%) prior to enzymatic hydrolysis. Cellulase enzyme was added directly to portions of the pretreated microcrystalline cellulose. The amount of protein solubilized by the water pretreatment was measured in the supernatant fraction using the bicinchoninic (BCA) protein assay (32). Appropriate blanks and controls were used to ensure that carbohydrate breakdown did not interfere with the BCA assay. Scanning Election Microscopy (SEM) was completed in the Materials Science Department at Purdue on a Jeol (JSM-T300) microscope.

Microbiological Methods. *P. ostreatus* (NRRL 2366) was obtained in the form of mycelial growth on a potato dextrose agar (PDA) slant from the Northern Regional Research Laboratory (NRRL) in Peoria, IL. Initial transfers were made by adding deionized distilled water (DDW), 1-2 mL to the original slant and then using this liquid to prepare streak plates on PDA agar. The plates were incubated several days at room temperature before new PDA slants were prepared from individual colonies. The new slants were incubated for two days at room temperature before being used to inoculate liquid nutrient media to produce submerged cultures.

Modified procedures of Kaneshiro (*33*), and Lindenfelser et al. (*34*), were used to inoculate either rapeseed or cowpea stems with the submerged *P. ostreatus* pellets. Flasks containing wetted stems (100 g H_2O/20 g stems) were sterilized and allowed to cool. Using aseptic conditions, 10 mL of the submerged culture media was added to each flask containing stems. Appropriate controls were also prepared. Incubation was for 30 days at 27°C; growth was visible after two days. During incubation, sterile DDW was added periodically to moisten the plant stems.

Results

Compositional Analysis. Proximate analysis data for rapeseed, cowpea and rice samples is presented in Table I. Individual carbohydrate components by ADF and NDF procedures are given in Table II.

The carbohydrate values calculated by difference (Table I) with some exceptions agree with values obtained through fiber analysis (Table II), and indicate the large amount of inedible material in the stems, seed pods and leaves. The values reported in Table II should be more accurate for determination of total CHO content in these materials.

Table I. Proximate Analysis of Inedible Plant Material (dry weight basis)

Plant/Portion	Carbohydrate	Protein	Fat	Ash	Total
Rapeseed					
Stems	73	15	1	11	100
Siliques	68	22	1	10	101
Leaves	42	45	2	11	100
Cowpea					
Stems	87	4	1	8	100
Leaves	46	34	4	16	100
Pods	66	27	2	5	100
Rice					
Leaves	51	35	2	12	100
Stems	61	21	1	17	100

Enzymatic Hydrolysis. Several commercially available enzyme preparations were screened for the ability to hydrolyze cellulose present in the stem, leaf or seed pod fractions of rapeseed, cowpea and rice plants (Table III). Cytolase CL was chosen for evaluation of effectiveness of pretreatments because overall glucose production was higher with this enzyme system.

Table II. Carbohydrate Composition of Inedible Plant Material[1]

Plant/Portion	Cellulose	Hemicellulose	Lignin	Total
Rapeseed				
Stems	38	10	18	66
Siliques	35	12	18	65
Leaves	14	15	5	34
Cowpea				
Stems	27	16	9	52
Pods	37	17	15	69
Leaves	11	12	5	28
Rice				
Stems	33	31	12	76
Hulls	34	19	10	63
Leaves	27	27	4	58

[1]Percentage dry weight basis ± 5%.

Water Pretreatment of Microcrystalline Cellulose. Enzymatic hydrolysis of microcrystalline cellulose (Avicel) following water pretreatment at 220°C without pH control is shown in Figure 1. The initial rate of the enzymatic hydrolysis may be enhanced although the final rate and conversion to glucose was equivalent to control samples. Other characteristics of the pretreated Avicel were a decrease in pH from 5.4 to 3.3 and considerable browning (degradation) of the Avicel. Avicel was also pretreated at 220°C, while maintaining the pH between 5.5 and 7.0 during the heating period (31). Results are presented in Figure 2.

Water Pretreatment of Plant Material. Hot water treatment of plant stems at 180°C and 200°C resulted in significant improvement of the enzymatic hydrolysis of the cellulose in the material (Figure 3). Both the rate and the extent of cellulose conversion to glucose were enhanced. Compositional, chemical and physical changes, all possibly related to this enhanced susceptibility to enzymatic action, were noticed in the plant material following the water pretreatment. Scanning electron microscopy (SEM) on control and heated rapeseed stems produced the photomicrographs presented in Figure 4.

Compositional changes in carbohydrate components following the water pretreatment are given in Table IV. Water pretreatment of rapeseed reduces the fraction of hemicellulose which remains, and is accompanied by an apparent increase in cellulose content. The fraction of lignin appears to be constant, thus indicating that in relative amounts, water removes some lignin. Approximately half the protein

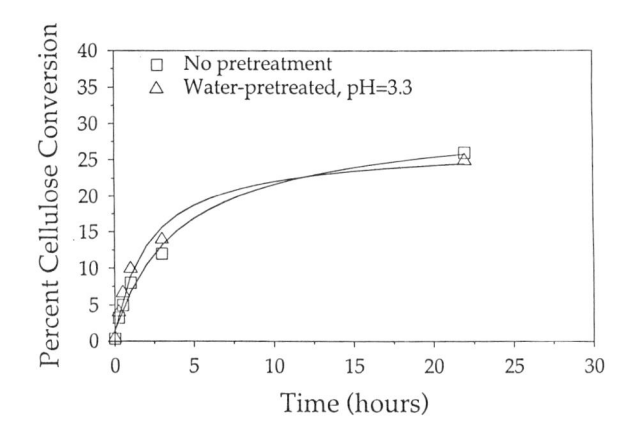

Figure 1. Enzymatic conversion of microcrystalline cellulose after water pretreatment without pH regulation. (Reproduced with permission from reference 31. Copyright 1993 Society of Automotive Engineers, Inc.)

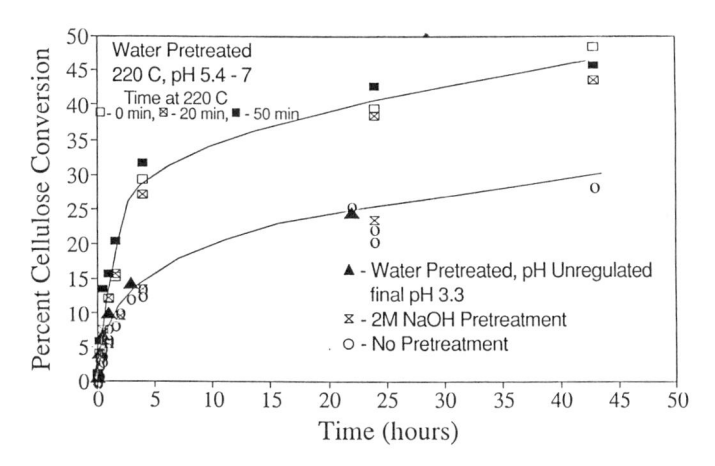

Figure 2. Enzymatic conversion of microcrystalline cellulose after water pretreatment at a maintained pH. (Reproduced with permission from reference 31. Copyright 1993 Society of Automotive Engineers, Inc.)

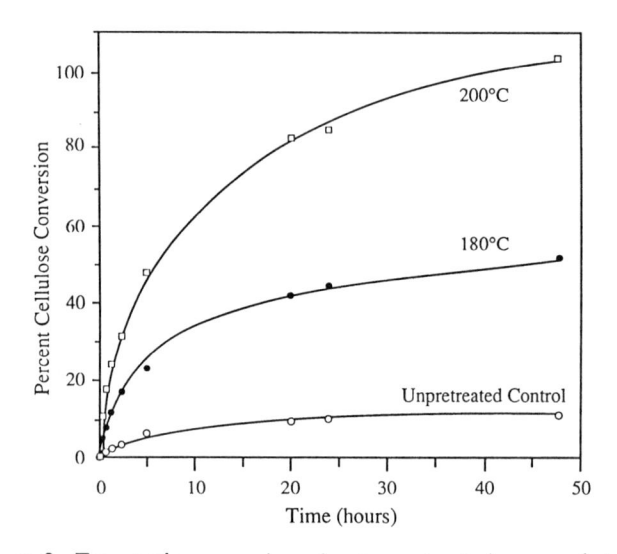

Figure 3. Enzymatic conversion of water pretreated rapeseed stems.

in the stem samples was solubilized as detected by the BCA protein assay in the supernatant portion of the heat treated material.

Table III. Percentage Cellulose Hydrolysis to Glucose Following a 24 hour Incubation with Enzymes

Plant Material	Cytolase CL	Multifect XL	Cytolase M103S	Rhozyme HP-150
Rapeseed				
Stems	23	15	22	9
Siliques	11	11	17	5
Rice				
Leaves	23	37	8	2
Stems	8	13	7	0
Hulls	23	29	6	15
Cowpea				
Stems	61	57	34	21
Pods	36	34	34	7
Leaves	49	20	---	3

Unlike the crystalline Avicel (Figure 2), the pH of the plant material did not decrease dramatically. In most cases the pH dropped about one pH unit (Table V).

Biological Pretreatment. Mycelial growth of *P. ostreatus* alone as a pretreatment did not increase the susceptibility of the cellulose to enzyme action, but when the growth was followed by the hot water treatment, a significant increase in glucose production was measured (Figures 5 and 6). Other associated changes following growth and heating to 180°C were a reduction in the hemicellulose content (Table IV); hemicellulose was removed during microbial growth. Liquid chromatography of the supernatant fraction indicated that little breakdown of hemicellulose to mono and disaccharide components was occurring (data not shown). The amount of xylose and arabinose present in the supernatant fraction following heating would correspond to a less than 1% degradation of hemicellulose. It is likely that the hemicellulose is present in soluble larger molecular weight fractions. Mycelial growth of the *P. ostreatus* resulted in a pH decrease in the plant material (Table V).

Discussion

Water Pretreatment of Microcrystalline Cellulose. In an effort to understand pretreatment effects on cellulose, a model was developed to predict cellulose and glucose degradation during aqueous heating (35). At a pH below 7.0, the degradation reactions are catalyzed by H+. Two of the primary degradation products, formic and levulinic acid, contribute significant quantities of hydrogen ions. Hence, as degradation progresses, the rate of degradation increases.

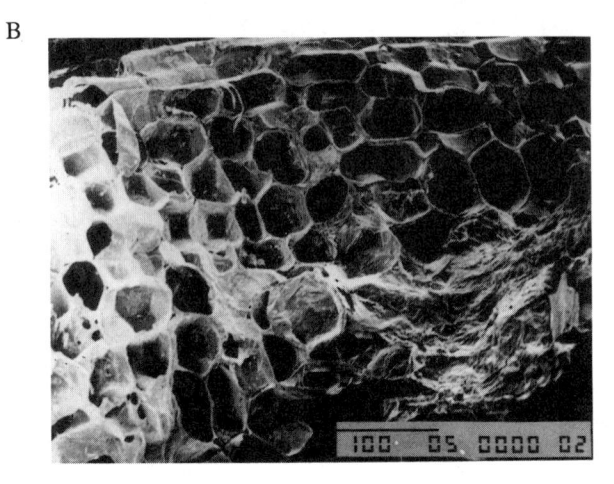

Figure 4. Photomicrographs of control and water pretreated rapeseed stems: (A) Control, (200x); (B) 180°C, (200x); (C) 200°C, (350x).

C

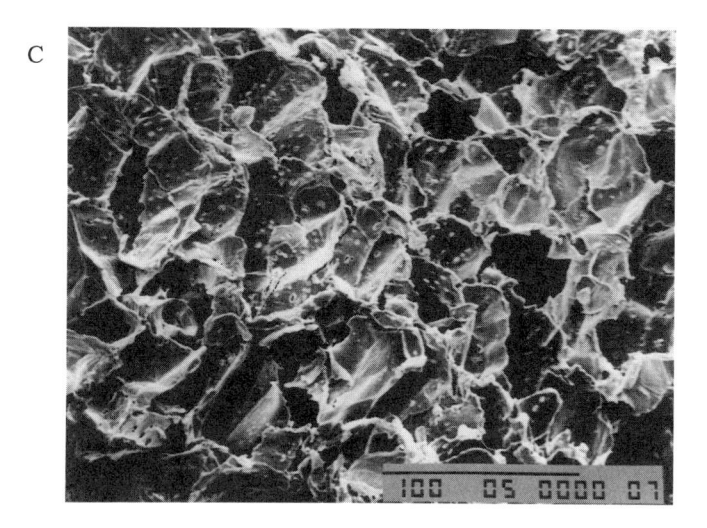

Figure 4. Continued.

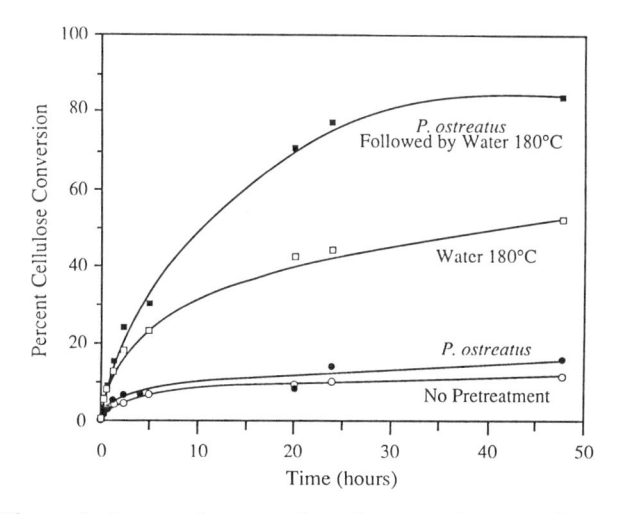

Figure 5. Enzymatic conversion of pretreated rapeseed stems.

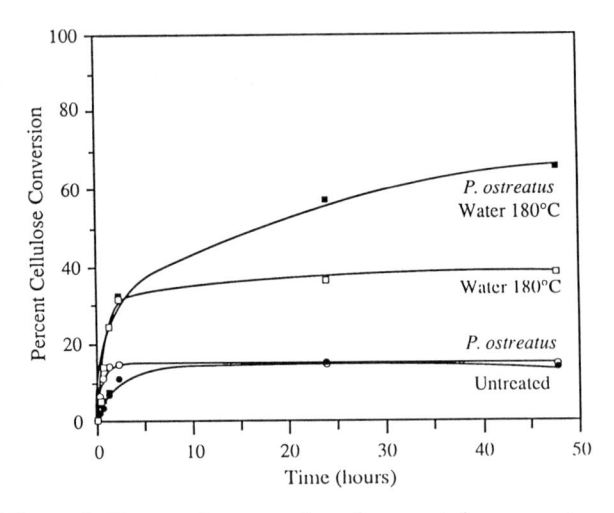

Figure 6. Enzymatic conversion of pretreated cowpea stems.

Maintaining the pH close to 7.0 should minimize the rate, and therefore extent, of degradation.

Table IV. Carbohydrate Composition of Rapeseed Stems Following Water Pretreatment[1]

Treatment	Hemicellulose	Cellulose	Lignin
Control	17	32	28
180°C	4	50	24
200°C	0	66	28
Growth of P. ostreatus	4	43	23
P. ostreatus and 180°C	7	53	28

[1] Percentage dry weight basis ± 5%

Table V. Effect of Pretreatment on pH

Material	Initial pH	Final pH
Rapeseed Stems 180°C	5.7	4.7
200°C	5.7	4.3
Soybean Hulls	6.6	4.8
Rice Stems and Leaves	5.8	5.6
Rapeseed Stems (P. ostreatus, 180°C)	5.3	4.0
Rapeseed Siliques (P. ostreatus, 180°C)	4.3	3.9

Our solution for controlling pH during pretreatment has been to continuously add small amounts of 0.5 M KOH solution when the Avicel is at high temperatures. By adding ~12 mL of KOH solution, the pH was maintained between 5.5 and 7 for the entire heat up profile and for a substantial time at 220°C (Figure 2). This resulted in a significant decrease in degradation and enhanced enzymatic hydrolysis as shown in Figure 2. This clearly shows that liquid water pretreatment at high temperatures can enhance the susceptibility of crystalline cellulose to hydrolysis. Further enhancements were not seen by increasing the time at 220°C or the pretreatment temperature (Figure 2). This is most likely due to the small amount of degradation that occurred during these conditions. Improved pH control may result in even greater susceptibility of Avicel.

Water Pretreatment of Plant Material

Enhanced Enzymatic Reactivity. The use of liquid water at 180°C to treat inedible plant material improved the subsequent enzymatic hydrolysis of the treated material (Figure 3). Complete conversion of the cellulose by enzymatic hydrolysis was observed for materials pretreated at 200°C with no hold time. Compared to crystalline cellulose, plant materials are complex and experience compositional, chemical and physical changes following water pretreatment, some of which may relate to the enhanced enzymatic reactivity.

Changes in Carbohydrate Composition. The primary compositional change which occurred as a result of the pretreatment was solubilization of most of the hemicellulose present (Table IV). The hydrolysis of the hemicellulose was assumed to be minimal because LC showed lack of xylose and furfural compounds. Hemicellulose is physically associated with cellulose, and hemicellulose removal is known to increase pore volume, thereby increasing enzyme accessibility and hydrolysis (9, 36). Mok and Antal (17) noted that hemicellulose can easily be removed from biomass by treatment with dilute acid, but they also found that this same goal can occur with water alone, achieving complete hemicellulose removal using water between 200-230°C.

The amount of lignin present in the rapeseed stems does not appear to change following heat treatment in water (Table IV). It has been reported that little delignification occurs until temperatures exceed 180°C (37). However, chemical bonds between cellulose and lignin may have been affected. Meshitsuka (38) states that at temperatures higher than the softening temperature of lignin (60-80°C), and close to that of cellulose (230-253°C), the melted lignin separates and then coagulates into cellulose free particulates. Lignin is also converted to a form which is extractable in alkali and some organic solvents. The temperatures used to pretreat the cellulose are well above the softening temperature of lignin, but non-alkaline; therefore, the lignin may have been rearranged or separated from the cellulose, but not solubilized and removed. This rearrangement of the hydrophobic lignin could possibly improve enzyme access to the internal pores of the cellulose.

Following the pretreatment, the material became enriched in cellulose due to removal of hemicellulose and protein (Table IV). This is desirable, providing a less heterogeneous cellulosic substrate than the original material. Meshitsuka (38) states that under proper conditions, only a small extent of cellulose hydrolysis occurs during the steam explosion process. Conditions in the present work were less severe than those of steam explosion; therefore, cellulose hydrolysis would not be expected.

pH Effect. Plant materials were able to buffer changes in pH (Table V), possibly due to their heterogeneous structure which includes proteins, hemicellulose, and salts. The central hypothesis of this research is that pressure cooking of cellulose in liquid water at controlled pH avoids acid formation and hydrolysis of the cellulose during the pretreatment. This effect has been described in buffered solvent pulping (37) where it was noted that the acids and bases present in pulping liquors are nonspecific catalysts, promoting both desirable and undesirable reactions, and that if pulping takes place at relatively neutral pH's, comparatively high temperatures can be used without damage. An advantage, then, of pretreating plant materials in water is their ability to buffer changes in pH.

Physical Changes. The SEM photomicrographs (Figure 4) indicate that changes are occurring on the surface of the stem material as a result of the heat treatment. Following heating at 180°C, the stems appear to contain more voids than controls. At the pretreatment temperature of 200°C, the photomicrographs show a similar pattern as the 180°C samples, but it appears that the surface is even more disrupted at the higher temperature.

Biological Pretreatment of Plant Material. Mycelial growth of *P. ostreatus* on plant stems did not result in enhanced glucose production following cellulase addition (Figures 5 and 6). Previous reported work has been somewhat varied as to the effectiveness of fungal pretreatments, due perhaps to culture and substrate choice, growth conditions, and enzymatic techniques used to measure cellulose digestion in *vivo* or in *vitro*. Biological delignification is seen as a competitive race between lignin reduction and carbohydrate reduction; as lignin is reduced, the rate of carbohydrate degradation increases (*39*). It has been suggested that only nitrogen limitation allows extensive degradation of pure lignin (*20*). In the present study, the amount of hemicellulose was significantly decreased following *P. ostreatus* growth, but amounts of cellulose and lignin were not reduced. It is possible that the nitrogen content of the herbaceous tissues, 1.0-5.0%, compared to woody tissues, 0.03-1.0%, (*22*) was not limiting during the time that the *P. ostreatus* was grown on the stems. The presence of nitrogen may in fact limit the extent of delignification occurring in the plant stems. It is interesting to note that although *P. ostreatus* growth reduces the hemicellulose content as much as the water pretreatment at 180°C, the reactivity of the cellulose is not improved over control samples. Obviously, hemicellulose reduction alone is not the only criteria for an effective pretreatment.

When fungal growth on rapeseed and cowpea stems was combined with water pretreatment, cellulose conversion was greatly enhanced (Figures 5 and 6). The reasons for this enhancement are not obvious. The presence of fungal mycelial can be seen in Figure 7A, and when this fungal treated stem material was heated, an increase in the number of voids was apparent (Figure 7B), similar to water treated stems (Figure 4B). Changes in the amounts of hemicellulose, cellulose, and lignin are similar in rapeseed stems heated to 180°C and in stems treated with fungus prior to heating to 180°C (Table 4). It has been suggested that cellulose pore size is a major limitation to cellulase action, and it has been noted that removal of hemicellulose increases pore volume (*36, 40*). The water treatment was effective in decreasing the amount of hemicellulose present and resulted in enhanced susceptibility to enzyme action, presumably by increasing available pores. It is possible that the success of the two step pretreatment is also related to an increase in pore volume. Markam and Bazin (*41*), note that regarding the mycelial growth habits of the filamentous fungi, the most important feature is the ability of a mycelial fungus to penetrate and ramify a substrate and to mobilize and transport the products of polymer degradation over considerable distances through the mycelial network. It seems plausible that this type of growth action combined with hot water treatment could result in an increase in pore volume, resulting in a highly reactive cellulose. Measurement of changes in pore volume relating to the pretreatments described is an important area to consider for future research.

Summary

Liquid water pretreatment was applied to rapeseed stems to improve the subsequent enzymatic hydrolysis of cellulose. Following water treatment (180°C), enzymatic conversion of cellulose to glucose in rapeseed stems ranged from 60-90% versus only 15-23% for untreated stems. Increasing the temperature of the pr~tr~~t~~~t from 180°C to 200°C allowed near complete conversion, but undesirable r

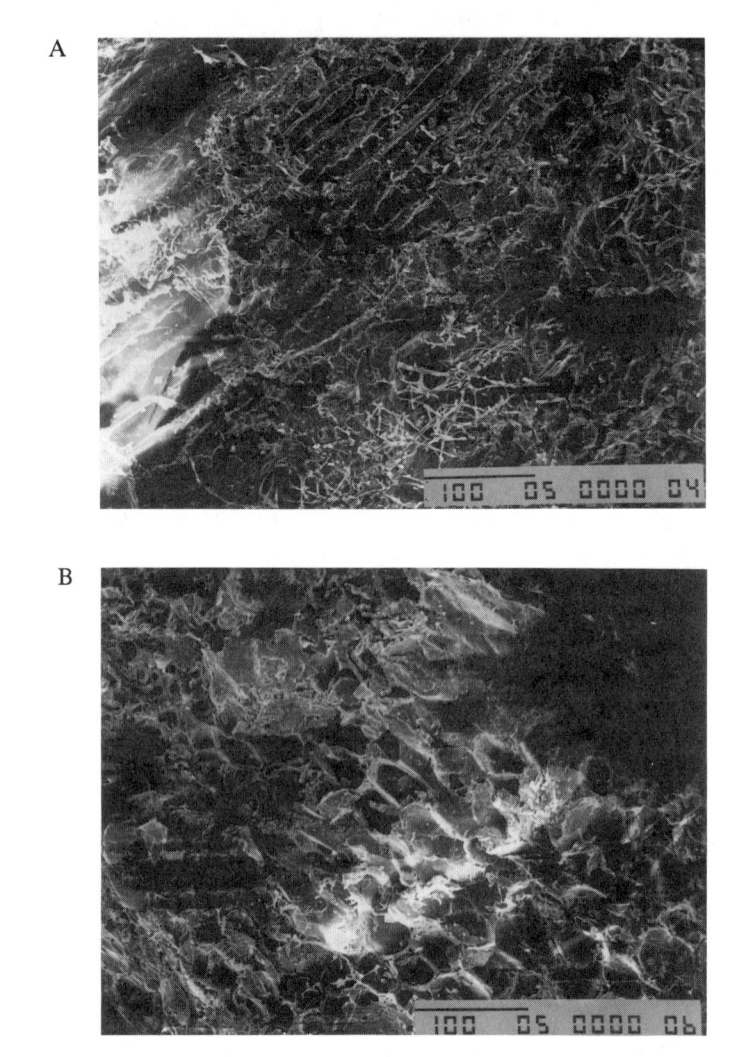

Figure 7. Photomicrographs of *Pleurotus ostreatus* treated rapeseed stems: (A) stems after mycelial growth, (200x); (B) stems after mycelial growth and heating at 200°C, (200x).

proteins and sugars occurred at the higher temperature. In an effort to increase cellulose conversion using lower pretreatment temperatures, a biological pretreatment which uses hemicellulose and/or lignin to form edible products through mycelial growth of the oyster mushroom, *Pleurotus ostreatus*, was placed prior to heating the stems at 180°C. *P. ostreatus* is capable of degrading lignin and hemicellulose as well as producing a nutritious mushroom. Although not a successful pretreatment alone, when *P. ostreatus* growth was followed by hot water treatment, cellulose conversion to glucose was significantly improved. This simple and safe scheme of fungal growth followed by heating in water appears to be an effective pretreatment system for bioregenerative waste recycle in a CELSS.

Acknowledgments

We thank Pat Jaynes, Mark Brewer, and Linda Anderson for experimental assistance. The material in this work was supported through the NASA NSCORT Center at Purdue University and USDA Grant CSR90-37233-5410.

Plant materials were provided through the laboratory of Dr. Cary Mitchell, whose research has been an invaluable source of information on characteristics of typical CELSS species. We also thank the laboratory of Dr. Suzanne Nielsen, whose many discussions and interactions in developing analytical protocols were appreciated. We thank Dr. John Patterson, whose laboratory and experience have been essential in the fiber analysis of cellulosic materials. We thank Genencor, Intl., for the gift of the enzymes. Parts of the literature survey were conducted under NREL Subcontract XAC-4-13511-01.

Literature Cited

(1) Drysdale, A., Sager, J., Wheeler, R., Fortson, R., and Chetirikin, P., **1993**, "CELSS Engineering Parameters," Proceedings of 1993 ICES Meeting, *SAE Technical Paper Series* 932130.
(2) Westgate, P. J., Kohlmann, K. L., Hendrickson, R., and Ladisch, M. R., **1992**, Bioprocessing in space, *Enzyme Microb. Technol., 14*:76.
(3) Ladisch, M. R., Ladisch, C. M., and Tsao, G. T., **1978**, Cellulose to sugars: New path gives quantitative yield, *Science, 201*:743.
(4) Ladisch, M. R., Waugh, L., Westgate, P., Kohlmann, K., Hendrickson, R., Yang, Y., and Ladisch, C., **1992**, Intercalation in the pretreatment of cellulose *in*: "*Harnessing Biotechnology for the 21st Century*," M. R. Ladisch and A. Bose, eds., Proceedings of the Ninth International Biotechnology Symposium and Exposition, American Chemical Society, Washington, DC.
(5) Nidetzky, B. and Steiner, W., **1993**, A new approach for modeling cellulase-cellulose adsorption and the kinetics of the enzymatic hydrolysis of microcrystalline cellulose, *Biotech. and Bioeng., 42*:469.
(6) Ladisch, M. R., **1989**, Hydrolysis, *in*: "*Biomass Handbook*," O. Kitani and C.W. Hall,eds., Gordon and Breach, New York.
(7) Jeffries, T. W., **1990**, Biodegradation of lignin-carbohydrate complexes, *Biodegradation , 160*:145.
(8) Ladisch, M. R., Lin, K. W., Voloch, M., and Tsao, G. T., **1983**, Process considerations in the enzymatic hydrolysis of biomass, *Enz. Microb. Technol., 5*(2):82.
(9) Marsden, W. L. and Gray, P. P., **1986**, Enzymatic hydrolysis of cellulose in lignocellulosic materials, *in*: "*CRC Critical Reviews*," *3*(3):235.
(10) Petersen, G. R. and Baresi L., **1990**, The conversion of lignocellulosics to fermentable sugars: A survey of current research and applications to CELSS, *in*: "*Advanced Environmental/Thermal Control and Life Support Systems*," SP-831, 901282, Society of Automotive Engineers, Inc., Warrendale, PA.

(11) Sjostrom, E., **1981**, *"Wood Chemistry Fundamentals and Applications,"* Academic Press, New York.
(12) Bobleter, O., Niesner, R., and Rohr, M., **1976**. The hydrothermal degradation of cellulosic matter to sugars and their fermentative conversion to protein, *J. Appl. Polymer Sci., 20*:2083.
(13) Haw, J. F., Maciel, G. E., Linden, J. C., and Murphy, V. G., **1985**, Nuclear magnetic resonance study of autohydrolysis and organosolv-treated lodgepole pinewood using carbon-13 with cross polarization and magnetic-angle spinning, *Holzforschung, 39*:99.
(14) Hormeyer, H. F., Schwald, W. Bonn, G., and Bobleter, O., **1988**, Hydrothermolysis of birch wood as pretreatment for enzymatic saccharification, *Holzforschung, 42*(2):95.
(15) Walch, E., Zemann, A., Schinner, F., Bonn, G., and Bobleter, O., **1992**, Enzymatic saccharification of hemicellulose obtained from hydrothermally pretreated sugarcane bagasse and beech bark, *Biores. Technol., 39*:173.
(16) Brownell, H. H. and Saddler, J. N., **1987**, Steam pretreatment of lignocellulosic material for enhanced enzymatic hydrolysis, *Biotech and Bioeng., 29*:228.
(17) Mok, W. S-L. and Antal, M. J. Jr., **1992**, Uncatalyzed solvolysis of whole biomass hemicellulose by hot compressed liquid water, *Ind. Eng. Chem. Res., 31*: 1157.
(18) Beltrame, P. L., Carniti, P., Visciglio, A., Focher, B., and Marzett, A., **1992**, Fractionation and bioconversion of steam exploded wheat straw, *Biores. Technol., 39*:165.
(19) Heitz, M., Capek-Menard, E., Koeberle, P.G., Gagne, J., Chornet, E., Overend, R. P., Taylor, J. D., and Yu, E., **1991**, Fractionation of *Populus tremuloides* at the pilot plant scale: Optimization of steam pretreatment conditions using Stake II technology, *Biores. Technol., 35*:23.
(20) Leisola, M. S. A. and Fiechter, A., **1985**, New trends in lignin biodegradation, *in: "Advances in Biotechnological Processes 5,"* Alan Liss Inc.
(21) Kirk, T. K., and Farrell, R. L., **1987**, Enzymatic combustion: The microbial degradation of lignin, *Ann. Rev. Microbiol., 41*:465.
(22) Hudson, H. J., **1986**, Fungi as decomposers of wood, *in: "Fungal Biology,"* A. J. Willis and M. A. Sleigh, eds., Edward Arnold Ltd., London.
(23) Mitchell, C. A., **1993**, The role of bioregenerative life-support systems in a manned future in space, *Transactions Kansas Acad. of Sci., 96*:87.
(24) Frick, J., **1993**, "Evaluation of dwarf rapeseed (*Brassica napus*) as an oilseed crop for bioregenerative life support systems," M. S. Thesis, Purdue University, West Lafayette, IN.
(25) Ohler, T., **1994**, "Evaluation of cowpea (*Vigna unguiculata*) (L. WaLp) as a candidate species for inclusion in a CELSS," M. S. Thesis, Purdue University, West Lafayette, IN.
(26) Goering, H. K. and Van Soest, P. J., **1970**, Forage Fiber Analysis, *in: "Agricultural Handbook No. 379 "*, Agricultural Research Service, U. S. Department of Agriculture, Washington, DC, Jacket No. 387-598.
(27) Van Soest, P. J. and Wine, R. H., **1967**, Use of detergents in the analysis of fibrous foods. IV. Determination of plant cell-wall constituents, *J. Assoc. Off. Anal. Chem., 50*:50.
(28) Van Soest, P. J. and Wine, R. H., **1968**, Determination of lignin and cellulose in acid-detergent fiber with permanganate, *J. Assoc. Off Anal. Chem., 54*:780.
(29) Ghose, T. K., **1987**, Measurement of cellulase activities, *Pure and Appl. Chem., 59* (2):257.

(30) Lin, K. W., Jacobson, B. J., Pereira, A. N., and Ladisch, M. R., **1988**, Liquid chromatography of carbohydrate monomers and oligomers, *in*: "*Methods in Enzymology*", *160*:145.
(31) Kohlmann, K., Westgate, P. J., Weil, J., and Ladisch, M. R., **1993**, "Biological-based systems for waste processing." Proceedings of 1993 ICES Meeting, *SAE Technical Paper Series* 932251.
(32) Smith, P. K., Krohn, R. I., Hermanson, G. T., Mallia, A. K., Gartner, F. H., Provenzano, M. D., Fujimoto, E. K., Goeke, N. M., Olson, B. J., and Klenk, D. C., **1985**, Measurement of protein using bicinchoninic acid, *Analytical Biochem.*, *150*:76.
(33) Kaneshiro, T., **1977**, Lignocellulosic agricultural wastes degraded by *Pleurotus ostreatus*, *Dev. Ind. Microbiol.*, *18*:591.
(34) Lindenfelser, L. A., Detroy, R. W., Ramstack, J. M., and Worden, K. A., **1979**, Biological modification of the lignin and cellulose components of wheat straw by *Pleurotus ostreatus*, *Dev. Ind. Microbiol.*, *20*:541.
(35) Weil, J., **1993**, "Unified model for the hydrolytic effects during cellulose pretreatments," M. S. Thesis, Purdue University, West Lafayette, IN.
(36) Grethlein, H. E., **1985**, The effect of pore size distribution on the rate of enzymatic hydrolysis of cellulosic substrates, *Bio. Technol.*, *3*:155.
(37) Faass, G. S., Roberts, R. S., and Muzz, J. D., **1989**, Buffered solvent pulping, *Holzforschung*, *43*(4):245
(38) Meshitsuka, G., **1991**, Utilization of wood and cellulose for chemicals and energy, *in*: "*Wood and Cellulosic Chemistry*," D.N.S. Hon and N. Shiraishi, eds. Marcel Dekker, Inc., New York.
(39) Reid, I. D., **1983**, Effects of nitrogen supplements on degradation of aspen wood lignin and carbohydrate components by *Phanerochaete chrysosporium*, *Appl. and Environ. Microbiol.*, *45*(3):830.
(40) Grethlein, H. E. and Converse, A. O., **1991**, Common aspects of acid prehydrolysis and steam explosion for pretreating wood, *Bioresource Technol.*, *36*:77.
(41) Markam, P. and Bazin, M. J., **1991**, Decomposition of cellulose by fungi, *in*: "*Handbook of Applied Mycology*," D. Arona, B. Raj, K. Mukerji, and G. R. Knudson, eds., Marcel Dekker, Inc., New York.

RECEIVED September 28, 1995

Chapter 16

Comparison of Protein Contents of Cellulase Preparations in a Worldwide Round-Robin Assay

William S. Adney, A. Mohagheghi, Steven R. Thomas, and Michael E. Himmel

Alternative Fuels Division, Applied Biological Sciences Branch, National Renewable Energy Laboratory, 1617 Cole Boulevard, Golden, CO 80401-3393

A recent activity sponsored by the International Energy Agency (IEA) entitled, "Biotechnology for the Conversion of Lignocellulosics," coordinated 13 laboratories to analyze 6 standard cellulase enzyme preparations supplied by NOVO Industries. The objective of this activity was to compare values obtained for protein and solids contents, using strictly standardized protocols, in an international round-robin format. The National Renewable Energy Laboratory (NREL) organized this activity and distributed the enzyme preparations and procedures used in the study. Results from the study showed conclusively that values used to sell and rank cellulase preparations, such as specific activity, may be more than an order of magnitude in error depending on the protein assay used.

Many methods are used by researchers to estimate protein concentrations in complex biological samples, such as commercial enzyme preparations. The Lowry method (1) is one of the most referenced assay procedures in the biochemical literature and is considered the standard for protein determination. The Lowry method, as first described, has many recognized deficiencies, including the instability of the alkaline copper reagent and sensitivity to compounds known to interfere with the accuracy of the assay. These interfering compounds include those potentially found in commercial cellulase preparations, such as disulfides, glycerol, polyols, aminosugars, and salts (2-6). Early modifications of the original Lowry method were introduced to simplify protein determination in the presence of known interfering agents (7,8). A more recent procedure developed by Markwell et al. (9) addresses the issues of determining protein concentrations in membrane and lipoprotein samples by adding sodium dodecylsulfate (SDS) to the alkali reagent, which acts to solubilize and disperse the protein component. Increased copper tartrate concentrations were also reported to facilitate protein determinations in the presence of sucrose, disulfides, and ethylenediaminetetraacetic acid (EDTA) (10).

0097-6156/95/0618-0256$12.00/0
© 1995 American Chemical Society

Two other commonly used methods for determining protein concentrations are the Coomassie Brilliant Blue G-250 dye binding method, known as the Bradford Method (*11*) and, more recently, a method based on adduct formation with bicinchoninic acid (BCA) (*12*). As was noted for the case of the Lowry assay, the Bradford assay is sensitive to interference by ionic and non-ionic detergents (*13,14*) and other small biomolecules (*15*). The BCA assay has reported interference from Zwitterionic buffers and chelating agents (*16*), sulfhydryl compounds (*17*), and lipids (*18*). As with the biuret assay, compounds that chelate copper ions affect the BCA assay by depleting the reagent.

Reliable and accurate protein determination is required to conduct kinetics and biophysical chemistry on purified enzymes. Accurate determination of protein concentrations in commercial enzyme preparations is also crucial for determining their specific activities and, thus, market value. For this reason, the IEA activity entitled, "Biotechnology for the Conversion of Lignocellulosics," which targets comparison of protein determinations for cellulase enzyme preparations, was initiated in 1989 and organized by K. Grohmann. The second round of activity occurred in 1991-1992, with the results presented at the 1992 IEA meeting in Vancouver, BC. During the 1992-1993 activity, only 5 laboratories participated, using the modified Lowry procedure as outlined in this paper. From this limited data base, the results showed that the standard curve data were unreliable, the dry weight measurements were inaccurate, and the results for the ultra-filtered enzymes were probably more reliable. Following this meeting, it was concluded that more participants were needed, and the data should be collected in triplicate by each laboratory to establish a useful data base from which some general conclusions could be drawn.

The current study was organized by A. Mohagheghi at NREL and involved the participation of the 13 laboratories listed in Appendix 1, which tested the standard enzymes using the procedure outlined below. The solids content of the cellulase preparations was also determined as outlined in the methods by these laboratories. The Kjeldahl method for nitrogen analysis was used for the absolute determination of protein content for this study. A comparison of these different methods of quantifying protein is presented in this paper.

Materials and Methods

Cellulase Enzyme Preparations. Enzymes used in this study were prepared and supplied by NOVO Industries (A/S DK-2880 Bagsvaerd, Denmark) and included Novozym (batch DCN 0012), Celluclast 1.5L (batch CCN 3027), Pulpzym HA (batch CXN 0005), and the ultrafiltered concentrates (UFC) of these three preparations. Ultrafiltration was performed using 10 kD cutoff membranes. NREL distributed 100-mL aliquots of the enzymes in 125 mL polyethylene plastic bottles to all participants by express mail in cold storage.

Appendix 1

List of Participants

Name	Organization	Country
Mary Jim Beck	Tennessee Valley Authority	U.S.A.
Alvin O. Converse	Dartmouth College	U.S.A.
Elizabeth Garcia/ Sharon Shoemaker	University of California-Davis	U.S.A.
Franz Gruber	Technische Universität Wien	Austria
Marianne Hayn	Institut for Biochemie, University of Graz	Austria
Dietmar Haltrich/ Brend Nidetzky	Universität Für BodenKultur, Institut Für LebensMittel Technologie	Austria
Maria Juoly/ M.P. Coughlan	University College Galway	Ireland
Ali Mohagheghi/ William Adney	National Renewable Energy Laboratory	U.S.A.
Marja Paloheimo	Alko Ltd.	Finland
Tapani Reinikainen	VTT Biotechnology Laboratory	Finland
Henrik Stalbrand	University of Lund	Sweden
Jack Saddler/ Richard Hanz	University of British Columbia	Canada
Marja Turunen	Alko Ltd.	Finland

Modified Lowry Procedure. The modified Lowry method as described by Markwell et al. (9) and improved by Sargent (10) was used by all laboratories to determine protein during the round-robin evaluation of the enzyme preparations.

Preparation of the Lowry Protein Reagents. Reagents were prepared by each laboratory using a standard procedure. Reagent A: 2% Na_2CO_3, 0.4% NaOH, 0.16% Na-tartrate, 1.0% SDS; Reagent B: 4% $CuSO_4$, $5H_2O$; Reagent C: Mixture of A & B (100:1), prepared fresh daily; Reagent D: Folin-Ciocalteu 2N phenol, diluted 1:1 with distilled water, prepared fresh daily. The Folin-Ciocalteu reagent and bovine serum albumin (BSA) was obtained form Sigma Chemical (St. Louis, MO).

Modified Lowry Procedure. The procedure used by all laboratories is briefly described. One mL of protein or standard sample (10-100 µg) was first mixed with 3 mL of Reagent C and then incubated at room temperature for at least 10 min, after which 300 µL of Reagent D was added followed by vigorous mixing. This solution was then incubated at room temperature for 45 min, at which time the absorbance was read at 660 nm against the reagent blank in a 1-cm cell. These determinations were done in triplicate and the absorbance values reported.

Standard curves for each individual analysis were used to determine the respective protein concentrations of the cellulase preparations using a linear regression fit of the BSA standard. Participants were asked to prepare a standard stock solution of 100 µg/mL BSA in distilled water. Volumes of 0, 50, 100, 200, 300, 400, 500, and 750 µL of the stock solutions were brought to 1 mL final volume with distilled water. The standard curves and sample sets were done in triplicate by each laboratory. Concentrations of the enzyme preparations were then corrected for dilutions before comparison of assay results.

Other Protein Assay Methods. For comparison, some round-robin participants also used the Bradford dye-binding method (11) and the BCA method (12). Cellulase enzyme preparations were also subjected to Kjeldahl nitrogen analysis following the general method of Hiller et al. (19) and improved by the Association of Official Analytical Chemists (20). NREL participants also developed a chromatographic method for estimating protein contents in complex samples.

The Bradford Dye-Binding and BCA Assays. Commercial reagents were used for both the Bradford (BioRad, Richmond, CA)) and BCA (Pierce, Rockford, IL) methods. The standard protein used in all round-robin determinations was BSA. Methods described by the product inserts were used for both methods.

Kjeldahl Nitrogen Analysis. The protein content of the enzyme preparations was also estimated by determining the total nitrogen content by Kjeldahl nitrogen analysis. The following relationship was used to calculate protein content: protein (g) = nitrogen (g) x 6.25.

NREL Chromatographic Protein Estimation Method. Protein concentrations in the commercial enzyme preparations were estimated by uv absorbance following chromatographic fractionation. To eliminate absorbance caused by non-protein compounds, the samples were diluted 50-fold and subjected to analytical-scale size exclusion chromatography (SEC) in 20 mM acetate, 100 mM NaCl pH 5.0 buffer. A Pharmacia SMART System equipped with a Superdex 75 PC 3.2/30 SEC column was used to separate proteins from uv-absorbing small molecules and integrate their peak areas. This system was equipped with detection at 280, 260, and 215 nm using a Pharmacia µPeak monitor. Samples were injected in 20 µL volumes and the flow rate was maintained at 40 µL/min. Peak areas were determined using standard SMART System software. Only those peaks showing 280/260 nm ratios greater than 1 were assumed to be protein and included in the total protein calculation. The method of Kalckar (21), where 1.45 $OD_{280\ nm}$ - 0.74 $OD_{260\ nm}$ = mg protein/mL, was used to estimate protein concentration in selected chromatographic fractions. Kalckar's method permits the approximation that 1 optical density (OD) unit at 280 nm is characteristic of a 0.90 mg/mL protein (BSA) solution.

Solids Content Analysis. Percent total solids for the enzyme samples were determined by all laboratories that had enough enzyme to perform the analysis. The analysis was performed by placing 2 mL of enzyme solution in desiccated, cooled, pre-dried (80°C) aluminum weigh boats. The solutions were then dried for 24 h at 80°C, cooled in a desiccator, and weighed. Sample drying was repeated until a constant weight was obtained. Samples were run in triplicate and the solids content calculated as the weight of dried material minus the original weight of the aluminum weigh boat divided by the weight of starting enzyme.

Preparation of the Enzymes for Protein Assay. The commercial enzymes were diluted 200-fold by adding 1 g of enzyme to a 200-mL volumetric flask and then adjusting the volume to 200 mL using distilled water. Assays were performed on aliquots of 100, 200, and 300 µL of the diluted enzyme solutions. Participants were requested to assay each dilution in triplicate and to report all absorbance values. The ultrafiltered enzymes were treated identically, following an initial 500-fold dilution.

Results

Round-Robin Lowry Protein Determinations. Thirteen laboratories performed the modified Lowry protein determination on 6 enzyme preparations using the protocol described above. These results, corrected for the respective dilutions, are presented in Tables 1 and 2. In the case of one laboratory, the 200 and 300 µL sample loadings were much lower than the value reported by the same laboratory for the 100 µL sample. Because of this discrepancy, values from this laboratory were excluded from the data base. Contrary to instructions, another laboratory used 25, 50, and 100 µL loadings of enzyme to conduct the modified Lowry procedure. Results from this laboratory were compared using only the 100 µL value from the data base, although the lower loadings gave absorbance values within the range of the standard curve.

A summary of the round-robin results for the commercial preparations, excluding these individual laboratory differences, is presented in Table 1. The differences in mean values for individual protein loadings used in this study (100, 200, and 300 µL)

Table 1. Summary of Round-Robin Results for Lowry Protein Assay for Commercial Enzyme Preparations

Enzyme	Novozym			Celluclast			Pulpzym		
Enzyme volume (µL)	100	200	300	100	200	300	100	200	300
Mean (mg/mL)	142.63	144.18	137.65	143.60	138.50	131.70	71.63	68.17	70.39
Median (mg/mL)	147.57	143.49	132.64	147.18	139.39	128.45	70.43	67.29	66.89
Standard Deviation	18.68	13.33	22.12	20.67	11.74	21.24	10.91	6.39	14.42
Variance	348.97	177.63	489.20	427.13	137.82	451.00	119.01	40.79	207.95
Range	70.55	43.92	81.82	83.14	46.09	79.89	42.05	22.18	53.40
Minimum	84.70	126.70	118.74	84.09	117.51	110.95	47.79	59.52	56.19
Maximum	155.25	170.62	200.56	167.23	163.61	190.85	89.84	81.70	109.59
Total samples (n)	12	11	11	12	11	11	12	11	11

Table 2. Summary of Round-Robin Results for Lowry Protein Assay for Ultrafiltered Enzyme Preparations

UF-Enzyme	UF-Novozym			UF-Celluclast			UF-Pulpzym		
Enzyme volume (µL)	100	200	300	100	200	300	100	200	300
Mean (mg/mL)	177.58	167.77	163.66	236.00	222.19	214.80	167.86	159.88	158.47
Median (mg/mL)	175.72	173.03	163.87	243.72	235.94	216.03	176.48	171.48	163.89
Standard Deviation	50.27	38.69	28.61	84.32	55.37	35.96	47.41	33.54	20.13
Variance	2526.95	1497.01	818.42	7109.18	3065.83	1393.27	2247.73	1125.2	405.28
Range	174.77	134.21	103.10	306.42	179.87	120.26	159.87	104.64	61.43
Minimum	61.99	81.87	112.96	48.39	108.05	150.86	47.42	80.21	111.68
Maximum	236.76	216.08	216.06	354.81	287.92	271.12	207.30	184.85	173.11
Total samples (n)	9	8	8	9	8	8	9	8	8

were tested by oneway variance analysis using values reported at each protein loading. Significant differences could not be demonstrated for the different enzyme loadings used for any of the commercial enzyme determinations, indicating that although absorbance values reported for the 200 and 300 µL sample loadings were outside the range of the standard curve, no statistically significant difference in the absolute calculated protein value could be detected. The range and standard deviations for each loading is relatively large. The combined mean protein concentrations obtained for all levels of sample enzyme loadings were compared to individual values obtained by each laboratory as a percentage of deviation from the overall mean as presented in Figure 1. Deviations from the mean for individual laboratories varied, with the largest deviation of -23% from Laboratory 13. In general, laboratories deviated from the mean in the same direction for the 3 regular enzyme samples.

Nine laboratories performed the modified Lowry protein analysis on the ultrafiltered enzyme preparations. Laboratory 4 did not have enough enzyme and deviated from the standard procedure to compensate for enzyme deficiencies. Laboratories 8, 10, and 11 did not attempt the modified Lowry assay on the ultrafiltered enzyme samples. A summary of the results from the laboratories that performed the standardized modified Lowry assay on the ultrafiltered samples is presented in Table 2. In general, the dispersion of data about the mean, as indicated by the standard deviation between laboratories, was much greater for the ultrafiltered samples than for the commercial enzyme preparations. Differences in mean values for individual protein loadings were tested by analyzing variance in the same fashion as the commercial enzyme preparations. Again, significant differences could not be established for the different enzyme loadings used for the ultrafiltered enzyme determinations. This is not surprising, considering the exceptionally large values for the range, which in many cases were equal to the actual mean protein estimate. The standard deviations for the ultrafiltered samples were considerably larger than standard deviations for the commercial enzyme determinations. The individual laboratory variations from the overall mean for the ultrafilter enzyme preparations are presented in Figure 2.

Round-Robin Solids Content Analysis. Not all laboratories had enough sample to run triplicate solids determinations, and some laboratories did not report values for all enzyme samples. Data from the individual measurements for the total solids analysis were analyzed and are presented in Table 3. In general, the results from the total solids analysis were more closely grouped than the results from the modified Lowry protein determinations. These results indicate that no laboratory deviated from the overall sample mean by more than 5%, compared to deviations as high as 60% that were found in the modified Lowry protein determinations.

Comparison of Protein Methods. Four methods using different chemistries for determining protein concentrations were compared at NREL. These methods included the round-robin results using the modified Lowry protein assay, the Bradford dye binding method, the BCA method, the Kjeldahl nitrogen analysis method, and a new chromatographic fractionation-uv absorbance method. The results of this comparison are presented in Figure 3. The Bradford method appears to significantly underestimate the protein concentration for both the commercial and ultrafiltered enzyme preparations.

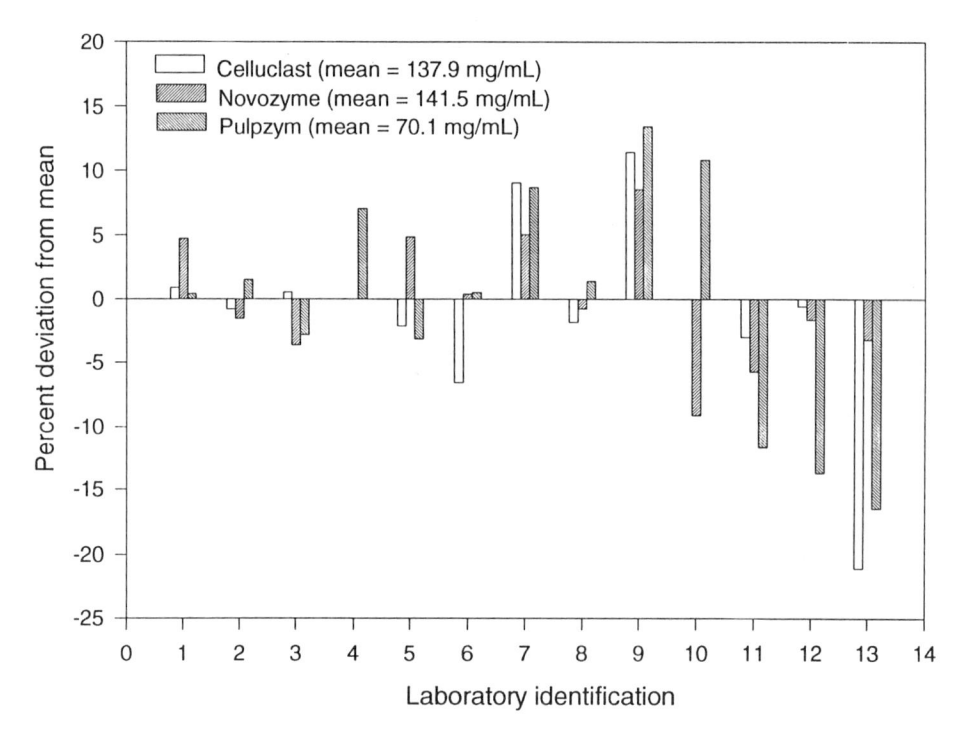

Figure 1. Comparison of the percent deviation from the mean of individual Lowry protein measurements of cellulase protein samples.

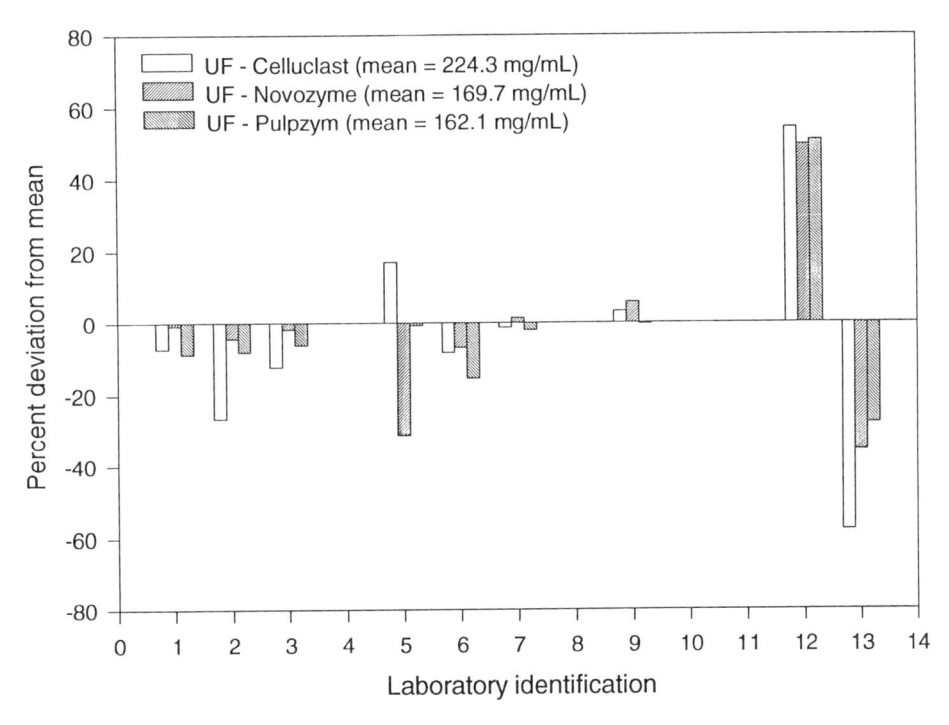

Figure 2. Comparison of the percent deviation from the mean of individual Lowry protein measurements of ultrafiltered protein samples.

Table 3. Summary of Round-Robin Solids Content Analysis Results

Enzyme	Novozym	Celluclast	Pulpzym	UF Novozym	Uf Celluclast	UF Pulpzym
Mean (% dry weight)	0.374	0.506	0.504	0.318	0.330	0.349
Standard Error	0.008	0.008	0.000	0.002	0.001	0.002
Median	0.363	0.519	0.504	0.318	0.330	0.352
Standard Deviation	0.049	0.044	0.002	0.010	0.006	0.008
Variance	0.002	0.002	0.000	0.000	0.000	0.000
Range	0.193	0.167	0.010	0.037	0.020	0.027
Minimum	0.339	0.363	0.501	0.302	0.320	0.330
Maximum	0.532	0.530	0.511	0.339	0.340	0.356
Total count (n)	34	35	36	26	25	26

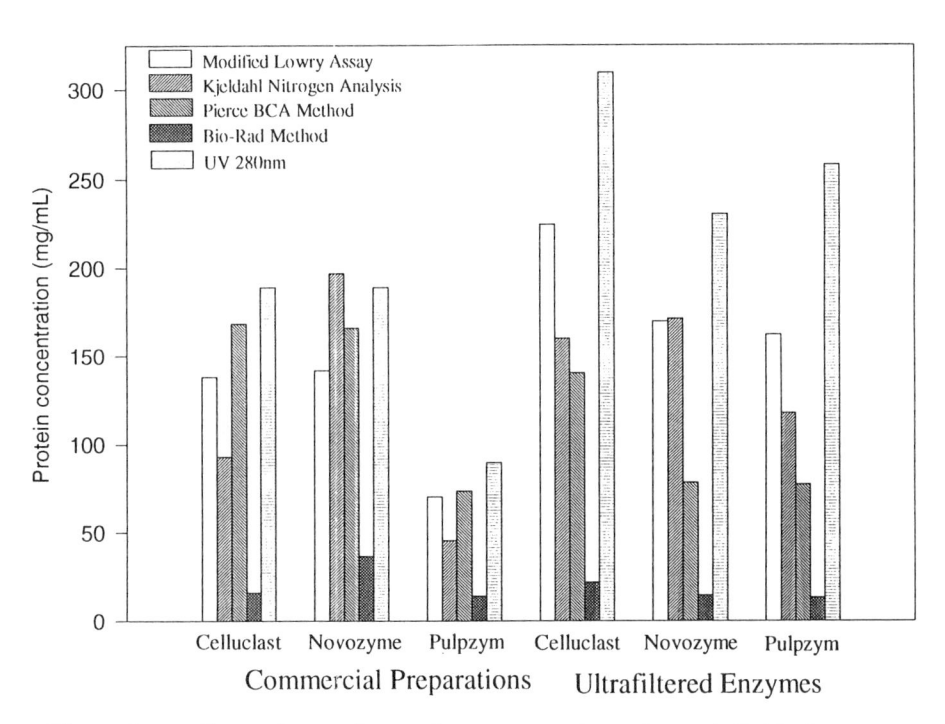

Figure 3. Comparison of protein concentrations from individual enzyme preparations using different methods.

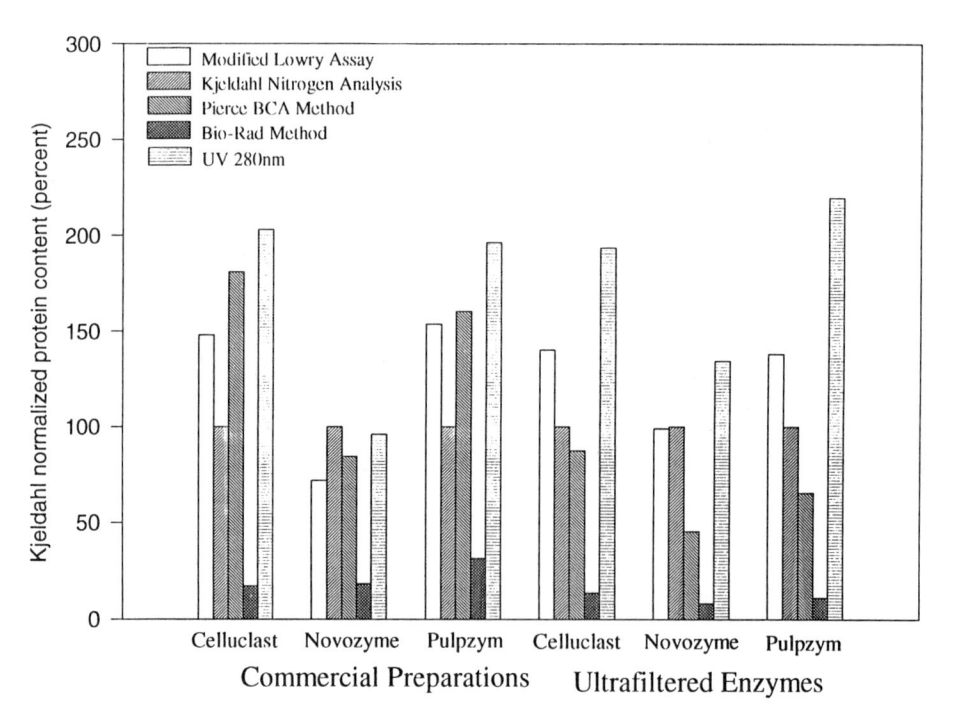

Figure 4. Comparison of protein concentrations normalized to Kjeldahl nitrogen protein values.

Figure 4 presents the data (shown in Figure 3) normalized to the Kjeldahl nitrogen values expressed as 100 percent. This data presentation may be defended because the Kjeldahl method may be the least sensitive of this group of assays to the effects of interfering compounds expected in concentrated (especially ultrafiltered) culture broths (*19*).

In addition, Laboratory 2 also reported that it compared the Lowry protein results to the Bradford method and obtained values 4 times lower using the Bradford method. This group suggested that this difference could be caused by the inadequate protein disaggregation with the Bradford method (i.e., SDS). Laboratory 13 also compared the modified Lowry and Bradford methods and found that the former gives much higher values than the latter by a 3.2:1 ratio.

Discussion

Assays that determine protein concentrations of cellulase enzyme samples for commercial application should be sensitive, accurate, precise, and easily applied. Interfering compounds, which may either suppress or enhance the color response of the protein for a given assay, must also be considered. For example, the BCA protein assay reportedly has greater tolerance for commonly interfering compounds than does the Lowry or the Bradford methods (*22,23*). A complication for selecting such assays for a specific protein determination, however, is that even the venerable Lowry assay still harbors some mechanistic unknowns (*24*). Most assay procedures use BSA as the standard protein against which the unknown proteins are measured; however, different methods exhibit unique levels of responses toward different proteins (*22*).

In this round-robin study, variability in the measurements of modified Lowry protein concentrations reported from all laboratories was greater for the ultrafiltered enzymes than for the commercial enzyme preparations. This result was puzzling, since ultrafiltration should remove low molecular weight compounds that could interfere with the accuracy of the measurement. In general, relatively large interlaboratory variations in protein concentrations were reported using the modified Lowry assay, whereas less variability was observed for the total solids analysis for these same enzyme preparations.

In the NREL portion of the study (shown in Figure 3), concentration values found for commercial protein preparations appeared to follow the trend: chromatographic uv method>BCA method>modified Lowry method>Bradford method. This trend is more easily seen in Figure 4. Results from the BCA assay, when compared to the modified Lowry protein values obtained in the round-robin assay, delivered statistically equivalent values for the commercial enzyme preparations, but consistently lower values when the ultrafiltered preparations were tested. In fact, the trend noted for the commercial preparations is largely lost in the case of the ultrafiltered enzymes. Because this study was based on real commercial preparations rather than artificially created mixtures, the actual protein values cannot be readily determined.

When the Bradford method was compared to the other methods used in this study, the protein concentration determined was lower by a factor of at least 5.

Values obtained during this study using the Bradford dye-binding method are now considered suspect because of the large variation in dye reactivity displayed by different proteins and documented by others (*15,22*). Unfortunately, this assay is widely used in industry because it is sensitive, simple, quick, and relatively few compounds have been reported to interfere.

Protein concentrations and enzyme catalytic activities are routinely used to describe, compare, and cost commercial enzymes. This study has shown that large differences in the estimation of protein concentrations can occur, depending on the specific type of protein assay used and how the assay is standardized. The impact of such variation, which in some instances may be as large as 90%, is obviously significant to industry. In the case of the cellulase preparations used in this study, reports of specific activities clearly depend on the type of method used to determine protein concentrations. If the Bradford method is used for determining protein concentration, for example, the specific activities derived for these enzyme preparations will be considerably higher than if the BCA or modified Lowry method is used. Because of this, we recommend the BCA, modified Lowry, or chromatographic fractionation methods for commercial enzyme preparations. The chromatographic method is admittedly cumbersome, however. Because of low profit margins in the biomass to ethanol industry, the issue of specific activity determination for various cellulase preparations is especially critical for their ultimate ranking.

Acknowledgments

This work was funded by the Biochemical Conversion Element of the Biofuels Program of the U.S. Department of Energy and the IEA. The authors would like to thank Dr. K. Grohmann of the USDA United States Citrus and Subtropical Products Research Laboratory for initially organizing this IEA activity. The authors also acknowledge M.J. Beck, A.O. Converse, E. Garcia, S.P. Shoemaker, F. Gruber, M. Hayn, D. Haltrich, B. Nidetzky, M. Juoly, M.P. Coughlan, M. Paloheimo, T. Reinikainen, H. Stalbrand, J. Saddler, R. Hanz, and M. Turunen for their input to this activity.

Literature Cited

1. Lowry, O.H.; Rosebrough, N.J.; Farr, A.L.; Randall, R.J. *J. Biol. Chem.* **1951**, *193*, 265-275.

2. Vallejo, C.C.; Lugunas, R. *Anal. Biochem.* **1970**, *36*, 207-212.

3. Berg, D.H. *Anal. Biochem.* **1971**, *42*, 505-508.

4. Lo, C.; Stelson, H. *Anal. Biochem.* **1972**, *45*, 331-336.

5. Bensadoun, A.; Weinstein, D. *Anal. Biochem.* **1976**, *70*, 241-250.

6. Peterson, G.L. *Anal. Biochem.* **1979**, *100*, 201-220.

7. Peterson, G.L. *Anal. Biochem.* **1977**, *83*, 346-356.

8. Hartree, E.E. *Anal. Biochem.* **1972**, *48*, 422-427.

9. Markwell, M.A.K.; Haas, S.M.; Tolbert, N.E.; Bieber, L.L. *Methods Enzymol.* **1981**, *72*, 296-303.

10. Sargent, M.G. *Anal. Biochem.* **1987**, *163*, 476-481.

11. Bradford, M.M. *Anal. Biochem.* **1976**, *72*, 248-254.

12. Smith, P.K.; Krohn, R.I.; Hermanson, G.T.; Mallia, A.K.; Gartner, F.H.; Provenzano, M.D.; Fujimoto, E.K.; Goeke, N.M.; Olson, B.J.; Klenk, D.C. *Anal. Biochem.* **1985**, *150*, 76-85.

13. Boccaccio, G.L.; Quesada-Allue, L.A. *An. Asoc. Quim. Argent.* **1989**, *77*, 79-88.

14. Carroll, K.; O'Kennedy, R. *Biochem. Soc. Trans.* **1988**, *16*, 382-383.

15. Compton, S.J.; Jones, C.G. *Anal. Biochem.* **1985**, *151*, 369-374.

16. Kaushal, V.; Barnes, L.D.; *Anal. Biochem.* **1986**, *157*, 291-294.

17. Hill, H.D.; Straka, J.G.; *Anal. Biochem.* **1988**, *170*, 203-208.

18. Kessler, R.; Franestil, D. *Anal. Biochem.* **1986**, *159*, 138-142.

19. Hiller, A.; Plazin, J.; Van Slyke, D.D. *J. Biol. Chem.* **1948**, *176*, 1401.

20. Official Methods of Analysis of the Association of Official Analytical Chemists, (Horwitz, W., ed.), 13th Ed., No. 2.057, AOAC: Washington, DC, 1980, p. 15.

21. Kalckar, H. *J. Biol. Chem.* **1947**, *167*, 461.

22. Davies, E.M. *Am. Biotech. Lab.* **1988**, 28-37.

23. Wiechelman, K.; Braun, R.; Fitzpatrick, J. *Anal. Biochem.* **1988**, *175*, 231-237.

24. Sengupta, S.; Chattopadhyay, M.K. *J. Pharm. Pharmacol.* **1993**, *45*, 80.

RECEIVED September 11, 1995

Chapter 17

Economic Fundamentals of Ethanol Production from Lignocellulosic Biomass

Charles E. Wyman

Alternative Fuels Division, Applied Biological Science Branch, National Renewable Energy Laboratory, 1617 Cole Boulevard, Golden, CO 80401–3393

> Information on the production potential for ethanol from lignocellulosic resources is summarized, and the benefits for reducing carbon dioxide accumulation are discussed. The fundamentals of ethanol production are examined in terms of the primary features influencing cost: feedstock cost, feedstock composition, product yield, energy use, and other operating costs. These costs are compared to the revenues that can be realized for fuel ethanol and co-product electricity, and the margin left to cover annualized capital costs is determined. Then the allowable capital cost for the plant is calculated, and analogies are made with existing technology for ethanol production from corn. From this analysis, it is shown that ethanol from cellulosic biomass can be a cost-competitive fuel.

Currently, petroleum provides the largest single source of energy in the United States (40%), transportation fuels are almost totally derived from petroleum (about 97%), and about two thirds of the petroleum used in the United States is for transportation. Because about half the oil used in the United States is imported, we are strategically vulnerable to disruptions in our supply of transportation fuels. In addition, at a cost of about $40 billion annually for oil imports, the largest fraction of our trade deficit is contributed by petroleum imports, and we are susceptible to significant economic dislocations if oil prices increase dramatically, as has occurred in the past *(1-2)*.

Transportation fuels are substantial contributors to urban air pollution *(3-6)*. About two-thirds of the carbon monoxide pollution in our cities is due to the transportation sector. In addition, approximately one-third of the ozone-forming compounds that cause smog in our cities is due to transportation. Because transportation fuels are derived from fossil sources and constitute a significant fraction of energy used in the United States, it is not surprising that about one-third of the carbon dioxide accumulation in the United States is due to the transportation sector.

Of course, carbon dioxide is a major greenhouse gas that could contribute to global climate change *(7)*.

Ethanol is a diverse, high-performance transportation fuel that has the potential to be produced on a large scale from plentiful sources of lignocellulosic biomass. Successful commercialization of this technology would substantially decrease or even eliminate our dependence on imported oil, thereby reducing the strategic vulnerability of the United States transportation sector, lowering the trade deficit dramatically, and creating substantial employment *(1)*. In addition, ethanol production from lignocellulosic biomass can reduce the accumulation of carbon dioxide by 90% or more, substantially decreasing the contribution to global climate change *(8)*. Technology for producing ethanol from biomass has been improved dramatically over the past decade or more so that ethanol could be competitive now in the United States for existing markets *(9-12)*, and opportunities have been identified to further reduce the cost of ethanol production to be competitive as a neat fuel with gasoline. Yet many still question the benefits of ethanol from lignocellulosic biomass and its potential to be economically competitive.

In this chapter, the current status of ethanol use to improve air quality will be reviewed to provide an update of the growing demand for this fuel. Then the technology will be briefly reviewed to acquaint the reader with the production of ethanol from lignocellulosic biomass and the impact on carbon dioxide buildup. Against this background, the economics of biomass ethanol production will be examined from a fundamental perspective and through comparison to the existing corn ethanol industry to demonstrate that ethanol could be made from lignocellulosic biomass at competitive prices. The goal of this approach is to clearly demonstrate the growing market for and benefits of ethanol production from biomass and establish in an unambiguous manner that ethanol can be made at competitive prices for advanced technology in well engineering processes.

Ethanol Use

Urban air pollution is due to evaporative and tailpipe vehicle emissions. Evaporative emissions occur during vehicle refueling and operation, as fuel components evaporate into the atmosphere. These emissions include various volatile hydrocarbons that cause ozone formation and smog. In addition, several components — such as benzene that constitute evaporative emissions — can be toxic. Tailpipe emissions, on the other hand, are emitted from the exhaust system of vehicles. Problematic examples include oxides of nitrogen, carbon monoxide, partial combustion products, and unburned hydrocarbons, which result from incomplete fuel combustion as well as high engine temperatures. These components contribute to carbon monoxide accumulation in cities as well as to ozone formation and smog.

Ethanol can be used in several ways as a fuel to help address air pollution. First, ethanol can be directly blended with gasoline as in the 10% mixtures now typically used in the United States (E10) or 22% blends used in Brazil (E22). Direct blends of ethanol with gasoline serve to extend gasoline by reducing the amount of gasoline required while boosting octane, thereby reducing the need for toxic octane boosters. Ethanol also provides oxygen for the fuel to promote more complete combustion. However, the vapor pressure of the resulting mixture increases when ethanol is directly blended with gasoline at low levels, causing concerns about evaporative emissions *(13-14)*.

Ethanol can also be reacted with isobutylene or other olefins to form ethers such as ethyl tertiary butyl ether (ETBE). When blended with gasoline, ETBE provides the same benefits as direct ethanol blends in terms of extending the gasoline supply, boosting octane, and providing oxygen. Additionally, ETBE actually reduces vapor pressure when mixed with gasoline. Thus, ozone formation and smog decrease with ETBE blends.

Finally, ethanol can be used as "pure" fuel in the form of hydrous ethanol containing 95% ethanol and 5% water (as in Brazil) or with small amounts of gasoline to promote cold starting. Mixtures of 95% ethanol with 5% gasoline are denoted as E95, while 85% ethanol with 15% gasoline is designated as E85 *(15)*. Neat ethanol fuel has a high octane, a high heat of vaporization, and other favorable properties that result in higher efficiency operation than gasoline for properly optimized engines. As a result, a 20%–30% increase in efficiency relative to gasoline is possible *(1)*. Neat ethanol also has low toxicity, low vapor pressure, and low photochemical reactivity, reducing the potential for smog formation and other environmental impacts. In the longer term, pure ethanol is more readily adaptable than gasoline to fuel cell applications. Fuel cells can achieve far higher efficiencies than internal combustion engines, while realizing tremendous advantages in reducing air pollution *(16)*.

To improve urban air quality, oxygenated gasoline has been required in 39 carbon monoxide non-attainment areas in the United States since 1993. This fuel must contain 2.7% oxygen during the winter months. This requirement includes a waiver for the higher vapor pressure that results when ethanol is blended with gasoline, because ethanol reduces carbon monoxide emissions that are of concern in the winter months, while smog formation associated with higher vapor pressure is not a serious problem during this period. Reformulated gasoline (RFG) will be required beginning in 1995 in nine ozone non-attainment areas within the United States. RFG must contain at least 2% oxygen year-round, but no vapor pressure waiver is provided for ethanol at this time. In addition, RFG must have a reduced aromatic content, especially of benzene *(6)*.

In December 1993, the U.S. Environmental Protection Agency (EPA) proposed a new rule that will require 30% of oxygenates in RFG to be derived from renewable sources, and the EPA enacted a phased in approach in June 1994 requiring 15% of oxygenates in RFG be from renewable sources in 1995, and 30% thereafter. Ethanol, with production now at about 3.8 billion L (1 billion gal) per year, is expected to be the primary fuel affected. The renewable oxygenate standard (ROS) was adopted to reduce oil imports and carbon dioxide accumulation, and create domestic employment. However, some controversy surrounds the requirement for renewable oxygenates in RFG. First, the increase in gasoline vapor pressure when ethanol is blended with gasoline is of concern. Some controversy also surrounds the amount of carbon dioxide that accumulates when ethanol is produced from corn, as is now the practice in the

United States. In addition, opponents question the use of existing tax incentives ($0.14/L ethanol, $0.54/gal federal) to encourage corn ethanol use.

The ROS would allow direct ethanol blends to count toward the renewable oxygenate requirement from September 16 to April 30 to reduce carbon monoxide and unburned hydrocarbon emissions. However, from May 1 to September 15, only ethanol in ETBE would count toward the standard. In this way, ETBE would reduce tailpipe emissions of carbon monoxide and unburned hydrocarbons, while reducing evaporative emissions that lead to smog formation during the summer months when smog formation is an issue. To date, implementation of the ROS has been held up by legal challenges.

Other approaches are also potentially viable for reducing smog formation for blends. First, the vapor pressure of the gasoline blending stock could be reduced to compensate for the increased vapor pressure for a low-level ethanol blend. In addition, the higher vapor pressure is due to non-ideal behavior, and ethanol has a far lower vapor pressure than gasoline; thus, as the amount of ethanol is increased beyond about 22%, the vapor pressure of the gasoline-ethanol mixture actually is reduced from that of the gasoline to which ethanol is added. As mentioned previously, ETBE blends reduce vapor pressure while mixtures of ETBE and ethanol could achieve vapor pressures equal to that of the gasoline blending stock. Thus, the higher vapor pressure exhibited for 10% blends is not an inherent limitation of ethanol.

Ethanol Production from Lignocellulosic Biomass

Currently, more than 11 billion L (3 billion gal) of ethanol are produced annually from cane sugar in Brazil, but sugar is too expensive in the United States to achieve economical conversion to ethanol. In the United States over 3.8 billion L (1 billion gal) of ethanol is produced annually from starch crops, mostly corn. However, the cost to produce ethanol from corn is still higher than to produce gasoline, and federal and state tax incentives are used to compensate for the higher price.

In addition to producing ethanol from starch and sugar crops, ethanol can be made from lignocellulosic biomass. Examples of existing sources of lignocellulosic biomass include agricultural and forestry residues, a major fraction of municipal solid waste (MSW), wastepaper, and various industrial wastestreams. Future sources of lignocellulosic biomass could be herbaceous (grasses) and woody crops grown to support ethanol production.

Figure 1 illustrates the process for enzymatic conversion of lignocellulosic biomass to ethanol. First the biomass is pretreated to reduce its size and open up the structure to facilitate conversion of this naturally resistant material into ethanol. Often the hemicellulose fraction (which comprises about 20% to 40% of the material) is broken down to form its component sugars such as xylose during a pretreatment step; these sugars are subsequently fermented to ethanol. Left behind is a solid residue of cellulose and lignin, a small portion of which is fed to a cellulase enzyme production step. These enzymes are then added to the bulk of the solid cellulose-containing material to break down the cellulose into glucose (hydrolysis), and an appropriate organism ferments the glucose into ethanol. Following conversion of the sugars from

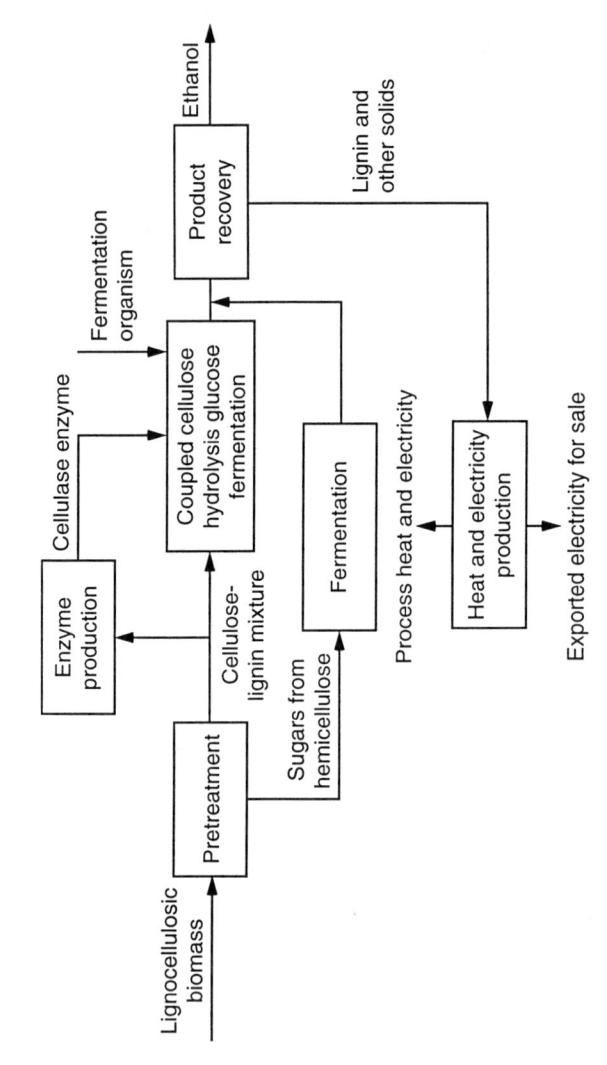

Figure 1. Processing lignocellulosic biomass to ethanol.

the cellulose and hemicellulose fractions into ethanol, the fermentation broth is sent to a purification step where ethanol is recovered for use as a fuel. The solid residue left following purification contains primarily the lignin fraction, representing about 15% to 20% of the original biomass substrate, that can be burned to provide the heat and electricity to power the entire conversion process as well as excess electricity that can be exported for sale *(11-12)*.

Because lignin can fuel the conversion process, and because low levels of fossil energy inputs are required to grow biomass, most carbon dioxide released during production and utilization of ethanol from lignocellulosic biomass is recaptured to grow new biomass to replace that harvested for ethanol production, and little if any net carbon dioxide accumulates. The result is a 90% or greater reduction in carbon dioxide accumulation compared to use of RFG *(8)*.

Economic Fundamentals

Lignocellulosic biomass provides a low-cost, abundant, domestic resource that could produce enough ethanol to displace a substantial fraction, if not all, gasoline used in the United States. However, the cost of conversion to ethanol has historically been too high because of the recalcitrant nature of lignocellulosic materials. Over the past 14 years, substantial progress has been made in reducing the cost of ethanol production from lignocellulosic biomass from about $0.95/L ($3.60/gal) in 1980 *(9-10)* to $0.32/L ($1.22/gal) (11), and it can be competitive now, particularly for niche markets that use low-cost feedstocks or other cost-saving measures. Furthermore, additional opportunities have been identified to advance the technology so that ethanol produced from lignocellulosic biomass can compete with gasoline without special tax considerations.

Although detailed process designs and economic evaluations have been employed to estimate the cost of ethanol from biomass for current technology and identify targets for continued cost reductions, such studies are highly dependent on the process design chosen, and different studies estimate different ethanol production costs. However, consideration of fundamental economic principles can show that it is possible to achieve ethanol production at competitive costs. In this section, a simplified analysis will illustrate the economic merits of ethanol production from biomass and its excellent potential to be cost competitive in the open market.

Feedstock costs. The amount of ethanol derived from a given weight of biomass is a critical factor in establishing the economics of ethanol production, since the ethanol yield determines the potential revenue stream for the process and biomass represents a major cost element. The cellulose fraction is hydrolyzed to glucose and fermented to ethanol as shown in the following stoichiometric equation:

$$n(C_6H_{10}O_5) + n\,H_2O \underset{\text{enzyme}}{\rightarrow} n\,C_6H_{12}O_6 \underset{\text{yeast}}{\rightarrow} 2nC_2H_5OH + 2n\,CO_2 \qquad (1)$$

Similarly, the hemicellulosic fraction is hydrolyzed to xylose and other sugars for fermentation to ethanol according to the following relationship:

$$3n(C_5H_8O_4) + 3n\ H_2O \xrightarrow{\text{acid}} 3n\ C_5H_{10}O_5 \xrightarrow{\text{yeast}} 5nC_2H_5OH + 5n\ CO_2 \qquad (2)$$

For the purposes of this analysis, the ratio of cellulose to hemicellulose is assumed to be 2:1 and 80% of the remaining material is assumed to be lignin. These ratios are based on studies by others *(12)*, but the ratio of cellulose to hemicellulose is not critical to the analysis, since the weight yield is only slightly higher for hemicellulose than for cellulose. On the other hand, as we will see, the amount of lignin affects the amount of excess heat or electricity that can be sold, but this effect is not expected to greatly change the results of the analysis presented here.

The volumetric ethanol yield is calculated from the overall carbohydrate fraction (cellulose and hemicellulose), the ratio of these components, the stoichiometry for conversion to ethanol, and the fractional yield of ethanol obtained from the carbohydrates, as shown in equation *(3)*.

$$Y = 1260xC[0.568f_c + 0.581(1-f_c)] \qquad (3)$$

in which Y is the volumetric ethanol yield in liters/tonne, x is the weight fraction of the feedstock carbohydrates (cellulose and hemicellulose) converted to ethanol, C is the weight fraction of carbohydrates in the feedstock, f_c is the fraction of the feedstock carbohydrates that are cellulose, and $(1-f_c)$ is the fraction of the feedstock carbohydrates that are hemicellulose. This relationship is illustrated in Figure 2 in terms of the volume of ethanol produced per weight of feedstock as a function of the fraction of the carbohydrates converted to ethanol for varying carbohydrate content, i.e., total cellulose and hemicellulose in the feedstocks. This figure clearly shows how dramatically the volumetric ethanol yield changes with the fraction of carbohydrate contained in the feedstock and its conversion to ethanol.

From an economic perspective, the key parameter is the cost of the feedstock for a given volume of ethanol produced. This value can be determined by dividing the feedstock costs per unit weight by the volumetric ethanol yield as:

$$\frac{\text{Feedstock cost}}{\text{Volume ethanol}} = \frac{\text{Feedstock cost/weight of feedstock}}{\text{Ethanol volume/weight of feedstock}} \qquad (4)$$

Once again, we can vary the carbohydrate content and percentage yield to determine their influence on the feedstock cost per volume of ethanol produced. Figure 3 shows the result for a feedstock costing $37/tonne ($34/ton). Figure 4 then compares the feedstock cost per volume for a feedstock containing 70% carbohydrates as a function of the yield to ethanol for feedstocks costing $37/tonne and $46/tonne ($42/ton). As expected, the volumetric feedstock costs are directly proportional to the cost of the feedstock and greatly influenced by the carbohydrate content and ethanol yield from carbohydrates.

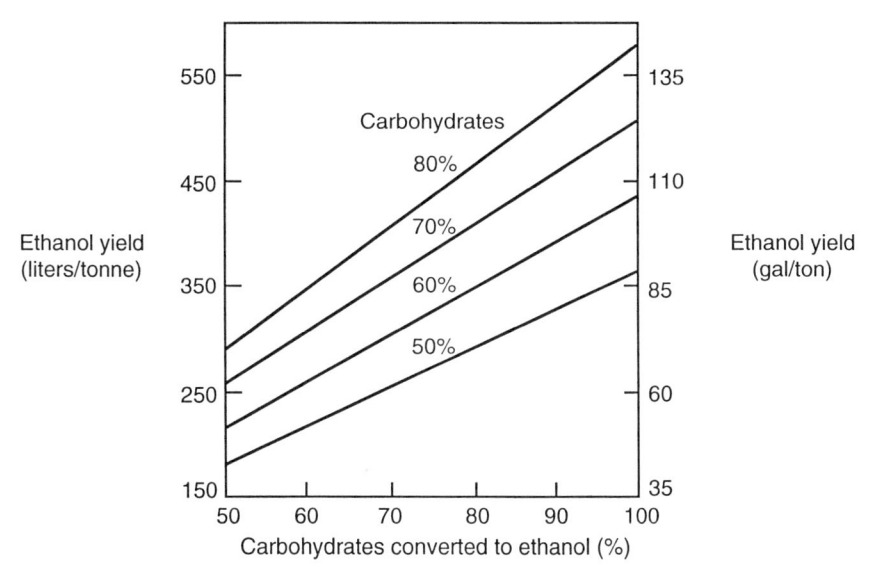

Figure 2. Volumetric ethanol yield per weight of feedstock as a function of the percentage of the total carbohydrates converted to ethanol (yield) and the carbohydrate content of the feedstock.

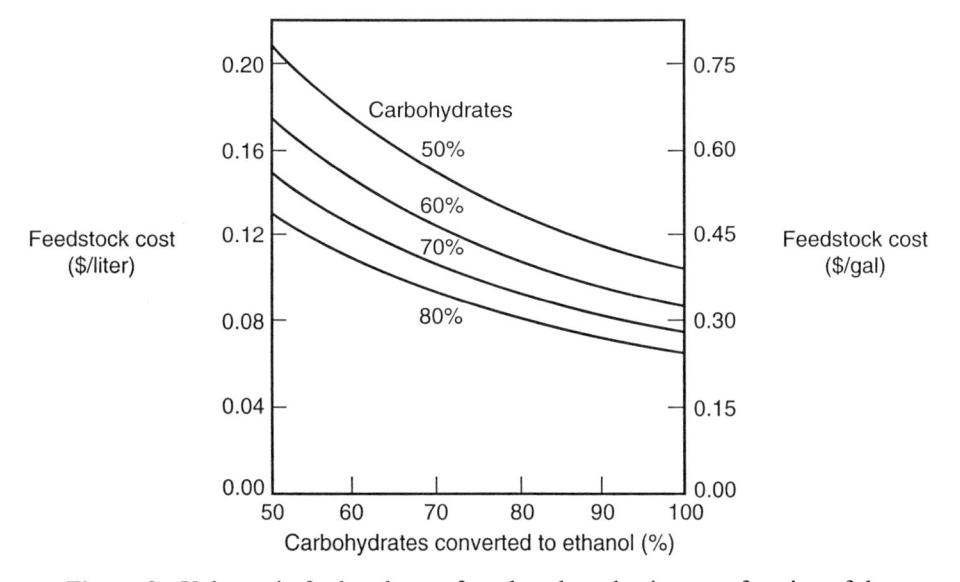

Figure 3. Volumetric feedstock cost for ethanol production as a function of the percentage of the total carbohydrates converted to ethanol for varying carbohydrate content and a feedstock costing $37/tonne ($34/ton).

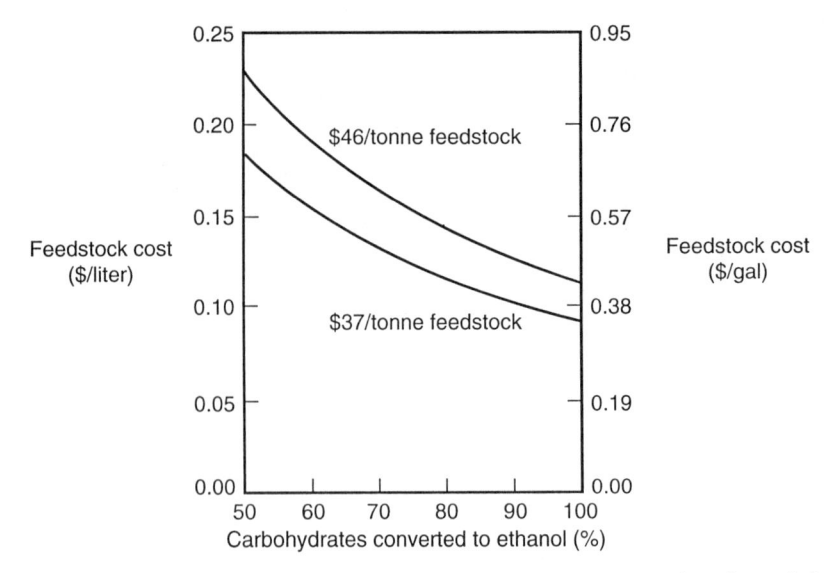

Figure 4. Comparison of the volumetric feedstock cost as a function of the percentage of the total carbohydrates converted to ethanol for varying feedstock costs and a feedstock containing 70% carbohydrates.

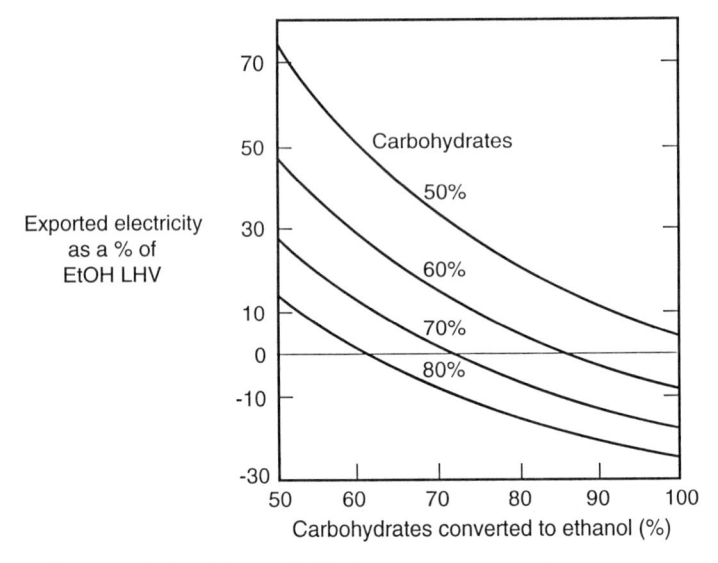

Figure 5. Estimated electricity available for export expressed as a percentage of the lower heating value (LHV) of ethanol for a plant requiring 22 MJ/L of ethanol produced (80,000 Btu/gal) for process heat and electricity requirements. Exported electricity is estimated for varying feedstock carbohydrate content as a function of the percentage converted to ethanol.

Process revenues. Two products are assumed to be manufactured for revenue generation from the ethanol process: ethanol and excess exportable electricity. As described previously, all materials not converted to ethanol are assumed to be burned to produce heat and electricity. The solid material that is burned following ethanol recovery is assumed for the purposes of this analysis to contain 50% moisture. Furthermore, it is assumed that the process requires either 22 MJ/L (80,000 BTU/gal) of ethanol produced or 11 MJ/L (40,000 BTU/gal) for process heat and electricity; the 11 MJ/L figure is typical of efficient modern corn ethanol plants. It is further assumed that the electricity is produced with a 33% efficiency and surplus electricity is sold.

Figure 5 shows the amount of excess electricity exported as a percentage of the lower heating value of the ethanol produced as a function of the material's carbohydrate content and the percentage of that carbohydrate converted to ethanol for a process requiring 22 MJ/L (80,000 BTU/gal) of ethanol produced. Shown in Figure 6 is similar information for a process that requires 11 MJ/L (40,000 BTU/gal) of process heat and electricity. From these figures, substantial amounts of electricity can be exported for a process with energy requirements similar to those of a modern corn ethanol plant, while little if any electricity export is possible for a less efficient plant. In fact, the less efficient plant benefits from a lower carbohydrate content so that more lignin is available to produce heat and electricity for the process.

For the purposes of this analysis, it is assumed that ethanol is sold at $0.18/L ($0.67/gal) and electricity is sold at $0.03/kWh. As shown in Figure 7, the ethanol selling price of $0.18/L for conventional gasoline corresponds to an oil cost of $25/bbl for ethanol achieving a range of 80% that of gasoline based on its superior properties that result in higher efficiency use.

The revenue can be determined from these unit prices for electricity and ethanol. Figure 8 shows the revenues per volume of ethanol produced for either ethanol alone or ethanol plus electricity sales for a feedstock costing $37/tonne ($34/ton) and a 70% carbohydrate content. Also shown is the feedstock cost per volume of ethanol produced as a function of the percentage of the carbohydrates converted to ethanol. The margin between the feedstock cost and the revenue from sale of ethanol or ethanol plus electricity is available to recover the remaining costs and realize a return on investment for the ethanol conversion process. From this figure, it can be seen that coproduct revenues in the form of electricity are very important for low ethanol yields from biomass, and become predictably less important as the yield improves. Furthermore, it can be seen that a significant margin is available to cover other costs of production (COP) and realize a return on capital.

For commodity products, the feedstock cost often represents 80% to 90% of the overall COP. From this, we could estimate the COP from the feedstock cost based on a factored cost estimate. For the purposes of this analysis, the feedstock cost is assumed to represent two-thirds of the overall production cost because a solid substrate is used in the conversion process, generally with higher conversion costs. Figure 8 presents the factored COP for a 70% carbohydrate feedstock costing $37/tonne ($34/ton) as a function of the fractional conversion of the carbohydrates to ethanol. As mentioned earlier, also shown are the revenues from the sale of ethanol or of ethanol plus electricity as well as the feedstock cost. Based on this analysis, the cost of ethanol

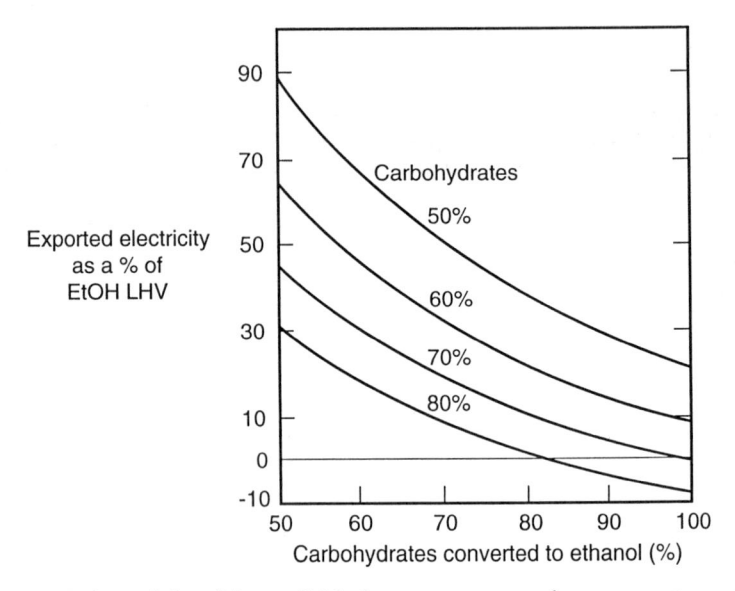

Figure 6. Estimated electricity available for export expressed as a percentage of the lower heating value (LHV) of ethanol for a plant requiring 11 MJ/L of ethanol produced (40,000 Btu/gal) for process heat and electricity. Exported electricity is estimated as a function of ethanol yield from carbohydrates.

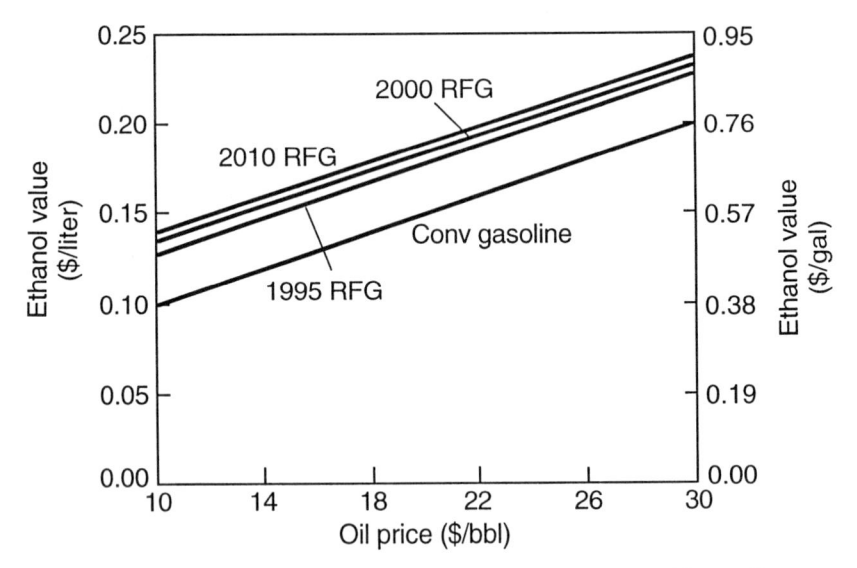

Figure 7. The value of ethanol as a neat fuel in competition with gasoline as a function of petroleum price for conventional gasoline and RFG designed to meet 1995, 2000, and 2010 EPA requirements.

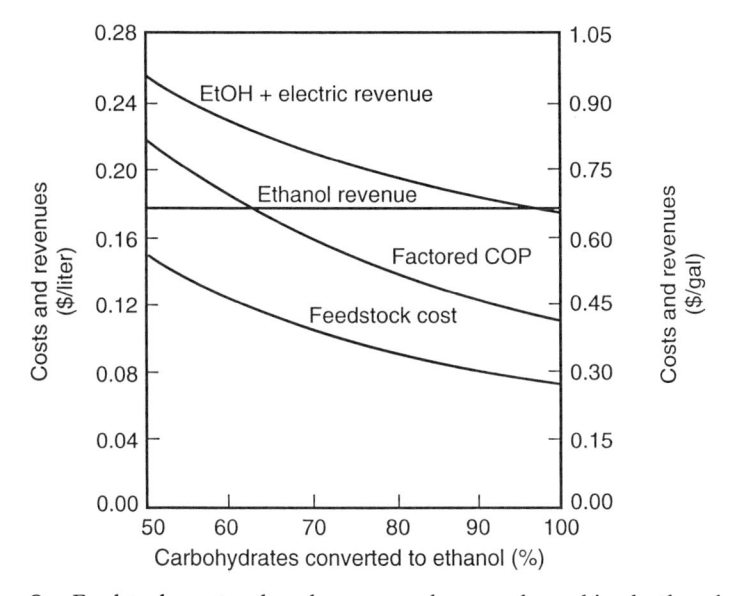

Figure 8. Feedstock cost, ethanol revenue alone, and combined ethanol and exported electricity revenue as a function of the percentage of carbohydrates converted to ethanol for a feedstock containing 70% cellulose and hemicellulose and costing $37/tonne ($34/ton). Also shown is the factored COP estimate based on feedstock cost alone.

production is projected to be significantly lower than the revenues derived from ethanol or ethanol plus electricity, particularly at high ethanol yields. Although simplistic, the analysis shows that production of ethanol (even though it is a low value product) could be economical for lignocellulosic materials if produced at the low processing costs typical of commodity products. Efficient, well-engineered, high-yield technology is needed to achieve this goal.

Unavoidable cost of production estimate. This analysis can be taken a step further by estimating costs felt to be unavoidable for ethanol production and determining the margin left to cover remaining costs that could be reduced through R&D. Table I summarizes the typical cost elements considered in estimating the required selling price for a product to cover all such costs as well as achieve a reasonable return on investment. These elements include the cost of feedstock as determined previously as well as costs for nutrients and other chemicals used in processing. In addition, labor costs for plant operation as well as associated direct overhead and general plant overhead expenses must be calculated. Similarly, maintenance costs and general plant overhead related to maintenance are estimated. Insurance and property tax expenses are included as well. Finally, the annualized cost of capital is calculated.

For the unavoidable cost analysis, feedstock costs are calculated as a function of yield and carbohydrate content, as discussed previously. Labor costs are determined for a 1745 tonne/d (1920 ton/d) ethanol plant that is assumed to require a total of eight operators at $29,800/yr each, an operating foreman at $34,000/yr, and an operating supervisor at $40,000/yr. These estimates are believed to be the minimum crew required to successfully operate a plant based upon operation of larger scale plants (on the order of 940 million L/yr or 250 million gal/yr) for corn ethanol production. The minimum costs possible for chemicals and nutrients is assumed as 3% of that for the feedstock. Current costs are projected to be higher than this, but as new pretreatment approaches and nutrient requirements and reuse are better defined, it may be possible to approach this level. Other costs are estimated by standard methods *(11-12)*. Maintenance costs are determined as 3% of the total fixed investment. Direct overhead is then calculated as 45% of the labor and supervision, while general plant overhead is calculated as 65% of labor and maintenance costs.

Table I Cost elements used to estimate selling price

- Feedstock
- Chemicals and nutrients
- Utilities
- Labor/Supervision
- Maintenance
- Direct overhead
- General overhead
- Insurance, property taxes
- Annual capital charge

The estimated minimum cost of nutrients and chemicals, operating labor, and direct and general plant overhead can be added to the feedstock costs as shown in Figure 9. The margin between the sum of all these costs and the revenue from either ethanol sales or ethanol plus electricity sales is available to cover the cost associated with capital recovery, including return on capital. As before, these costs are determined per volume of ethanol produced as a function of the ethanol yield from the carbohydrates for a 70% carbohydrate feedstock costing $37/tonne ($34/ton).

Allowable fixed capital investment. Having determined the minimum unavoidable costs for ethanol production, we can now determine the maximum capital investment that can be justified for this plant. Typically, the purchased capital cost is first estimated based on material and energy balances. Then the total installed capital cost is determined as a multiple of the purchased capital cost. In this case, the multiplier is taken as 2.85 based on other estimates *(12)*; although this value may seem low compared to chemical processes, it has proven reasonable for solid biomass such as corn when suspended in water at low temperatures and pressures. The fixed capital investment is calculated as 2% over the total installed capital cost. Finally, the total fixed investment is the sum of the start-up cost plus fixed capital investment, calculated as 5% above the fixed capital investment. These calculations are summarized in Table II *(11-12)*.

The capital cost can be annualized by multiplying the total capital investment by some factor, in this case 0.2 *(12)*. This factor is determined according to standard methods from the parameters in Table II. The plant is constructed over a three-year period, with 30% of the plant completed in the first year, 50% in the second year, and 20% in the third year. Furthermore, the plant is assumed to operate for 15 years with capacity at 60% of nameplate in the first year, 80% in the second year, and 100% thereafter. Straight-line depreciation is over a five-year period for equipment inside the battery limits and over a 15-year period for equipment outside the battery limits. Income tax is 37%, and no sales expenses are included in this calculation *(12)*. Typically, corn ethanol plants are built more rapidly and achieve full (and often over) capacity in less time.

Table II   Factors applied to estimate total fixed investment and annualized capital cost

- Total capital outlay for ethanol plant
    - Purchased capital cost (PCC)
    - Total installed capital cost (TIC) = 2.85 • PCC
    - Fixed capital investment (FCI) = 1.02 • TIC
    - Startup costs + FCI = Total Fixed Investment (TFI) = 1.05• FCI

- Capital cost annualized by multiplying the total capital investment by 0.20 to account for:
    - 3 years of construction with 30% 1st year, 50% 2nd year, 20% 3rd year
    - 15 years of operation

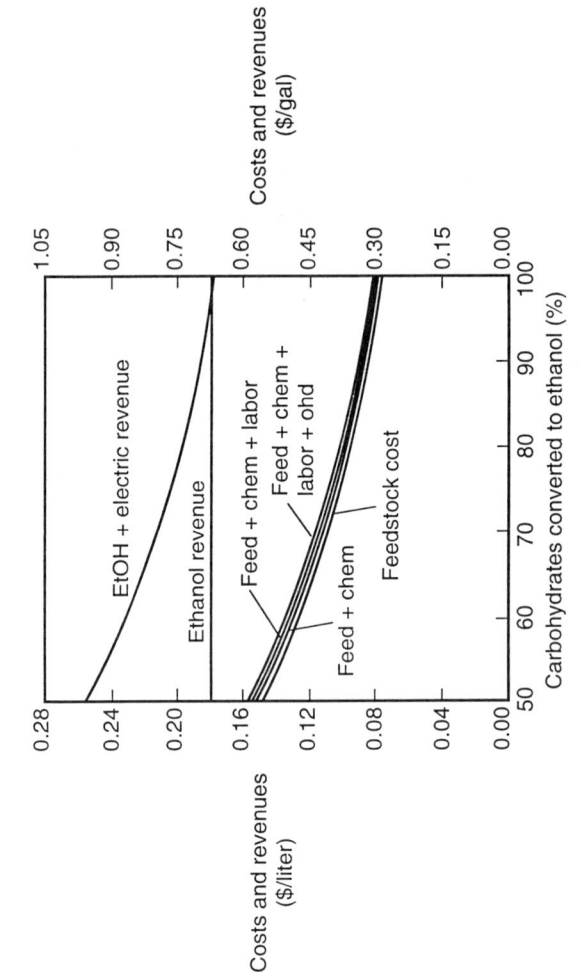

Figure 9. Revenue from sale of ethanol only and ethanol plus exported electricity as a function of the percentage of carbohydrates converted to ethanol for a feedstock containing 70% cellulose and hemicellulose and costing $37/tonne ($34/ton). Also shown are the estimated minimum costs for feedstock alone; feedstock with other chemicals; feedstocks, chemicals, and labor; and feedstock, chemicals, labor, and overhead.

- Income tax at 37%
- No sales expenses
- Capacity at 60% 1st year, 80% 2nd year, 100% thereafter.
- Straight-line depreciation over 5 years for ISBL, 15 years OSBL.

In addition to the annualized capital recovery factor, standard cost estimating methods also factor other costs from the capital investment. As mentioned previously, maintenance costs are determined at 3% of the total fixed investment. General plant overhead is then calculated as 65% of the maintenance cost. Insurance and property taxes are estimated at 1.5% of the total fixed investment.

Taking into consideration all costs estimated from the fixed capital investment, an allowable fixed capital investment (AFCI) can be calculated as:

$$AFCI = \frac{(Revenues - Costs)}{(0.20)(1.05) + 0.0678} \tag{5}$$

Figure 10 summarizes the allowable fixed capital investment based on this equation for feedstock costs of $37/tonne ($34/ton) and $46/tonne ($42/ton) as a function of the ethanol yield from the carbohydrate fraction of a biomass feedstock containing 70% carbohydrates. From this figure, a capital investment on the order of $0.25–$0.30/annual L ($0.95–$1.10/annual gal) of capital investment is allowable for a feedstock cost of $46/tonne ($42/ton). However, at a lower feedstock cost of $37/tonne ($34/ton), a capital cost on the order of $0.35/annual L ($1.33/annual gal) can be accommodated for conversion of biomass into ethanol. For comparison, it is worth noting that similar modern (dry mill) large-scale corn ethanol plants typically cost on the order of $0.29/L ($1.11/annual gal) of capacity *(17)* and the costs continue to drop as improvements are made. To achieve such low capital costs for a biomass ethanol plant certainly requires advanced technology, but these goals can be realized as improvements continue to be made to increase yields, consolidate steps, and speed rates *(see for example reference 18)*. Large-scale plants are also more likely to realize this goal.

It must be remembered that neat ethanol would likely be used for markets that can benefit from ethanol's favorable properties in reducing air pollution. In such cases, the competition would really be RFG *(8)*, and Figure 7 also includes the expected RFG price as a function of the price of oil. As before, this assumes that ethanol can achieve 80% of the range of gasoline because of its ability to achieve a 25% advantage in efficiency compared to gasoline. For this case, ethanol would be valued at about $0.21/L ($0.81/gal) to compete with RFG in such markets for an oil price of $25/bbl.

Figure 11 shows that we could afford to pay nearly $0.50/annual L of capacity ($1.90/annual gal) for an ethanol plant that could sell gasoline at a price competitive with RFG and purchase feedstock for $37/tonne ($34/ton). This allowable fixed capital investment is certainly less than for typical corn ethanol plant capital investments today. Thus, ethanol from biomass will be competitive for RFG markets if the technology is advanced for ethanol production from lignocellulosic biomass through continued R&D.

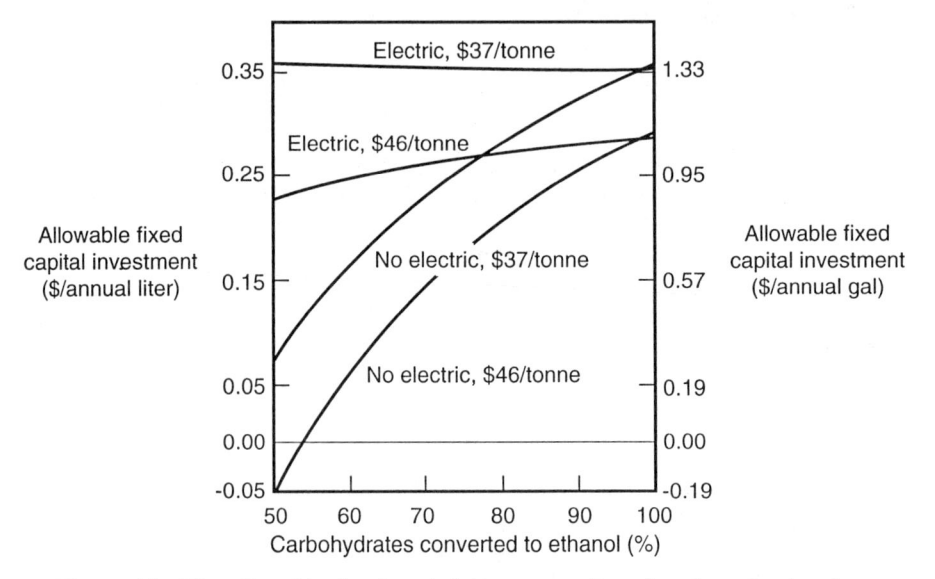

Figure 10. The allowable fixed capital investment for ethanol production from lignocellulosic biomass for varying carbohydrate conversion to ethanol for sale of ethanol only and of ethanol plus exported electricity. The feedstock is assumed to contain 70% cellulose and hemicellulose and cost either $37/tonne ($34/ton) or $46/tonne ($42/ton).

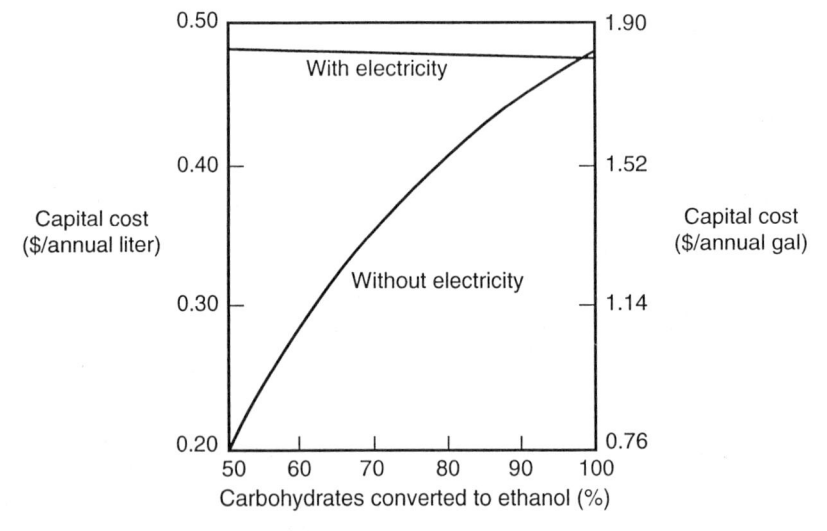

Figure 11. The allowable fixed capital investment as in Figure 10, but for an ethanol selling price of $0.21/L ($0.80/gal) and a feedstock cost of $37/tonne ($34/ton).

In closing, it is worth noting that ethanol today is primarily used for direct blending with gasoline in the United States. For such cases, ethanol typically sells in the range of $0.29/L to $0.32/L ($1.10–$1.30/gal). Such a price would allow a substantially higher capital investment in a biomass ethanol plant than shown here.

Conclusions

From this analysis, we can conclude that the production of low cost ethanol is possible from lignocellulosic biomass. Of course, high ethanol yields are vital to achieve economic viability. The latter is not surprising, and it is certainly true for any low-value, high-volume product to be competitive. It is also clear that a high carbohydrate content lowers the unit feedstock cost. Furthermore, it is important to use low-cost feedstocks for ethanol production to be competitive with conventional fuels. It is important to be sure value is obtained from the unconverted fractions if we are to achieve economic competitiveness, particularly for low ethanol yields. For example, this study assumes that unconverted fractions were burned to produce heat and electricity to power the process and to generate additional revenues from electricity exports. Greater electricity revenues result from more efficient plants, resulting in better economics; lower process energy use is possible compared to that assumed in this analysis. Other products could also be produced from the unconverted fraction if they generate appropriate revenues.

It is important to minimize all costs to maximize the margin available to cover capital recovery charges. Thus, the costs for nutrients and other chemicals must be minimized within the process. In addition, an efficient operation with a minimal operating staff is also important. The latter can also be achieved for a larger scale plant that realizes economies of scale. It must be realized that ethanol offers more than energy content, and if ethanol is valued for its ability to combat urban air pollution as a neat fuel or for direct blends in comparison to RFG, a higher fixed capital investment can still be profitable.

Overall, if high product yields, low operating costs, and reasonable coproduct markets are realized, the margin between revenues and operating costs is sufficient to provide a reasonable return on capital investment for a biomass ethanol plant similar in cost to a corn ethanol process. Through continued R&D, highly efficient, well-engineered processes should result that achieve this goal. Such cost-competitive lignocellulosic biomass-to-ethanol production would reduce oil imports, improve urban air quality, and curtail the buildup of greenhouse gases that lead to global climate change. On the other hand, this analysis suggests that an estimate of a high cost for ethanol production from biomass should not be interpreted as meaning that ethanol cannot be produced at low cost from lignocellulosic biomass but as evidence of a process design that does not achieve high product yields, low operating costs, adequate coproduct markets, and/or low capital costs.

Acknowledgements

The work reported in this paper is made possible through the support of the Ethanol from Biomass Project of the Biochemical Conversion Element of the Department of Energy Biofuels Program.

Literature Cited

1. Lynd, L.R., Cushman, J.J., Nichols, R.J., and Wyman C.E. *Science*, **1991**, *251*, 1318.
2. U.S. Department of Energy *The U.S. Petroleum Industry: Past as Prologue 1970-1992*; Energy Information Administration: Washington, DC, 1993. (DOE/E1A-0572).
3. U.S. Congress, Office of Technology Assessment *Replacing Gasoline: Alternative Fuels for Light-Duty Vehicles*; U.S. Government Printing Office: Washington, DC, 1990.
4. Sperling, D. "An Incentive-Based Transition to Alternative Transportation Fuels;" Energy and the Environment in the 21st Century, Proceedings of the Conference Held at the Massachusetts Institute of Technology: Cambridge, MA, March 26-28, 1990.
5. DeLuchi, M.A., Sperling D., and Johnson, R.A. *A Comparative Analysis of Future Transportation Fuels*; University of California at Berkeley, The Institute of Transportation Studies: Berkeley, CA, 1992. (Research Report UCB-ITS-RR-87-13).
6. Federal Register. *57 FR 4408*, February 5, 1992.
7. Intergovernment Panel on Climate Change. *Climate Change: The IPCC Scientific Assessment*; Houghton, J.T., Jenkins, G.J., and Ephraums, J.J., Eds.; Cambridge University Press: Cambridge, 1990.
8. Tyson K.S., Riley, C.J., and Humphreys, K.K. *Fuel Cycle Evaluations of Biomass-Ethanol and Reformulated Gasoline*; National Renewable Energy Laboratory: Golden, CO, 1993. (NREL/TP-463-4950).
9. Wright, J.D. *Energy Progress.* **1988**, *8(2)*, 71-78.
10. Wright, J.D. *Chem. Eng. Prog.* **1988**, *84(8)*, 62.
11. Hinman, N.D., Schell, D.J., Riley, C.J., Bergeron, P.W., and Walter, P.J. *Appl. Biochem. Biotech.* **1992**, *34/35*, 639.
12. U.S. Department of Energy *Assessment of Costs and Benefits of Flexible and Alternative Fuel Use in the U.S. Transportation Sector; Technical Report Eleven: Evaluation of a Potential Wood-to-Ethanol Process*; Office of Domestic and International Energy Policy: Washington, DC, January, 1993. (DOE/EP-0004).
13. Anderson, E. *Chemical and Engineering News* 1992, *October 12*, 8.
14. Anderson, E. *Chemical and Engineering News* 1992, *November 2*, 7.
15. Wyman, C.E., and Hinman, N.D. *Appl. Biochem. Biotech.* **1990**, *24/25*, 735.
16. Wyman, C.E., Bain, R.L., Hinman N.D., and Stevens, D.J. In *Renewable Energy: Sources for Fuels and Electricity;* Johansson, T.B., Kelly, H., Reddy, A.K.N., and Williams, R.H., Eds.; Island Press: Washington, DC, 1992, pp. 865-924.
17. Wood, P.R. *Fuel Reformulation* **1993**, July/August, 56.
18. Zhang, M., Eddy, C., Deanda, K., Finkelstein, M., and Picataggio, S. *Science,* **1995**, *267*, 240.

RECEIVED September 11, 1995

Chapter 18

Cellulose Degradation by Ruminal Microbes: Physiological and Hydrolytic Diversity Among Ruminal Cellulolytic Bacteria

Paul J. Weimer[1,2] and Christine L. Odt[1]

[1]U.S. Dairy Forage Research Center, Agricultural Research Service, U.S. Department of Agriculture, Madison, WI 53706
[2]Department of Bacteriology, University of Wisconsin, Madison, WI 53706

>Cellulose fermentation by ruminal microorganisms provides a major contribution to the nutrition of forage-fed ruminant animals. Ruminal cellulose degradation is mediated primarily by a few anaerobic bacterial species that avidly attach to forage particles in the rumen and degrade the cellulose components by cell-bound complexes of cellulolytic enzymes. The rate of cellulose degradation is largely a function of the available surface area of the cellulose. Crystallinity of the cellulose appears to be relatively unimportant in ruminal cellulose digestion, but evaluation of crystallinity effects is complicated by artifacts associated crystallinity measurements. Individual species vary considerably with respect to the attachment process and to utilization of different cellulose allomorphs, suggesting important differences in the mechanisms of cellulose digestion at the enzymatic level. These differences, combined with differences in microbial growth efficiency and product formation, provide both stability and flexibility to fiber digestion in the rumen.

The rumen is the definitive digestive organ in the ruminants, a specialized group of mammals (Order Artiodactales) that includes many familiar wild animals (e.g., deer, antelope and bison) as well as some of the more important farm animals (cattle, sheep, and goats). In essence, the rumen is a large fermentation vat in which feeds are fermented by anaerobic microorganisms to produce volatile fatty acids (used by the ruminant as oxidizable energy sources and anabolic precursors) and microbial cell mass (used by the ruminant as a protein source). The combined ruminal volume of the world's 1.1×10^9 domesticated cattle, 1.0×10^9 sheep, and 0.4×10^9 goats (*1*) is on the order of 10^{11} liters. Thus, the ruminal fermentation may be regarded as the world's largest and most important commercial fermentation process, and attempts to improve its efficiency and productivity can be viewed as a very worthwhile endeavor.

This chapter not subject to U.S. copyright
Published 1995 American Chemical Society

The primary role of ruminants in agriculture is to convert relatively recalcitrant cellulosic materials to useful endproducts (e.g., milk or meat). The role of the rumen microflora in accomplishing this process is facilitated by the relatively long (15-50 h) retention time for fibrous solids and by the many complex interactions among different microbial trophic groups (2). Ruminal digestion of plant fiber is currently thought to be limited primarily by the complex structure of the plant cell wall, particularly the physical protection provided by lignin, and by covalent linkages between lignin and/or phenolic acids and certain cell wall polysaccharides (e.g., arabinoxylans) (3). Cellulose itself appears to be rather easily digestible if the extensive matrix interactions among plant cell wall biopolymers can be disrupted (3,4). However, ruminal cellulose digestion can hardly be considered as a monolithic process, because of the great variety of ruminal microbial species known to be capable of cellulose hydrolysis. Although the taxonomic and physiological diversity among different cellulolytic groups (protozoa, fungi, and bacteria) has long been recognized, the diversity of the cellulolytic process within individual groups has received relatively little attention. Among the bacteria, generally considered to be the major cellulolytic group within the rumen (2), this diversity presents the potential for alternative mechanisms of cellulose hydrolysis and for complex ecological interactions among the different cellulolytic species. Moreover, the diversity among cellulolytic bacteria and its ecological expression have profound effects on animal nutrition.

Physiological Diversity Among the Ruminal Cellulolytic Bacteria

The ruminal cellulolytic bacteria may be functionally divided into two groups: those that adhere to plant fiber and those that do not. The nonadherent species are generally considered to be of little importance in ruminal cellulolysis, based on their relatively low numbers in vivo and their poor cellulolytic activity in vitro. Within the other group, comprising the adherent forms, three major species have been identified: *Fibrobacter succinogenes, Ruminococcus flavefaciens*, and *R. albus*. These three species share several common properties: all are strictly anaerobic (i.e., will not grow in media containing O_2 or having an elevated redox potential), do not form a protective resting stage, and are nonmotile. In addition, all three are highly specialized catabolically in that they can use only cellulose (or in some cases xylan) and its hydrolytic products as growth substrates. Our interest in these species arose from the fact that they produce different ratios of fermentation endproducts (Table I), and these differences could, in principle, have major effects on animal response. Moreover, differences in certain intrinsic properties of these species (e.g., Gram reaction) suggest structural and biochemical differences whose manipulation could provide a possible means of regulating their populations in rumino. For example, selective in rumino enhancement of *F. succinogenes* over the ruminococci might be desirable for beef cattle or for heifers (young cows prior to their first breeding and lactation). This enhancement would favor production of succinate, which in rumino is converted by other bacteria to propionate, which in turn is absorbed by the animal and used for gluconeogenesis that is essential for animal live weight gain. On the other hand, selective enhancement of the ruminococci might be desirable in lactating cows to provide the acetate needed for lipogenesis in the mammary gland, in order to produce milk with a sufficient content of fat for end uses such as cheese production.

Table I. Characteristics of adherent species of ruminal cellulolytic bacteria

Parameter	Fibrobacter succinogenes	Ruminococcus flavefaciens	Ruminococcus albus
Gram reaction	-	+	+
Growth substrates			
Cellulose	+	+	+
Cellodextrins	+	+	+
Glucose	+	-/+ [a]	-/+ [a]
Xylan	- [b]	+	+
Fermentation products [c]			
Succinate	M	m	-
Acetate	m	M	M
Formate	m	v [d]	m
Lactate	-	m/- [a]	m
Ethanol	-	-	M
Hydrogen	-	v [d]	M

[a] varies with strain
[b] xylan hydrolyzed, but hydrolytic products not utilized
[c] M = major product (>1 mol/mol anhydroglucose consumed); m = minor product (<1 mol/mol anhydroglucose consumed), (-) = not produced
[d] v = amount varies strongly with growth rate

One of the major goals of our work has been to quantitatively characterize the cellulose fermentation by these three species. In particular, we have been interested in determining the kinetics of cellulose hydrolysis (kinetic order, lag times, and rate constants) and certain fundamental growth parameters (maximum growth rate, growth yields, and maintenance coefficients). Moreover, we were interested in determining how endproduct formation varied with growth rate and with environmental factors such as pH. In the process of characterizing the physiology of these species, it became apparent that the adherent ruminal cellulolytic species displayed cellulolytic behavior that differed markedly from that of other, more intensively-studied nonruminal microbes such as the aerobic fungus *Trichoderma reesei*, or even that paradigm of bacterial cellulolysis, the thermophilic anaerobe *Clostridium thermocellum*. Further exploration of these capabilites have reinforced our view that the adherent ruminal bacteria represent a unique group of cellulolytic microbes that may be useful in answering some current questions on microbial cellulose degradation.

Diversity among Ruminal Cellulolytic Bacteria in Adherence to Cellulose

Unlike cellulolytic fungi or even anaerobic cellulolytic bacteria such as *C. thermocellum*, the ruminal cellulolytic bacteria retain virtually all of their cellulolytic enzyme systems at the cell surface throughout their incubation in pure culture. Significant rates of cellulose digestion are observed only if the cells are in physical contact with the cellulose. The typical analytical classes of cellulolytic enzymes (e.g., Avicelase, CMCase) are present in very small amounts in cell-free supernatants of batch cultures. Moreover, these enzyme activities are below detectable limits in cellulose-limited continuous cultures despite very high rates of cellulose removal and a lack of uncolonized cellulose fibers onto which free enzymes could adsorb. Localization of the cellulolytic enzymes at the bacterial cell surface presumably gives the cellulolytic microbe a competitive edge in utilizing the cellodextrin products of cellulose hydrolysis. Moreover, the cell surface localization allows the enzymes to function in a proteolytic environment so intense that 60-80 per cent of the forage protein is degraded by the rumen microflora before it can enter the abomasum for hydrolysis and subsequent intestinal absorption by the animal (5). Because the catalytic process occurs at the cell/cellulose interface, examination of the physical interaction between cell and cellulose may provide clues regarding the cellulolytic process.

R. flavefaciens normally grows as individual, spherical cells. This roughly spherical morphology is retained regardless of growth conditions, although this species can form short chains of cells during rapid growth on cellobiose (2) *R. flavefaciens* attaches to cellulose via a glycocalyx that forms a fibrous network with other cells (Figure 1). As originally reported by Kudo et al. (6), this network of adherent cells resists detachment by mechanical means, and is only poorly disrupted by treatment with methylcellulose, a soluble cellulose ether. *R. albus* also forms a glycocalyx that includes both a discrete layer ~100 nm thick surrounding individual cells, as well as an extensive network of fibers between cells (7,8).

Unlike the ruminococci, *Fibrobacter succinogenes* displays considerable morphological plasticity. Cells grown in batch culture (where growth rate is limited not by cellulose availability but by intrinsic properties of the organism's cellulolytic system) display a rather nondescript short-rod morphology (Figure 2). By contrast, when the organism is grown in cellulose-limited chemostats (where growth rate is maintained at suboptimal rates by limiting the rate at which cellulose is fed to the reactor), the cells display a flattened "fish scale" morphology that may optimize its contact with the cellulose particles (Figure 2). Examination of these cultures under the light microscope reveals that cellulose particles are normally fully colonized. However, particles prepared for scanning electron microscopy often show incomplete colonization (Figure 3). This suggests that sample preparation results in mechanical

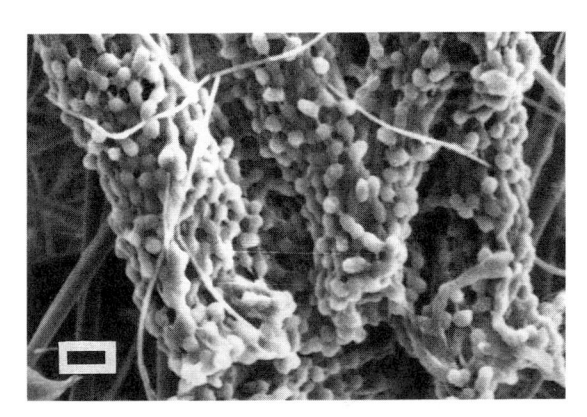

Figure 1. Scanning electron photomicrograph of *R. flavefaciens* FD-1 cells adhering to a particle of Avicel PH101 microcrystalline cellulose. Note the ramifying network of fibers connecting individual cells and resulting in nearly complete coverage of the cellulose. Samples were prepared by critical point drying and Au sputter coating, and were visualized with a JEOL JSM-35CF microscope at an accelerating voltage of 10 kV. Bar represents 2 µm.

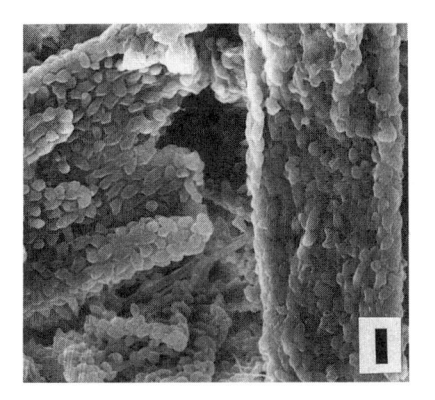

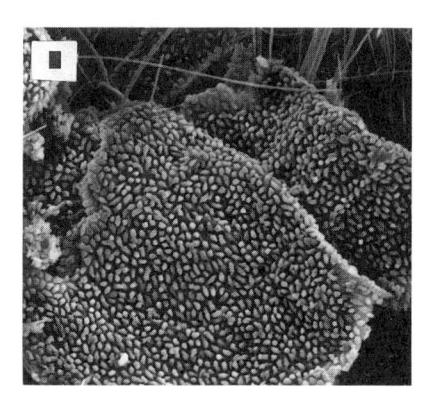

Figure 2. Scanning electron photomicrographs of *F. succinogenes* S85 grown in batch culture that contained an excess of cellulose (left panel) and in cellulose-limited continuous culture (right panel). Sigmacell 20 microcrystalline cellulose was used as the cellulose substrate. Accelerating voltage 5 kV. Bars represent 2 µm.

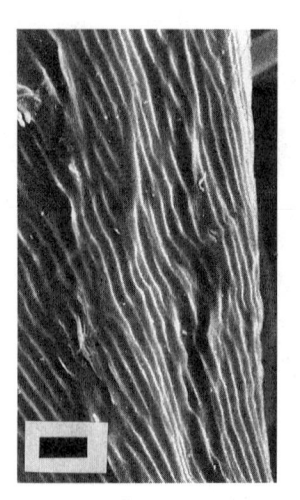

Figure 3. Scanning electron photomicrographs of *F. succinogenes* S85 and Sigma CF-1 crystalline cellulose. Accelerating voltage 5 kV. Left panel: Adherent cells. Center panel: Parallel grooves on cellulose fiber revealed by treatment of colonized fibers with methylcellulose to remove adherent cells. Right panel: Sigma CF-1 control (fiber not subjected to colonization or treatment). Bars represent 2 µm.

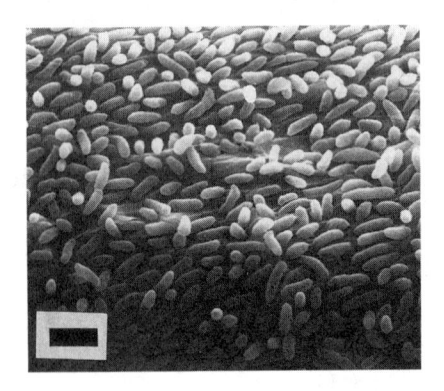

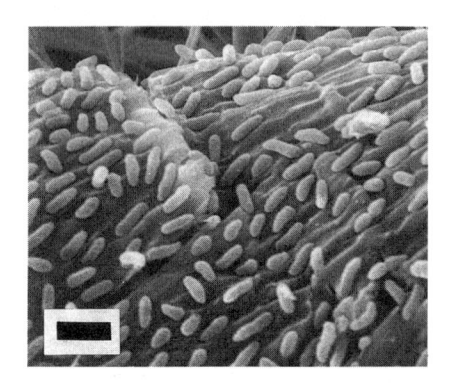

Figure 4. Scanning electron photomicrographs of *F. succinogenes* S85 attached to cellulose III_I (left panel) and cellulose IV_I (right panel). Accelerating voltage 5 kV. Bars represent 2 µm.

removal of cells from the fiber -- an interpretation in accord with the reportedly thin, tenuous (~10 nm-thick) glycocalyx produced by this species (6).

Of particular interest is the fact that cellulose digestion by *F. succinogenes* results in the formation of parallel grooves on the fiber surface whose width (~0.7μm) approximates the width of the bacterial cells (Figure 3). This phenomenon was first reported by Kudo et al. (6), who observed that treatment of the colonized particles with methylcellulose resulted in removal of attached cells; they proposed that the parallel grooves revealed upon cell removal were formed as a result of binding to, and subsequent degradation of, cellulose along a crystallographic plane.

We have examined this phenomenon of ordered binding further by growing *F. succinogenes* on celluloses having different unit cell geometries (allomorphic forms) and different degrees of crystallinity, and examining the cellulose particles by scanning electron microscopy before and after treatment with methylcellulose. Growth of this species on non-native allomorphs results in an ordered alignment of cells on the fibers (Figure 4), and the formation of these same types of parallel grooves as are observed on native crystalline cellulose. By contrast, cells grown on amorphous cellulose appear to be randomly oriented on the cellulose surface, and their removal with methylcellulose leaves an array of randomly-oriented cavities corresponding to the individual rod-shaped cells (Figure 5).

Further evidence for orientation of cells along a crystallographic plane is provided by comparison of cellulose fibers from different sources. Examination of many different cotton fibers recovered from *F. succinogenes* cultures has revealed that the angle of the grooves relative to the fiber axis is relatively constant for a particular fiber, but varies a great deal from fiber to fiber, spanning a range of from 3-46º (mean ± standard error = 22 ± 10º). By contrast, growth on the bast fibers of ramie (*Boehmeria nivea*) results in formation of grooves that are always virtually parallel to the fiber axis (Figure 6). These differences in groove angle correlate with the known variation in the angle of microfibrillar orientation relative to the longitudinal fiber axis of celluloses from these different sources (9).

The regular orientation of *F. succinogenes* cells along a crystallographic plane suggests that these organisms possess some sort of recognition factors that are themselves arrayed in an ordered fashion at the cell surface. We have not as yet obtained sufficient microscopic resolution of the *F. succinogenes* cell surface to permit visualization of any order (e.g., ordered protuberant structures) at the surface, even after treatment with the nonspecific anion-binding reagent, cationized ferritin, that is used to visualize protuberant structures on the surface of other anaerobic cellulolytic bacteria (10). However, we do know that the surface chemistry of *F. succinogenes* differs considerably from that of the ruminococci and other cellulolytic bacteria. In addition to its Gram-negative cell wall and its thin glycocalyx, *F. succinogenes* cells are not aggregated by the lectin BS-1, which aggregates numerous other species of anaerobic cellulolytic bacteria, including *R. albus* (10) and *R. flavefaciens*.

The differences in surface chemistry among different species are reflected in the different effects of physicochemical environment on the adherence of each species to

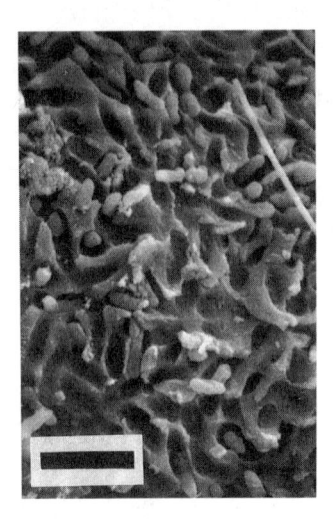

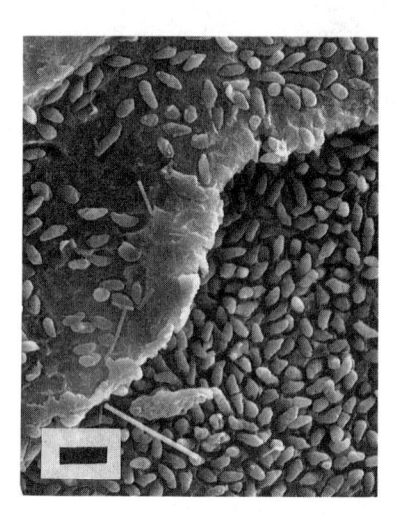

Figure 5. Scanning electron photomicrographs of *F. succinogenes* S85 attached to amorphous cellulose (left panel) and of randomly-ordered pits remaining after treatment of colonized particles with methylcellulose. Accelerating voltage 5 kV. Bars represent 2 µm.

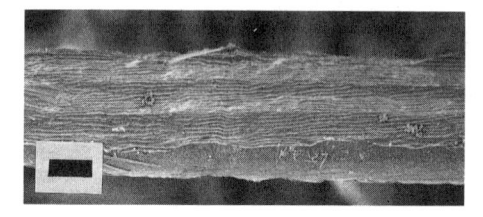

Figure 6. Scanning electron photomicrograph of a ramie fiber after colonization and partial digestion by *F. succinogenes* S85. The cells were removed from the fiber by treatment with methylcellulose. Microscopy was performed using a Hitachi S900 field-emission scanning electron microscope operated at 2 kV. Bar represents 10 µm.

cellulose. The adhesion of *F. succinogenes* is inhibited by heat, pronase, trypsin, and certain metabolic inhibitors (e.g., the ATPase inhibitor dicyclocarboxydiimide), suggesting that adherence requires intact functional surface proteins and the expenditure of metabolic energy (*11*). By contrast, adherence of the ruminococci appears to be a more passive process. Adhesion of *R. albus* is not inhibited by the presence of O_2, indicating that cells need not be viable to bind cellulose (*12*). Adhesion of *R. flavefaciens* is not inhibited by proteases or metabolic inhibitors, and may involve bridging of cells and glycocalyx material via divalent cations (*13*).

The common strategy of (and indeed, the requirement for) glycocalyx-mediated bacterial adherence presents technical difficulties in elucidating the biochemistry of ruminal cellulose digestion. Moreover, the involvement of a glycocalyx introduces the fundamental question of what occurs "under the blanket" where cellular structure contacts the surface of the cellulose fiber. The simplest scenario involves release of cellulases into a minute fluid space adjacent to the fiber surface, followed by cellulose hydrolysis and the bacterial uptake of the cellodextrin products. This scenario is teleologically attractive in that this confined space effectively concentrates the cellulases and sequesters both the cellulases (from ruminal proteases) and the cellodextrin products (from non-cellulolytic, cellodextrin-fermenting microbial competitors). Release of enzymes into this minute space would also permit migration of enzyme down the cellulose chain as hydrolysis proceeds.

On the other hand, release of free cellulases from the cell or glycocalyx surface is difficult to reconcile with several observations. Unless release of enzyme were asymmetric (i.e., occurred at only at the "fiber-facing" surface), we would expect cellulase to be excreted into the culture medium. This is not observed even in cellulose-limited chemostats that lack uncolonized cellulose fibers that could bind up freely-excreted enzymes. In the case of *F. succinogenes*, it is not clear how release of cellulases would simultaneously result in both oriented adherence of cells to cellulose and oriented degradation of cellulose relative to the microfibrillar axis.

Diversity with respect to hydrolytic capability

Early studies with *F. succinogenes* had concluded that this species was superior to other ruminal species in its ability to hydrolyze of crystalline cellulose (*14*). These studies were performed in batch culture at fixed incubation times, with no effort made to measure actual rates of degradation. However, studies with cellulose-fed continuous cultures have revealed that the rate constants for cellulose hydrolysis and the maximum specific growth rates of both *R. flavefaciens* (0.10 h^{-1})[*15*] and *R. albus* (0.08 h^{-1}) [*16*] on microcrystalline cellulose are equal to or greater than those of *F. succinogenes* (0.076 h^{-1}) [*17*]. The relative similarities in rate constants would seem to suggest that cellulose hydrolysis among the different species is similar in many respects, and that the different species have adapted to a rate of cellulose hydrolysis that may approach some sort of maximum within the constraints of an adherent cellulolytic lifestyle. Studies with cellulose-limited chemostats, in which exponential growth can be maintained at suboptimal rates, indicate that *R. albus* can maintain a rate constant for cellulose hydrolysis near 0.11 h^{-1} regardless of growth rate (*16*), while both *R. flavefaciens* and *F. succinogenes* display decreased hydrolytic rate constants with decreasing growth rate (*15,17*) Although other explanations are possible, the reduction in hydrolytic rate with decreasing growth rate may be due to a reduction of cellulase synthesis in response to diversion of an increasing proportion of energy for maintenance functions.

In contrast to their similarities in the hydrolysis of native crystalline cellulose, these species differ considerably in their ability to degrade different cellulose allomorphs (*18*). While both *F. succinogenes* and *R. flavefaciens* degrade cellulose at rates which follow the general pattern of amorphous > IV_I ~ III_I > III_{II} > I > II, some differences between the species are readily apparent. *R. flavefaciens* displays a much greater range of rates on the different forms, and is incapable of degrading cellulose II. These interspecific differences suggest the possibility of fundamental mechanistic differences in the cellulolytic system of each species. Among these allomorphs the differences in the sizes of their unit cells are very small, but differences in degree of interchain hydrogen bonding (and perhaps crystallite size) may be substantial. Thus differences in degradation rates (which were measured by weight loss) may reflect a combination of both enzymatic hydrolysis and nonenzymatic disintegration of hydrolyzed fragments. This nonenzymatic fragmentation may be an important determinant of the rate of the overall degradative process and thus warrants further investigation (*19*).

The more rapid degradation of amorphous cellulose than of crystalline cellulose has been widely observed in nonruminal systems and should thus be expected to hold in the rumen as well. However, we believe that the relationship between crystallinity and degradability has been oversimplified and is worthy of re-examination. Studies of the effect of cellulose crystallinity on its biodegradation have typically employed one of two approaches. The first has involved comparing the degradation of several different celluloses whose degree of crystallinity has been determined by one or more techniques (usually x-ray diffraction). Measurements of degradation have ranged from simple fixed-time assays to more elaborate kinetic studies. However, in almost all of these cases, other structural features (e.g., surface area or pore structure) that may impact degradation either were not measured, or have been measured under rather unrealistic conditions (e.g., surface areas measured on dried material whose fine structure has been severely altered by the drying process). A second approach has involved measuring crystallinity of the cellulose remaining after enzymatic or microbial degradation, and comparing these values with those of the same material before degradation. This approach assumes that the treatments used to recover the cellulose do not alter its crystallinity, an assumption that has not been systematically verified.

Our studies have indicated that, as expected, truly amorphous (non-reverting) celluloses are degraded relatively rapidly by ruminal bacteria. Introduction of a certain level of order (in the form of interchain hydrogen bonding to produce a crystalline lattice) decreases the rate of degradation. However, once a threshold level of crystallinity is reached (relative crystallinity index [RCI] of ~60, as measured by the x-ray diffraction method of Segal et al. [20]), the degradation rate no longer decreases with increasing crystallinity, but is instead limited by the surface area available for microbial colonization (21). There are three potential explanations for these crystallinity /degradability relationships. Firstly, this threshold value may represent an authentic average level of order at which the rate-limiting step in cellulose hydrolysis shifts from the rate of penetration into the matrix by the catalytic agent (viz. the microbial cells) to the rate of hydrolysis of surface chains on the cellulose fiber. This would seem unlikely for the ruminal bacteria, due to the large effective size of the catalytic unit (viz., the cell itself) relative to the microfibrils. Secondly, the hydrolysis of crystalline cellulose may require a prior enzymatic or nonenzymatic conversion of the crystalline region to an amorphous one, at least on a local molecular scale. Different crystallinity/degradation responses observed for different microbial species might simply reflect shifts in the relative rates of decrystallization versus hydrolysis. This hypothesis is difficult to test because of the lack of meaningful assays for any purported "decrystallase" ("hydrogen bondase") activity. Thirdly, the crystallinity measurements themselves may not accurately reflect the extent or distribution of hydrogen bonding within the fibers. This possibility can be tested by measuring relative crystallinity indices in original material versus values obtained following different treatments normally used for the recovery of cellulose from fermentation broths.

When such measurements are made on celluloses of intermediate crystallinity (e.g., Sigmacell 100), it is clear that any treatment (even resuspension in water and subsequent drying) causes substantial increases in the measured value of RCI (Table II). This apparent recrystallization is not observed with highly crystalline celluloses materials (e.g., Sigmacell 50). Recently, more detailed studies have shown that increases in RCI of cellulose can accompany simple wetting of samples without subsequent re-drying (Weimer, P.J; Hackney, J.M.; French, A.D., *Biotechnol. Bioeng.*, in press). Thus, claims that amorphous regions of cellulose are preferentially removed during microbial degradation must be evaluated cautiously if the sole evidence for such claims is an increased RCI of cellulose recovered from microbial cultures. This is particularly true for the ruminal bacteria, whose cellulases are usually localized at the cell surface and thus cannot readily penetrate even the most porous of the amorphous regions of cellulose fibers.

Concluding remarks

The differences in fermentation endproducts among different species of adherent ruminal cellulolytic bacteria has given impetus to a closer examination of the differences in the cellulolytic behavior of these organisms. It appears that the overwhelming localization of the cellulolytic enzymes at the cell surface has fostered a different relationship between cellulose structure and cellulose degradation than is observed in some other groups of cellulolytic microbes. Because individual cellulolytic proteins from several of these ruminal microbial species show strong

Table II. Effect of treatments on measured values of relative crystallinity index (RCI) of celluloses

Treatment [a]	Relative Crystallinity Index [b]	
	Sigmacell 50	Sigmacell 100
None [c]	94.9	78.9
Rewetting [d]	94.4	90.9
Alkaline hydrogen peroxide (AHP) [e]	91.9	85.1
Neutral detergent (ND) [f]	95.2	90.0
Acid detergent (AD) [f]	91.6	89.2
ND/AD/AHP/Acid chlorite [g]	93.5	93.5

[a] Except where indicated, all treatments were followed by lyophilization
[b] Determined by x-ray diffraction
[c] Untreated material as received from the vendor (Sigma, St. Louis, MO); not lyophilized
[d] Resuspended in water prior to lyophilization
[e] 5% H_2O_2, pH 11.5, room temperature, 6 h
[f] Methods of Goering and Van Soest (22)
[g] Sequential treatments with rinsing between each; acid chlorite treatment at 70 C for 6 h.

homology to those of nonruminal species, it is probable that the basic hydrolytic mechanisms will also show strong similarities. However, the localization and orientation of individual components of the ruminal cellulolytic systems -- as evidenced by the degradation of cellulose fibers in a uniquely oriented fashion by *F. succinogenes* -- might well differ from those of nonruminal bacteria, and represent an interesting alternative experimental system. Moreover, comparison of the properties of the cellulosome-type structures from different ruminal species may provide an explanation of the different surface properties of these species, as well as the differences in their ability to hydrolyze different cellulose allomorphs.

Many of the recent advances in our understanding of cellulolysis by both the aerobic fungi and *Clostridium thermocellum* have come from genetic studies, primarily through cloning of genes into *Escherichia coli*. Similar approaches have begun to yield information for the ruminal cellulolytic bacteria as well (23). However, the complexity of the ruminal cellulolytic system, particularly with regard to the essential role of cell adherence, suggests that an understanding of ruminal cellulolysis would be facilitated by direct genetic studies with these ruminal species. Unfortunately, development of a genetic technology for the predominant ruminal cellulolytic species remains an elusive goal. The development of a genetic system for these organisms would permit the integration of physiological and biochemical studies to help elucidate the cellulolytic mechanisms of this interesting microbial group.

Acknowledgments

We thank A.D. French and C. McCombs for the x-ray diffraction measurements. Fig. 6 was prepared by Y. Chen through the Integrated Microscopy Resource at the University of Wisconsin-Madison, funded by NIH Biomedical Technology Grant RR 00570.

Literature Cited

1. United States Department of Agriculture, *Agricultural Statistics - 1992*, U.S. Government Printing Office, Washington, DC, 1992.
2. Hungate, R.E. *The Rumen and its Microbes;* Academic Press, New York, NY, 1966.
3. Jung, H.G; Deetz, D.A., In *Forage Cell Wall Structure and Digestibility;* Jung, H.G.; Buxton, D.R.; Hatfield, R.D.; Ralph, J., Eds., Amer. Soc. Agron., Madison, WI, 1993, p.315-346.
4. Van Soest, P.J., *Fed. Proc.* **1973**, *32*, 1804-1808.
5. National Research Council. *Nutrient Requirements for Dairy Cattle;* 6th revised ed., National Academy Press, Washington, DC, 1988.
6. Kudo, H.; Cheng, K.-J.; Costerton, J.W. *Can. J. Microbiol.* **1987**, *33*, 267-271.
7. Cheng, K.-J.; Akin, D.E.; Costerton, J.W. *Fed. Proc.* **1977**, *36*, 193-197.
8. Patterson, H., Irvin, J., Costerton, J.W., Cheng, K.-J. *J. Bacteriol.* **1975**, *122*, 278-287.
9. Radhakrishnan, T.; Patil, N.B. *Text. Res. J.* **1968**, *38*, 209-210.
10. Lamed, R.; Naimark, J.; Morgenstern, E.; Bayer, E.A. *J. Bacteriol.* **1987**, *169*, 3792-3800.
11. Gong, J; Forsberg, C.W. *Appl. Environ. Microbiol.* **1989**, *55*, 3039-3044.
12. Morris, E.J.; Cole, O.J. *J. Gen. Microbiol.* **1987**, *133*, 1023-1032.
13. Roger, V.; Fonty, G.; Bony, S.K.; Gouet, P. *Appl. Environ. Microbiol.* **1990**, *56*, 3081-3087.
14. Halliwell, G.; Bryant, M.P. *J. Gen. Microbiol.* **1963**, *32*, 441-448.
15. Shi, Y.; Weimer, P.J. *Appl. Environ. Microbiol.* **1992**, *58*, 2583-2591.
16. Pavlostathis, S.G.; Miller, T.L.; Wolin, M.J. *Appl. Environ. Microbiol.* **1988**, *54*, 2660-2663.
17. Weimer, P.J. *Arch. Microbiol.* **1993**, *160*, 288-294.
18. Weimer, P.J.; French, A.D.; Calamari, T.A., Jr. *Appl. Environ. Microbiol.* **1991**, *57*, 3101-3108.
19. Walker, L.P.; Wilson, D.B.; Irwin, D.C. *Enz. Microb. Technol.* **1990**, *12*, 378-386.
20. Segal, L.; Creely, J.J.; Martin, A.E., Jr.; Conrad, C.M. *Text. Res. J.* **1959**, *29*, 786-794.
21. Weimer, P.J.; Lopez-Guisa, J.M.; French, A.D. *Appl. Environ. Microbiol.* **1990**, *56*, 2421-2429.

22. Goering, H.K.; Van Soest, P.J. *Forage Fiber Analysis;* Agriculture Handbook No. 379,United States Department of Agriculture, Washington, DC, 1970.
23. White, B.A.; Mackie, R.I.; Doerner, K.C., In *Forage Cell Wall Structure and Digestibility;* Jung, H.G.; Buxton, D.R.; Hatfield, R.D.; Ralph, J., Eds., Amer. Soc. Agron., Madison, WI, 1993, p.455-484.

RECEIVED August 17, 1995

Chapter 19

Induction of Xylanase and Cellulase in *Schizophyllum commune*

Dietmar Haltrich[1], Brigitte Sebesta[2], and Walter Steiner[2]

[1]Division of Biochemial Engineering, Institute of Food Technology, University of Agriculture, Peter-Jordan-Strasse 82, A−1190 Vienna, Austria
[2]Institute of Biotechnology, SFB Biokatalyse, University of Technology Graz, Petersgasse 12, A−8010 Graz, Austria

Various mono-, oligo-, and polysaccharides were screened for their ability to induce the synthesis of xylanase and cellulase (endoglucanase) activity in the basidiomycete *Schizophyllum commune*. Constitutive synthesis of low activities was observed on easily metabolizable compounds or even in the absence of an added inducer. Formation of both xylanase and cellulase was induced by cellulose and cellulose-rich substrates, of which bacterial cellulose was by far the best inducer, as well as by the disaccharide cellobiose. Furthermore, compounds structurally related to the latter, including sophorose, lactose, and 4-*O*-β-galactopyranosyl-D-mannopyranose, which was identified as a new inducer of xylanase and cellulase, provoked the synthesis of both enzymes significantly. Xylan isolated from birchwood, xylose, and β-methyl-D-xyloside, a structural analogue of xylobiose, only resulted in the constitutive enzyme levels formed. These results indicate that the synthesis of xylanase and cellulase is likely to be under common regulatory control in *S. commune*. Two of the compounds found to provoke production of xylanase, lactose and cellulose, were used as substrates for laboratory fermentations.

D-Xylans are the most abundant noncellulosic polysaccharides in hardwood and annual plants, where they account for 20 – 35% of the total dry weight. In softwoods they are found in lesser quantities, comprising approximately 8% of the tissue dry weight. The basic structure of xylans is a main chain of (1→4)-linked β-D-xylopyranosyl residues. Typically, these linear chains carry short side chains to a varying extent, whereas pure, unsubstituted xylans are extremely rare. The chemical composition of the substituent side groups and the frequency of their occurrence depend on the origin of the xylan as well as on the extraction procedures used for its isolation (*1*).

Due to the heterogeneity of the xylans, xylanolytic enzyme systems include several hydrolytic enzymes, the best known of these are endo-β-1,4-xylanases, which randomly attack the main chain of xylans, and β-xylosidases, which hydrolyze xylo-oligosaccharides to D-xylose. In addition to these, several accessory enzyme activities are necessary for debranching the variously substituted xylans (2).

The basidiomycete *Schizophyllum commune* is a common organism that grows under extremely variable conditions throughout the temperate and tropic zones (3). It is mostly found on the wood of deciduous trees, only rarely on coniferous wood, and occasionally on the roots and stems of herbaceous plants.

As a wood rotting fungus *S. commune* produces a number of lignocellulolytic enzymes, some of which are well characterized. *S. commune* is a particularly good producer of xylanases. Xylanase A, one of several extracellular xylanases formed, has been purified and characterized. Under appropriate culture conditions it is the predominantly excreted enzyme by *S. commune*. It has a reported molecular weight of approximately 22,000, an isoelectric point of 4.0–4.5, and a pH-optimum of 5.0 (4–6). Its specific activity is remarkably high and has been found to be in the range of 1.2 – 1.5·10^3 IU/mg (4,7). Purified xylanase A showed no appreciable cellulase, β-glucosidase, or β-xylosidase activity. It had no detectable activity on xylobiose and only low activity on xylotriose or xylotetraose, while it rapidly cleaved higher xylo-oligosaccharides (8). After an extended hydrolysis of larchwood xylan it yielded mainly xylose and xylobiose in an approximate ratio of 1 : 3 (9). The total amino acid sequence of xylanase A has been reported recently (6).

In addition to xylanase, auxiliary enzymes necessary for the complete hydrolysis of substituted xylans have been identified in *S. commune*. β-Xylosidase activity as part of the xylanolytic system is concurrently produced with xylanase (10) as is α-arabinosidase activity (10,11). Both enzyme activities have not been characterized in detail. α-Glucuronidase also consists in multiple forms in *S. commune* and its major form has been isolated (11). Unlike α-glucuronidases from other fungal sources, this enzyme is active on high molecular weight glucuronoxylans.

Several esterases have been reported to be produced in *S. commune*. These include acetyl xylan esterase (12,13) and ferulic acid esterase (11,14). Both enzymes have been partially purified and both enzymes show cooperativity with xylanase in the enzymatic degradation of substituted xylans (14,15). Recently an esterase liberating acetyl side groups from native softwood galactoglucomannan has been reported to be produced by *S. commune* (16).

In addition to an apparently complete xylanolytic enzyme system, *S. commune* produces cellulases as well as mannanases. The cellulases are excreted in multiple forms and include endoglucanases, cellobiohydrolases, and β-glucosidases (17–19). The formation of both xylanase and cellulase by *S. commune* has been investigated in some detail and several growth media have been developed pertaining to increased enzyme production (10,20). It is interesting to note that for the enhanced production of these two groups of enzymes cellulose or cellulose-rich substrates have been used.

The purpose of the present study was to investigate the induction of xylanase and endoglucanase in *S. commune*. Special attention was paid to the question, whether the synthesis of these two types of glycanases is under common or separate control, that is, whether both enzyme activities can be induced by the same low molecular weight compounds.

Experimental

Inducers and Chemicals. Bacterial cellulose was obtained from standing cultures of *Acetobacter pasteurianus* DSM 2004 (Deutsche Sammlung von Mikroorganismen, Braunschweig, FRG) grown according to Hestrin and Schramm (*21*). All other chemicals were analytical grade reagents purchased from Merck (Darmstadt, FRG) or obtained as follows: 4-*O*-β-galactopyranosyl-D-mannopyranose (β-1,4-gal*p*-man*p*), lactobionic acid, lactulose, D-lyxose, β-methyl-D-xylopyranoside, pectin from apple, and D-ribose were from Sigma (St. Louis, MO, USA); cellobiose, L-sorbose, and xylan from oat spelts were from Fluka (Buchs, Switzerland); xylan from beechwood was from Lenzing AG (Lenzing, Austria); sophorose was from Roth (Karlsruhe, FRG); xylan from birchwood was from Fluka and from Roth; carboxymethylcellulose was from Hercule France S.A. (Rueil Malmaison, France).

Microbial Strain, Mycelial Preparation, and Induction Experiments. A wild strain of *Schizophyllum commune* was used throughout this work. It was isolated from decomposing jute in Bangladesh and has been identified by the Centraalbureau voor Schimmelcultures (CBS, Baarn, The Netherlands). It was maintained on glucose-maltose-Sabouraud agar with subculturing every 4 to 6 weeks.

Mycelial biomass was produced in the following medium containing: glucose, 1.0% (w/v); yeast extract, 1.0%; NH_4NO_3, 0.14%; KH_2PO_4, 0.13%; $MgSO_4 \cdot 7H_2O$, 0.1%. The pH-value was adjusted to 7.0 prior to sterilization; tap water was used for media preparation. This medium without a carbon source and yeast extract is referred to as basal medium. Each flask was inoculated by adding a piece (approx. 1 cm^2) of the actively growing part of a 7–8-day-old fungal colony on Sabouraud agar. For obtaining a homogeneous preparation of the mycelium, media were homogenized with a laboratory homogenizer (Ultra-Turrax T25, Janke & Kunkel, Staufen i.Br., FRG) at 9,500 rpm for 15 sec after inoculation. The inoculated flasks were continuously shaken on an orbital shaker at 150 rpm, temperature and relative humidity were maintained at 30 °C and 60%, respectively. The mycelium of *S. commune* was harvested in the late exponential phase of growth by filtering through nylon cloth and successively washed with cold basal medium. Then it was suspended in the basal medium, to which yeast extract (0.1% w/v) and the respective inducers (2.4–9.0 mM or 0.06–0.30% w/v for polysaccharides) were added, so that the concentration of biomass was 1.0–1.5 mg/mL (dry weight). The flasks were incubated at 30 °C under continuous agitation (150 rpm) for various lengths of time. Blanks were prepared in the same way, except that no inducer was added.

Bioprocess Experiments. Fermentations were carried out in 12-L laboratory fermenters (Braun Biostat-V, B. Braun, Melsungen, FRG) with working volumes of 9–10 L, each equipped with three Rushton turbine impellers (each with six flat blades) or with a draft tube and two marine propellers. Polypropylenglycol P2000 (Fluka) was used as antifoam agent. The temperature was constant at 30 °C; the pH-value was not controlled. Agitation was gradually increased from initially 250 rpm to 550 rpm to ensure good mixing, even when the rheological behavior of the fermentation broth changed drastically during the course of the cultivation. Culture media employed included the basal medium supplemented with 5.5% lactose and 5.0% yeast extract or a medium optimized for xylanase production (*10*) containing 7.3% cellulose and 5.5% yeast extract added to the basal medium.

Enzyme Activity Assays. All enzyme activity assays were carried out in 50 mM sodium citrate buffer, pH 5.0. *Xylanase activity* (1,4-β-D-xylan xylanohydrolase, EC 3.2.1.8) was assayed using a 1% (w/v) solution of birchwood xylan (Roth) as a substrate (*22*). *Cellulase activity* (endo-1,4-β-glucanase, CMCase, EC 3.2.1.4) was determined according to IUPAC recommendations (*23*) employing carboxymethylcellulose as a substrate. The release of reducing sugars was measured using the dinitrosalicylic acid method. One unit of enzyme activity is defined as 1 µmol of xylose or glucose equivalents produced per minute under the given conditions.

β-Galactosidase activity (β-D-galactoside galactohydrolase, EC 3.2.1.23) was measured according to a modified method of Herr et al. (*24*) using a 0.4% (w/v) solution of *p*-nitrophenyl-β-D-galactopyranoside in citrate buffer as the substrate. Activities were expressed on the basis of liberation of *p*-nitrophenol.

Other Analyses. The carbohydrate content of lignocellulosic substrates was determined by sulfuric acid hydrolysis as described by Wilke et al. (*25*). The monomeric sugars thus obtained were then determined by high-pressure liquid chromatography (HPLC) using an HPX-87H column (Bio-Rad, Richmond, CA, USA).

Sugar concentrations in the fermentation samples were assayed enzymatically. Glucose was determined using the hexokinase method (Dipro, Wiener Neudorf, Austria); lactose and galactose were measured using a combination test kit (Boehringer Mannheim, Vienna, Austria).

Results

In order to determine the effect of the nature of a certain compound on the synthesis of xylanase and endoglucanase activities by *Schizophyllum commune*, glucose-grown mycelium was incubated in the basal medium supplemented with various mono-, oligo-, and polysaccharides. In these experiments several polymers that are present in lignocellulosic material were tested as were oligosaccharides, which are derived from these glycans by the action of the respective glycan-degrading enzymes, or compounds structurally related to these oligosaccharides. Some carbohydrates, e.g. arabinose or glucuronic acid, were also included since these are found as substituent side groups in naturally occurring xylans and might have an inducing effect on the synthesis of xylanolytic enzymes. In addition to a control medium without a carbon source added, several easily metabolizable aldoses and polyols were also included as blanks. Figure 1 shows, that most of the monosaccharides tested, such as glucose, xylose or arabinose as well as the polyols xylitol or glycerol, did not induce xylanase formation when compared to the constitutive level of xylanase activity obtained for the control medium which did not contain a carbon source.

The various cellulosic substrates employed as inducers (carboxymethylcellulose, Avicel, bacterial cellulose) all showed a strong inducing effect as did several disaccharides (cellobiose, 4-*O*-β-galactopyranosyl-D-mannopyranose, lactose, sophorose). Bacterial cellulose was used in these induction experiments since it does not contain xylan (*26*). This is in contrast to commercially available cellulose, which is purified

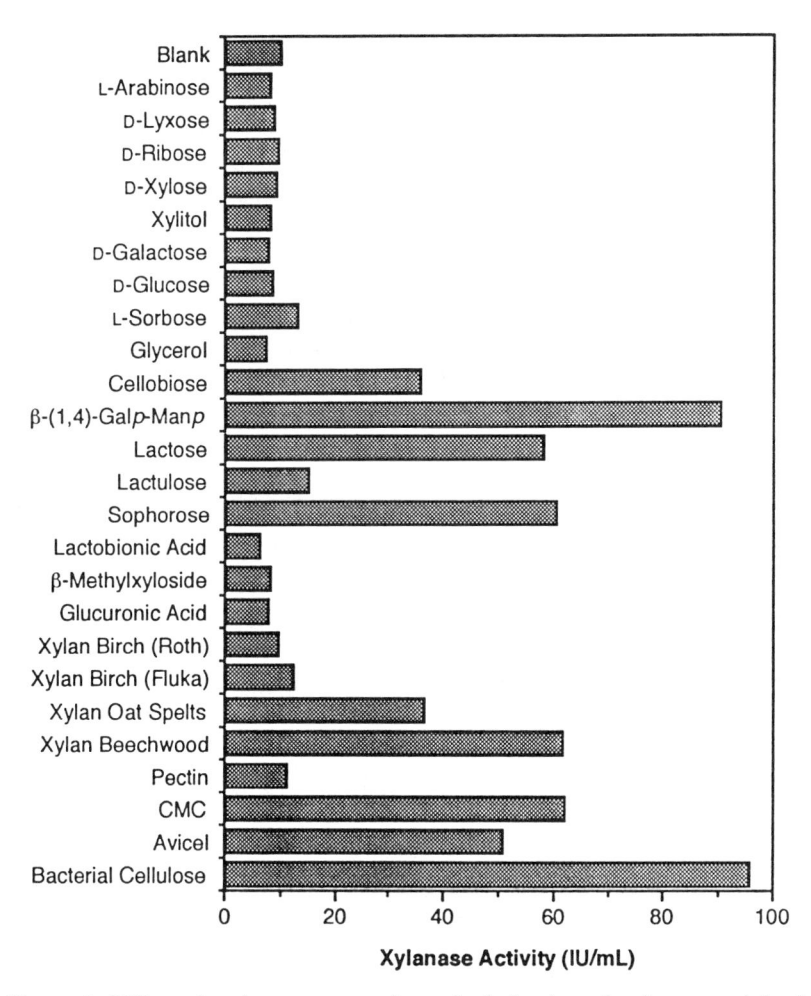

Figure 1. Effect of various compounds on the induction of xylanase activity in washed, glucose-grown mycelium of *Schizophyllum commune* after 25 h of incubation. Concentration of mycelium was 1.5 mg/mL (dry weight). Concentration of carbohydrates was 6.0 mM except for sophorose (4.5 mM) and 4-O-β-galactopyranosyl-D-mannopyranose (β-1,4-gal$p$-man$p$; 2.4 mM). Concentration of polysaccharides was 1.9 mg/mL.

from lignocellulose and which contains appreciable amounts of xylan as a contaminant (*27*). This small amount of xylan present in the growth medium can induce the formation of xylanase, a fact that has been proven for *Trichoderma* strains (*27,28*).

Several xylan preparations were used as possible inducers for xylanase in *S. commune*. Interestingly, xylan from birchwood, which is a soluble, deacetylated glucuronoxylan, did not induce the synthesis of xylanase and resulted in the formation of only the constitutive levels of this enzyme activity (9.5 and 12.3 IU/mL). On the contrary, an insoluble, unsubstituted xylan from beechwood proved to be a good inducer of xylanase activity (61.8 IU/mL) as did xylan from oat spelts, which resulted in the synthesis of intermediate levels of xylanase (36.3 IU/mL). This difference presumably results from the greatly varying cellulose content of these different preparations employed. Whereas xylan from birchwood is relatively pure and contains less than 0.5% glucan (*29*), the xylan from beechwood used in this study contained approximately 7% glucan. Xylan from oat spelts also comprises appreciable amounts of glucan (*27*). β-Methyl-D-xyloside, a nonmetabolizable structural analogue of xylobiose known to induce xylanase in both yeasts and fungi (*30,31*), did not induce the synthesis of xylanase in *S. commune*. These results obtained in the induction experiments are in good agreement with the xylanase production reported for 7-day incubations (*32*).

Cellulase activity is induced by the same compounds that also result in a significant increase of xylanase (Figure 2). The different cellulosic substrates used as inducers showed a very strong inducing effect as did the glucan-rich xylan preparation from beechwood and several disaccharides such as cellobiose or compounds with related structures including lactose or β-1,4-gal*p*-man*p*, whereas all other carbohydrates tested gave only very low levels of cellulase. Bacterial cellulose was by far the best inducer for both xylanase and cellulase activities.

The time course of the induction of xylanase by lactose and Avicel, each at a concentration of 3.0 mg/mL, is shown in Figure 3. A control experiment contained glucose which was shown to result only in constitutive or even repressed enzyme levels formed. After an induction period of approximately 12 h xylanase activity could be detected in the culture filtrate. The formation of xylanase leveled off after approximately 24 h with the depletion of the more easily metabolizable inducer lactose, whereas it continued to increase on Avicel throughout the incubation period of 76 h. The reason for this effect is believed to be due to the prolonged availability of inducing molecules (presumably cellobiose or another oligosaccharide derived from cellulose by the action of cellulolytic enzymes) which are slowly liberated and thus resulted in an extended presence of the inducer. A similar mechanism was also suggested to explain the induction of xylanase in the yeast *Cryptococcus albidus* by xylan-derived xylobiose (*33*).

The effect of various concentrations of lactose and Avicel on the induction of xylanase is listed in Table I. The inducing effect increased with rising concentrations of the inducers. This effect was especially pronounced with elevated concentrations of Avicel where a 2.5-fold increase in the concentration (from 1.2 to 3.0 mg/mL) led to an almost 5-fold increase in xylanase activity formed during an incubation period of 76 h.

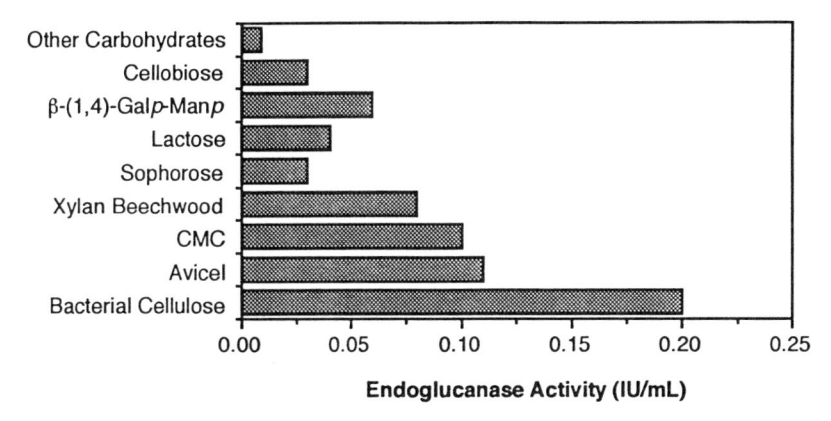

Figure 2. Effect of various compounds on the induction of cellulase activity in *Schizophyllum commune* after an incubation period of 25 h. Cellulase activity was assayed using carboxymethylcellulose as the substrate. Concentrations of cells as well as of the respective inducers were as described in Figure 1. The average of all other carbohydrates tested is shown in one bar.

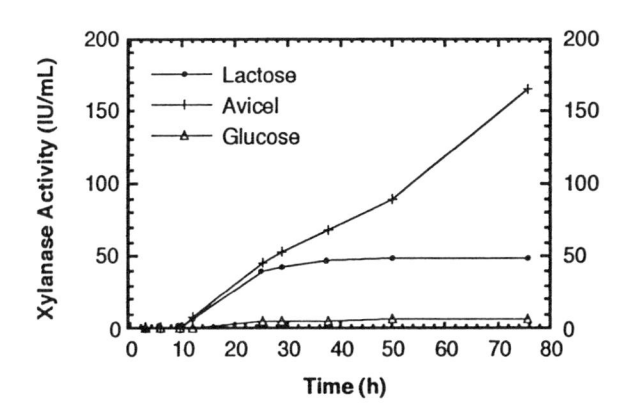

Figure 3. Time course of the xylanase synthesis in washed, glucose-grown mycelium of *Schizophyllum commune* using different inducers. Concentrations of mycelium was 1.0 mg/mL (dry weight). Concentration of carbohydrates was 3 mg/mL.

Table I. Effect of various concentrations of lactose and Avicel on the induction of xylanase activity in *S. commune*[a]

Inducer	Xylanase Activity (IU/mL)				
	6 h	12 h	24 h	37.5 h	76 h
Glucose, 1.2 mg/mL	0.5	0.8	4.8	4.8	5.1
Lactose, 0.6 mg/mL	0.6	1.8	8.8	10.0	10.3
Lactose, 1.2 mg/mL	0.7	2.4	20.3	25.3	26.7
Lactose, 3.0 mg/mL	0.8	6.4	39.7	47.1	48.9
Avicel, 0.6 mg/mL	0.6	1.8	7.1	11.3	17.5
Avicel, 1.2 mg/mL	0.6	2.4	12.9	21.7	34.0
Avicel, 3.0 mg/mL	0.6	2.4	45.3	68.8	165.0

[a] Concentration of mycelium was 1.0 mg/mL (dry weight)

To further assess the results obtained in the induction experiments, laboratory fermentations of *S. commune* employing lactose and cellulose as growth substrates were performed. The time course of a cultivation using 5.5% lactose as a carbon source is shown in Figure 4. The inoculum (8.5% v/v) was a 6-day-old shake flask preculture (cell dry weight 11.8 g/L) grown on the fermentation medium. Concentrations of lactose, glucose, and galactose as well as the activities for xylanase and β-galactosidase were followed during the course of the cultivation. Xylanase activity of approximately 230 IU/mL was reached after 135 h of growth (Figure 4), corresponding to a volumetric productivity of 1,700 IU/L·h. The release of glucose and galactose from lactose, due to the action of β-galactosidase (Figure 4), may limit xylanase production because neither aldose induced xylanase synthesis in the induction experiments. By applying an appropriate strategy of feeding the substrate lactose, such as in a fed-batch or continuous cultivation, much higher xylanase yields can be expected in accordance to results obtained for the production of cellulases by the fungus *Trichoderma reesei* employing lactose as a substrate (34).

In comparison to a cultivation employing the soluble substrate lactose, the time course of a cultivation on a medium containing cellulose (Avicel, 7.3%) as a substrate is shown in Figure 5. A 9-day-old preculture grown on the same medium was used as an inoculum for this fermentation (10% v/v). After 108.5 h of growth, fresh medium (1.0 L) was added to the fermentation to make up for losses caused by sample removal. This addition was necessary to maintain axial flow around the draft tube. The aeration rate was gradually increased during the first 70 h and then held constant at 0.4 vvm during the remainder of the fermentation period. A higher aeration rate was not possible, since it resulted in air-shrouding of the impellers. The maximum value for xylanase activity of approximately 4,800 IU/mL was obtained after 220 h, corresponding to a productivity of 22,000 IU/L·h. When cellulose was employed as a substrate, the pH-value started to rise significantly to a final value of 7.15 after an

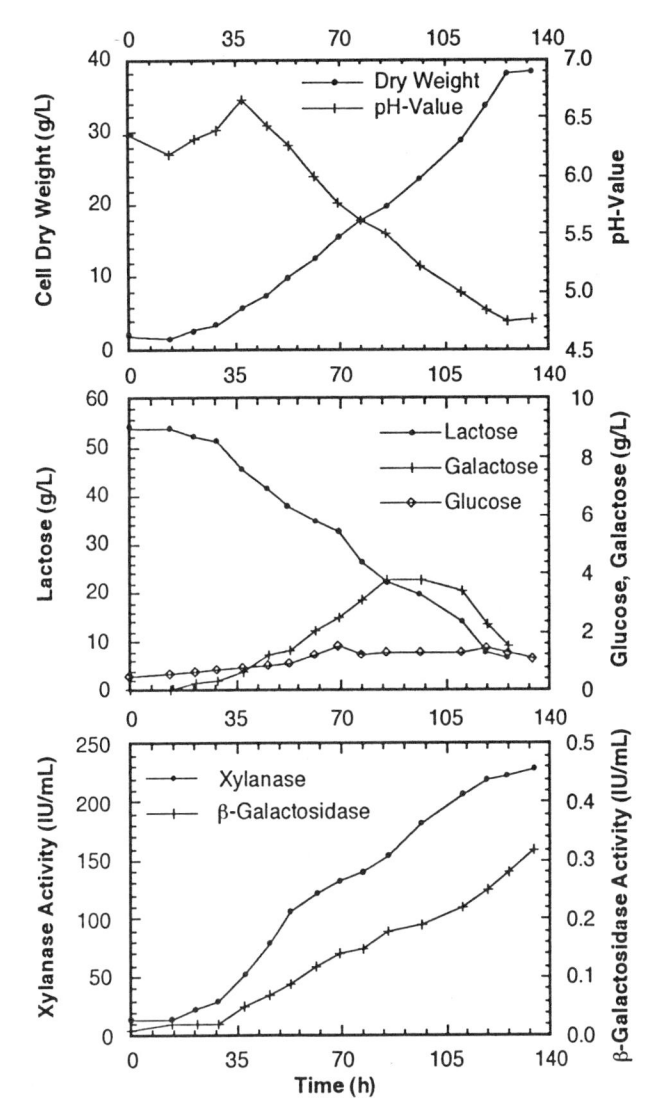

Figure 4. Time course of a batch cultivation of *Schizophyllum commune* in a 12-L laboratory fermenter (9 L working volume) on a medium containing 5.5% lactose as a carbon source. The pH-value was not controlled.

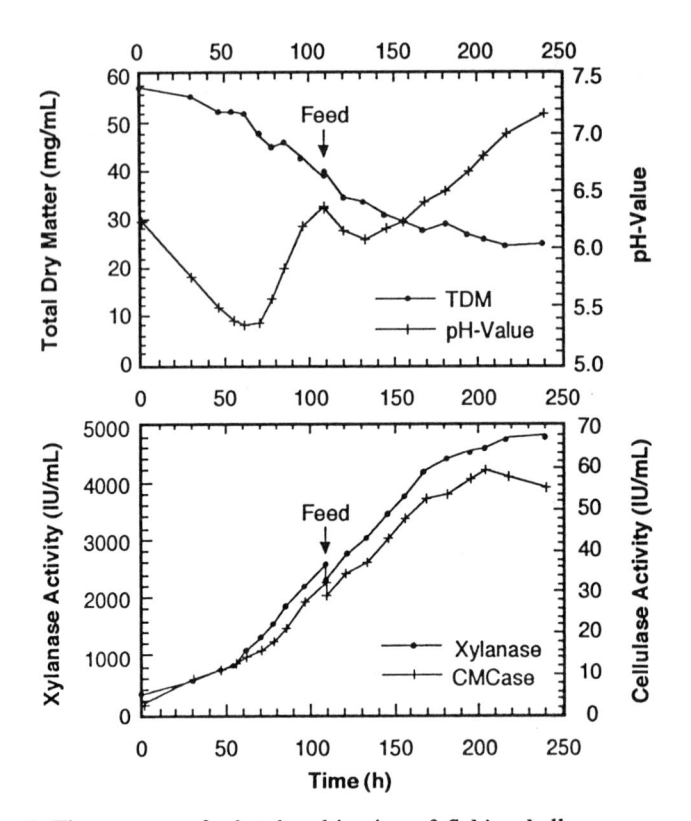

Figure 5. Time course of a batch cultivation of *Schizophyllum commune* on a medium containing 7.3% cellulose as a substrate. The cultivation was carried out in a 12-L laboratory fermenter (9.5 L working volume) equipped with a draft tube and two axial propellers. Fresh medium was added after 108.5 h to compensate for losses caused by the taking of samples, thus maintaining the axial flow around the draft tube. The pH-value was not controlled. Aeration was gradually increased to 0.4 vvm, agitation was increased from initially 250 rev/min to 550 rev/min (Adapted from ref. *10*).

initial decrease; this short sharp decrease followed by a constant increase was also observed when fresh substrate was added to the cultivation. Contrarily, the pH-value decreased during the cultivation on lactose from an initial value of 6.35 to reach a value of 4.75.

Discussion

In *Schizophyllum commune* formation of extracellular xylanase and cellulase activities is inducible as can be concluded from both growth and induction experiments. For the cellulase biosynthesis in this organism it has been demonstrated that, at least for an initial phase, induction occurs at the level of transcription (*18*). The low activity of both enzymes formed under conditions of starvation or in the presence of slowly utilizable carbon sources can be regarded as a basal, constitutively formed level of enzyme activity. Growth on cellulose or cellulose-rich substrates greatly stimulated the formation of both enzymes. Interestingly, cellulose induced not only synthesis of cellulase but also of xylanase activity. A strict connection of xylanase and cellulase production in *S. commune* was also observed for a number of other carbohydrates used as inducers or substrates, such as cellobiose, lactose, sophorose, or L-sorbose (*13,32*).

The formation of xylanases along with cellulases on media containing cellulose as the sole carbon source has been described for several *Trichoderma* spp. (*13,28,35,36*); for *Aspergillus* spp. (*37,38*); for *Sclerotium rolfsii (39)*, *Thermoascus aurantiacus (40)*, *Talaromyces emersonii (41)*, *Thielavia terrestris (42)*, *Penicillium pinophilum (43)*, and *Fusarium oxysporum (44)*. However, there are distinct differences between these strains and *S. commune*. Most of these strains yielded higher levels of xylanase when grown on xylan than when cultivated on cellulose at a similar concentration. However, this was not found in all of these organisms (*35*). Often xylan as a substrate specifically induced production of xylanase with no or only very low cellulase activities (*13,28,36,37*), while on the contrary cellulose induced both xylanase and cellulase activities. For *T. reesei* QM 9414 and *T. harzianum* this production of xylanase when grown on cellulose was shown to be caused by xylan which is present as an impurity in commercially available cellulose (*27,28,36*). When these strains were grown on cellulose substrates with a reduced xylan content, xylanase production decreased. It could be shown that this remnant xylan still present in Avicel does not induce the formation of xylanase by *S. commune,* since Avicel which was treated with xylanase followed by an alkaline extraction induced both higher xylanase as well as cellulase activities than untreated Avicel when employed as a substrate (*32*). Furthermore, bacterial cellulose obtained from *Acetobacter pasteurianus,* which is free of xylan, showed a very strong inducing effect on the formation of both xylanase and cellulase in *S. commune*. Whereas cellulose strongly induced the synthesis of xylanase in *S. commune*, formation of this enzyme activity is not promoted by components that are generally found to induce this enzyme activity in fungi or yeasts. These typical inducers include xylan or hemicellulose-rich material, xylooligosaccharides, pentoses such as xylose or arabinose, and various synthetic alkyl and aryl-β-D-xylosides (*27,45,46*). In growth and induction experiments xylanase formation was not induced in *S. commune* when the relatively pure birchwood xylan, pentoses, or β-methyl-D-xyloside, a structural analogue of xylobiose, were employed as an inducing substance. In contrast to these results, xylan from beechwood when used as an inducer or carbon source resulted in the formation of considerable xylanase and cellulase activities. However, these high enzyme activities seem to be caused by the high cellulose content of the latter xylan preparation.

From these results it was concluded that the synthesis of xylanase and cellulase are under a common regulatory control in *S. commune*. Induction of cellulase, xylanase as well as mannanase under the control of a single common regulator gene has also been proposed for another wood rotting fungus, *Polyporus (Bjerkandera) adustus* (47). A possible natural inducer for both enzyme activities in *S. commune* might be cellobiose, a small molecule that can be easily taken up by the cell and which is derived from insoluble cellulose. Cellobiose as a substrate induces xylanase activity markedly. A stronger inducing effect of this disaccharide may be inhibited by the action of β-glucosidase which is excreted into the growth medium in considerable amounts. As a matter of fact, it has been shown that thiocellobiose, which is not hydrolyzed by the extracellular enzymes of *S. commune*, acts as a gratuitous inducer of both cellulase and xylanase (48). Both xylanase and cellulase activities were also induced by lactose (4-*O*-β-D-galactopyranosyl-D-glucopyranose) and 4-*O*-β-D-galactopyranosyl-D-mannopyranose which are both structurally related to cellobiose (4-*O*-β-D-glucopyranosyl-D-glucopyranose). Lactobionic acid (4-*O*-β-D-galactopyranosyl-D-gluconic acid) and lactulose (4-*O*-β-D-galactopyranosyl-D-fructofuranose), both having structures derived from lactose but lacking the two β-1,4-linked pyranosyl rings, did not or only slightly induce xylanase and cellulase activities.

This common control in the regulation of the synthesis of xylanases and cellulases in *S. commune* might include other enzyme activities necessary for the degradation of hemicellulosic and cellulosic material as well. For both α-arabinosidase and α-glucuronidase it could be shown that production was higher on media containing cellulose, while growth on wheat bran, a substrate rich in xylan containing α-arabinosyl and α-glucuronosyl substituents, resulted in lower enzyme activities (11). Biely et al. (13) observed markedly higher production of acetyl xylan esterase activity by *S. commune* grown on cellulose than when grown on xylan or acetyl xylan, which only led to the synthesis of the constitutive level of this enzyme activity. Similar results were also obtained by MacKenzie and Bilous (14). Addition of only a small amount of cellulose to a medium containing xylan was sufficient to initiate the production of acetyl xylan esterase. A common regulatory control of acetyl xylan esterase with cellulases and xylanases has also been suggested by Biely et al. (49). In *T. reesei* both acetyl xylan and cellulose induced the formation of acetyl xylan esterase (13), whereas in several *Aspergillus* spp. this enzyme activity was considerably induced by oat spelt xylan and only to a lesser extent by cellulose (50).

Production of ferulic acid esterase by *S. commune* was highest when the organism was grown on cellulose, whereas cultivation on xylan or wheat bran, which contains ferulic acid esterified to arabinoxylan, resulted in much lower enzyme activities. In several *Aspergillus* spp., such as *A. niger* or *A. oryzae*, formation of ferulic acid esterase could be induced by different ferulic acid-containing carbon sources (11,51). Formation of another esterase, namely acetyl glucomannan esterase, by *S. commune* when grown on media containing cellulose has also been described (16), however, production of this particular enzyme on media containing different carbon sources has not been reported.

Formation of mannanase activity was greatly stimulated when *S. commune* was grown on cellulose. Production of this enzyme during growth on Avicel was approximately 10-fold higher than on a medium employing galactomannan from guar seeds as a carbon source (*32*). Substituting varying parts of the microcrystalline cellulose in the medium with guar gum did not specifically induce the formation of mannanase as has been described for other organisms (*47*).

Literature Cited

1. Stephen, A. M. In *The Polysaccharides;* Aspinall, G. O., Ed.; Academic Press: New York, London, 1983, Vol. 3; pp. 97–193.
2. Poutanen, K.; Tenkanen, M.; Korte, H.; Puls, J. *ACS Symp. Ser.* **1991**, *460*, 426–436.
3. Wessels, J. G. H. *Wentia.* **1965**, *13*, 1–113.
4. Jurasek, L.; Paice, M. G. *Meth. Enzymol.* **1988**, *160*, 659–662.
5. Ujiie, M.; Roy, C.; Yaguchi, M. *Appl. Environ. Microbiol.* **1991**, *57*, 1860–1862.
6. Oku, T.; Roy, C.; Watson, D. C.; Wakarchuk, W.; Campbell, R.; Yaguchi, M.; Jurasek, L.; Paice, M. G. *FEBS Lett.* **1993**, *334*, 296–300.
7. Wakarchuk, W.; Methot, N.; Lanthier, P.; Sung, W.; Seligy, V.; Yaguchi, M.; To, R.; Campbell, R.; Rose, D. *Progress Biotechnol.* **1992**, *7*, 439–442.
8. Bray, M. R.; Clarke, A. J. *Progress Biotechnol.* **1992**, *7*, 423–428.
9. Paice, M. G.; Jurasek, L.; Carpenter, M. R.; Smillie, L. B. *Appl. Environ. Microbiol.* **1978**, *36*, 802–808.
10. Haltrich, D.; Preiß, M.; Steiner, W. *Enzyme Microb. Technol.* **1993**, *15*, 854–860.
11. Johnson, K. G.; Silva, M. C.; MacKenzie, C. R.; Schneider, H.; Fontana, J. D. *Appl. Biochem. Biotechnol.* **1989**, *20/21*, 245–258.
12. Biely, P.; Puls, J.; Schneider, H. *FEBS Lett.* **1985**, *186*, 80–84.
13. Biely, P.; MacKenzie, C. R.; Schneider, H. *Can. J. Microbiol.* **1988**, *34*, 767–772.
14. MacKenzie, C. R.; Bilous, D. *Appl. Environ. Microbiol.* **1988**, *54*, 1170–1173.
15. Biely, P.; MacKenzie, C. R.; Puls, J.; Schneider, H. *Bio/Technology.* **1986**, *4*, 731–733.
16. Tenkanen, M.; Puls, J.; Rättö, M.; Viikari, L. *Appl. Microbiol. Biotechnol.* **1993**, *39*, 159–165.
17. Paice, M. G.; Desrochers, M.; Rho, D.; Jurasek, L.; Roy, C.; Rollin, C. F.; de Miguel, E.; Yaguchi, M. *Bio/Technology.* **1984**, *2*, 535–539.
18. Willick, G. E.; Seligy, V. E. *Eur. J. Biochem.* **1985**, *151*, 89–96.
19. Lo, A. C.; Barbier, J.-R.; Willick, G. E. *Eur. J. Biochem.* **1990**, *192*, 175–181.
20. Desrochers, M.; Jurasek, L.; Paice, M. G. *Dev. Ind. Microbiol.* **1981**, *22*, 675–684.
21. Hestrin, S.; Schramm, M. *Biochem. J.* **1954**, *58*, 345–352.
22. Bailey, M. J.; Biely, P.; Poutanen, K. *J. Biotechnol.* **1992**, *23*, 257–270.
23. Ghose, T. K. *Pure Appl. Chem.* **1987**, *59*, 257–268.

24. Herr, D.; Baumer, F.; Dellweg, H. *Appl. Microbiol. Biotechnol.* **1978**, *5*, 29–36.
25. Wilke, C. R.; Yang, R. D.; Sciamanna, A. F.; Freitas, R. P. *Biotechnol. Bioeng.* **1981**, *23*, 163–183.
26. Savidge, R. A.; Colvin, J. R. *Can. J. Microbiol.* **1985**, *31*, 1019–1023.
27. Senior, D. J.; Mayers, P. R.; Saddler, J. N. *ACS Symp. Ser.* **1989**, *399*, 641–654.
28. Hrmová, M.; Biely, P.; Vrsanská, M. *Arch. Microbiol.* **1986**, *144*, 307–311.
29. Rapp, P.; Wagner, F. *Appl. Environ. Microbiol.* **1986**, *51*, 746–752.
30. Morosoli, R.; Durand, S.; Letendre, E. D. *FEMS Microbiol. Lett.* **1987**, *48*, 261–266.
31. Gomes, D. J.; Gomes, J.; Steiner, W. *J. Biotechnol.* **1994**, *33*, 87–94.
32. Haltrich, D.; Steiner, W. *Enzyme Microb. Technol.* **1994**, *16*, 229–235.
33. Biely, P.; Krátky, Z.; Vrsanská, M.; Urmanicová, D. *Eur. J. Biochem.* **1980**, *108*, 323–329.
34. Persson, I.; Tjerneld, F.; Hahn-Hägerdal, B. *Process Biochem.* **1991**, *26*, 65–74.
35. Royer, J. C.; Nakas, J. P. *Enzyme Microb. Technol.* **1989**, *11*, 405–410.
36. Senior, D. J.; Mayers, P. R.; Saddler, J. N. *Appl. Microbiol. Biotechnol.* **1989**, *32*, 137–142.
37. Hrmová, M.; Biely, P.; Vrsanská, M. *Enzyme Microb. Technol.* **1989**, *11*, 610–616.
38. Bailey, M. J.; Poutanen, K. *Appl. Microbiol. Biotechnol.* **1989**, *30*, 5–10.
39. Lachke, A. H.; Deshpande, M. V. *FEMS Microbiol. Rev.* **1988**, *54*, 177–194.
40. Yu, E. K. C.; Tan, L. U. L.; Chan, M. K.-H.; Deschatelets, L.; Saddler, J. N. *Enzyme Microb. Technol.* **1987**, *9*, 16–24.
41. Tuohy, M. G.; Coughlan, M. P. *Bioresource Technol.* **1992**, *39*, 131–137.
42. Gilbert, M.; Breuil, C.; Saddler, J. N. *Bioresource Technol.* **1992**, *39*, 147–154.
43. Brown, J. A.; Collin, S. A.; Wood, T. M. *Enzyme Microb. Technol.* **1987**, *9*, 355–360.
44. Yoshida, N.; Fukushima, T.; Saito, H.; Shimosaka, M.; Okazaki, M. *Agric. Biol. Chem.* **1989**, *53*, 1829–1836.
45. Biely, P. *Trends Biotechnol.* **1985**, *3*, 286–290.
46. Wong, K. K. Y.; Saddler, J. N. *Progress Biotechnol.* **1992**, *7*, 171–186.
47. Eriksson, K. E.; Goodell, E. W. *Can. J. Microbiol.* **1974**, *20*, 371–378.
48. Rho, D.; Desrochers, M.; Jurasek, L.; Driguez, H.; Defaye, J. *J. Bacteriol.* **1982**, *149*, 47–53.
49. Biely, P.; MacKenzie, C. R.; Schneider, H. *Meth. Enzymol.* **1988**, *160*, 700–707.
50. Khan, A. W.; Lamb, K. A.; Overend, R. P. *Enzyme Microb. Technol.* **1990**, *12*, 127–131.
51. Tenkanen, M.; Schuseil, J.; Puls, J.; Poutanen, K. *J. Biotechnol.* **1991**, *18*, 69–84.

RECEIVED September 11, 1995

Chapter 20

Simultaneous Production of Xylanase and Mannanase by Several Hemicellulolytic Fungi

G. M. Gübitz and Walter Steiner

Enzyme Technology Laboratory, Institute of Biotechnology, Technical University Graz, Petersgasse 12, A–8010 Graz, Austria

> The simultaneous production of xylanase and mannanase by the hemicellulolytic fungi *Thermomyces lanuginosus, Sclerotium rolfsii* and *Schizophyllum commune* was studied. Substrate-optimization experiments were carried out by using statistical design methods to obtain a maximal mannanase activity under the constraint of a certain high level of xylanase activity. To increase the level of mannanase activity different mannan-rich carbon sources such as konjak powder, locust bean gum, copra meal powder and others, were added to the culture media formerly optimized for xylanase production. The highest level of mannanase activity (3,3 µkat/ml) was detected in the culture filtrate of *S. rolfsii* when grown on a mixture of α-cellulose and konjak powder as carbon source. The secreted mannanases were characterized by their molecular weights, pI values and pH and temperature optima and stabilities.

Beside cellulose and lignin, hemicelluloses are the main components of wood and annual plants (*1*). The most abundant hemicelluloses in wood are hetero-1,4-ß-D-xylans with a backbone consisting of ß-1,4-linked xylopyranose units and hetero-1,4-ß-D-glucomannans with a backbone composed of ß-1,4-linked mannopyranose and glucopyranose units (*2*). The breakdown of these hemicelluloses requires the synergistic action of xylan- and mannan-degrading enzyme systems. Endo-1,4-ß-xylanases, ß-xylosidases, endo-1,4-ß-mannanases, ß-mannosidases and ß-glucosidases cleave the main chain of the mannans and xylans, while other enzymes cleave the side chain substituents. In addition, hemicellulolytic enzymes are known to facilitate the chemical bleaching of both hardwood and softwood pulps. The breakdown of xylans in hardwood pulps by endo-1,4-ß-xylanases and side group-cleaving enzymes leads to a significant reduction in the amount of chemicals required for bleaching.

Softwoods contain arabinoglucuronoxylan and two different types of acetylated galactoglucomannans consisting of glucose, mannose and galactose in the ratios 1:3:1 and 1:4:0.1, respectively. For the complete hydrolysis of arabinoglucuronoxylans endo-1,4-ß-xylanase, ß-xylosidase, α-L-arabinofuranosidase and α-glucuronidase are needed. For the breakdown of galactoglucomannans, the enzymes endo-1,4-ß-mannanase, ß-mannosidase, ß-glucosidase and α-galactosidase are required.

Endo-1,4-ß-xylanases have been reported to occur in a wide spectrum of organisms including fungi, bacteria from marine and terrestrial environments, rumen bacteria and protozoa, insects, snails, marine algae and germinating seeds of terrestrial plants (3). Endo-1,4-ß-mannanases have been reported to be produced by bacteria (4-6), snails (7), and several fungi such as *Trichoderma harzianum* (8), *Trichoderma reesei* (1), *Polyporus versicolor* (9), *S. commune* (9), *Thielavia terrestris* (10), *Aspergillus tamarii* (11) and *Tyromyces* palustris (12). Out of these hemicellulolytic fungi, *Sclerotium rolfsii*, *Schizopyllum commune* and *Thermomyces lanuginosus* were chosen to study the simultaneous production of xylanases and mannanases. The production of xylanases by these fungi has previously been optimized in our laboratory (13-15).

The plant pathogen basidiomycete *S. rolfsii* is known to produce cellulolytic enzymes in high amounts (16). The production of endo-1,4-ß-xylanase(17), arabanase (18), polygalacturonase (19) and ß-D-xylosidase (20) by *S. rolfsii* has been described as well. ß-Glucosidase (21), α-L-arabinosidase (22), α-D-galactosidase (23) and an acid pectin esterase (24) from *S. rolfsii* have been purified. The wood-rotting fungus *S. commune* is known to be an excellent producer of lignocellulolytic enzymes. The formation of cellulase (25), endo-1,4-ß-xylanase (15), endo-1,4-ß-mannanase (9), acetyl mannan esterase (26) and acetyl xylan esterase (27) have been described and investigated. The thermophilic fungus *T. lanuginosus* produces mainly xylanase without any accompanying cellulase activity (13). Only little information is available in the literature about other lignocellulolytic enzymes produced by *T. lanuginosus*. High levels of both mannanase and xylanase activities in the culture filtrates could increase the potential of these fungi in the bio-bleaching of softwood pulps and other biotechnological applications.

Materials and Methods

Chemicals. Locust bean gum galactomannan (mannose galactose ratio of 4:1 (28)), and α-cellulose were obtained from Sigma (St. Louis, USA), *Amorphophallus konjac* glucomannan (mannose glucose ratio of 1,8:1) was obtained from Arkopharma (Carros, France) and guar gum galactomannan (mannose galactose ratio of 2:1) came from Meyhall (Kreuzlingen, Switzerland). Xylan from birch wood was obtained from Roth (Karlsruhe, Germany); xylan from beech wood from Lenzing AG (Lenzing, Austria). All other chemicals used were from Merck (Darmstadt, Germany). Complex carbon sources such as wheat bran, corn cobs, copra galactomannan (mannose galactose ratio of 14:1 (29)) were obtained locally.

Organisms and culture conditions. *T. lanuginosus, S. rolfsii* and *S. commune* have been isolated from decomposed jute stacks by I. Gomes at the Jute Research Institute, Dhaka, Bangladesh, and have been identified by the Centraalbureau voor Schimmelcultures (CBS, Baarn, The Netherlands) and are deposited at the culture collection at the Institute of Biotechnology, University of Technology, Graz, Austria. *T. lanuginosus* is deposited at the German type culture collection (DSM Braunschweig) under the number DSM 5826. Stock cultures of *T. lanuginosus* were maintained on potato dextrose agar (PDA) at 4°C and subcultured every month. *S. rolfsii* and *S. commune* were subcultured on Sabouraud agar, stored at 4°C and transferred every 4 to 6 weeks.

Medium components, concentrations (given in % w/v) and pH values varied in each experiment according to the optimization strategy and the fungal strain. For cultivation of *T. lanuginosus*, a medium adjusted to a pH of 6,5 and containing 2,85% yeast extract, 0,4% $(NH_4)_2SO_4$, 1,0% KH_2PO_4, 0,03% $MgSO_4·7H_2O$, 0,03% $FeSO_4$, and 0,03% $CaCl_2$ in addition to the carbon source was used. The medium for cultivation of *S. rolfsii* was adjusted to a pH of 5,0 and contained 0,12% KH_2PO_4, 0,15% $MgSO_4·7H_2O$, 0,06% KCl, 0,25% NH_4NO_3 and 6,74% bacto peptone plus the carbon source indicated below. To 100ml medium 300μl of a trace-element solution containing 0,1% $ZnSO_4·7H_2O$, 0,3% $MnCl_2·4H_2O$, 0,3% H_3BO_3, 0,2% $CoCl_2·6H_2O$, 0,01% $CuSO_4·5H_2O$, 0,02% $NiCl_2·6H_2O$ and 0,6% H_2SO_4 conc. were added. The culture medium for *S. commune*, adjusted to a pH of 7,0, contained 0,13% KH_2PO_4, 0,1% $MgSO_4·7H_2O$, 0,14% NH_4NO_3, 5,54% yeast extract, 100μl per 100ml medium of the trace-element solution described above and the carbon source. The pH were adjusted with NaOH or phosphoric acid. Tap water was used for all preparations.

All growth experiments were carried out as shake-flask cultures in 300-ml Erlenmeyer flasks containing 100ml of a given medium and using a piece (1cm^2) of an actively growing fungal colony on agar as an inoculum. The flasks were shaken on an orbital shaker with atmospheric humidity control at 60% relative humidity and at 150 rpm for 7 days at 50°C (*T. lanuginosus*), for 14 days at 30°C (*S. rolfsii*) and for 7 days at 30°C (*S. commune*). Cultures were centrifuged and the clear supernatant was used for the enzyme activity assays.

Electrophoresis and isoelectric focusing (IEF). Proteins in the culture filtrates were separated by SDS-PAGE under denaturing conditions using Biorad (Richmond, CA, USA) equipment. Electrophoresis was carried out according to the manufacturer's instructions using Coomassie Blue for protein staining. To determine the molecular weight of the separated proteins their bands were compared to those of molecular weight standards (range 14,4 kDa to 97,4 kDa) obtained from Biorad. For zymogramm analysis, a second gel was run simultaneously containing 0,02% w/v of either xylan from birch wood, galactomannan or carboxymethylcellulose. One gel was then stained with Coomassie Blue and the second gel was washed with a buffer containing 50mM disodium hydrogen phosphate, 12,5mM citric acid and 25% isopropanol, pH 6,3 to remove the SDS and stained with Congo Red for xylanase, mannanase and cellulase activity (30).

The pI values of the proteins in the ultrafiltrated culture filtrate were determined by isoelectric focusing using the IEF equipment from Biorad and the procedures recommended in the Biorad instruction manual. The different ampholytes (Biorad) used provided a pH range from 3 to 10, 3 to 5 and 4 to 6. Standard marker proteins (Biorad) with pI values from 4,45 to 9,6 were run at the same time. Agar replica gels containing 0,02% w/v of the above mentioned polysaccharides were stained with Congo Red for hemicellulase activities.

Enzyme assay. Endo-1,4-ß-xylanase was assayed according to Bailey et al. (31) using a 1%(w/v) solution of birch wood xylan (Roth, Karlsruhe, Germany) in 50mM sodium citrate buffer as a substrate set to the pH optimum of each enzyme. The diluted enzyme solution was incubated for 5 min. at 50°C. Released reducing sugars were assayed by adding 2-hydroxy-3,5-dinitrobenzoic acid reagent, boiling for 5 min., cooling and measuring the absorbance at 540 nm. Mannanase activity was determined similarly, using a 0.5%(w/v) solution of locust bean gum galactomannan. For the assay of glucomannan activity a 0,5% (w/v) solution of *Amorphophallus konjac* glucomannan was used. Filter paper activity was determined according to IUPAC recommendations (32) at a pH of 6,5.

Determination of pH and temperature stabilities. For studying the thermal stabilities of the mannanases the enzyme preparations were diluted 1:200 with 50mM sodium citrate buffer (pH 4,5). The buffered enzyme solutions were incubated on a water bath at constant temperature (50°C-80°C). Samples were drawn at certain intervals, cooled on ice immediately, and stored at 4°C. The pH stabilities were determined similarly diluting the enzyme preparations with 50mM sodium citrate buffer (pH3-pH6,5) and incubating the samples at 50°C. The residual enzyme activitivies were measured under standard conditions as described above at the pH optimum of each enzyme. For the determination of the thermal stabilities of cellulases from *S. rolfsii, S. commune* the samples were diluted with 50mM sodium citrate buffer (pH 4,5) and incubated at 60°C. The residual cellulase activities were determined at pH 6,5 as described above.

Results and Discussion

The starting point of the substrate optimization experiments for each fungus was the medium optimized for xylanase production in earlier investigations (13-15). In a first step, the influence of different mannans as carbon sources on the mannanase production was studied. The fungi were grown on several mannan-rich carbon sources such as konjak powder, locust bean gum, guar gum and coprameal powder and mixtures with other complex carbon sources. The concentrations of all components of the growth media, except the carbon sources, were kept constant. The results of these experiments are shown in Table Ia.

The highest mannanase activity was obtained in the culture filtrate of *S. rolfsii* grown on a mixture of konjak powder and α-cellulose (Table Ib). None of the fungi investigated produced pronounced mannanase activity on the mannans as sole carbon sources, while mixtures of mannan and cellulose or mannan and xylan increased the mannanase production in all tested strains (Table Ib). Konjak powder, locust bean gum, guar gum and coprameal were added in concentrations not exceeding 2% (w/v), since the viscosity of solutions of higher concentrations limits oxygen transfer and growth of organisms.

Table I. Effect of different carbon sources (a) and their combinations (b) on the mannanase production by *T. lanuginosus, S. commune* and *S. rolfsii*

a) Carbon source*	Concentration [% w/v] T. l.	S. r.	S. c.	Mannanase activity [nkat/ml] T. l.	S. r.	S. c.
Corn cobs (coarse)	3,10	4,00	4,00	5	236	54
Wheat bran	3,10	4,00	4,00	5	283	15
Rice husks	3,10	4,00	4,00	6	1165	150
Wheat straw (steamed)	3,10	4,00	4,00	2	1540	129
Barley husks	3,10	4,00	4,00	7	1475	198
Xylan (beech wood)	3,10	4,00	4,00	29	520	87
α-Cellulose	3,10	4,00	4,00	2	2078	97
Avicel	3,10	4,00	4,00	4	1328	452
Konjak powder	2,00	2,00	2,00	16	308	21
Guar gum	2,00	2,00	2,00	<<<	87	8
Copra meal	2,00	2,00	2,00	6	34	5
Locust bean gum	2,00	2,00	2,00	3	54	53
Mannose	3,10	4,00	4,00	5	420	60
b) Combinations:*						
Locust bean gum + Avicel	0,18 2,92	0,24 3,76	0,24 3,76	5	808	**546**
Konjak powder +Avicel	0,18 2,92	0,24 3,76	0,24 3,76	5	1420	514
Locust bean gum + α-Cellulose	0,18 2,92	0,24 3,76	0,24 3,76	14	1875	165
Konjak powder + α-Cellulose	0,18 2,92	0,24 3,76	0,24 3,76	17	**3367**	142
Locust bean gum + Xylan (beech wood)	0,94 2,16	1,21 2,79	1,21 2,79	27	931	313
Konjak powder + Xylan (beech wood)	0,94 2,16	1,21 2,79	1,21 2,79	**30**	1800	62

*Carbon sources were added to the basal media. All culture conditions are given in Materials and Methods.

It has been reported that the addition of locust bean gum galactomannan to culture media increased the mannanase production in organisms, such as *Streptomyces lividans* (33), *Aspergillus tamarii* (11), *Trichoderma harzianum* (8) and several thermophilic fungi (34). Konjac glucomannan induced the formation of mannanase in marine bacteria (35), in an alkalophilic *Bacillus sp.* (36) and in *Trichoderma reesei* (1). Compared to galactomannan, the addition of glucomannan to the culture media of *S. rolfsii* and *T. lanuginosus* yielded higher mannanase activities. No relation between type of mannan (galactomannan or glucomannan) added to the media for potential induction and type of formed mannanase activity (galactomannanase or glucomannanase) could be detected for the examined fungi, as it has been shown for *Trichoderma* harzianum (8). The production of mannanases along with cellulases on cellulose as carbon source has been reported for *Trichoderma reesei* (1), *Polyporus versicolor* (37) and other fungi (38). Using cellulose as the sole carbon source, the examined strains *S. rolfsii* and *S. commune* produced high mannanase activities. Cellulose is also reported to be an inducer for both xylanase and cellulase synthesis in S. commune and several other fungi (14).

To determine the effect of other media components, experiments using factorial designs were carried out. When compared to various concentrations of KH_2PO_4, $MgSO_4$ and several inorganic and organic nitrogen sources, the carbon sources were by far the most important factors for mannanase production (data not shown). The concentrations of the carbon components in the culture media (mixtures of mannan and cellulose or mannan and xylan) were optimized for each fungus by using a central composite design (14) (Table II).

The objective of all optimization experiments was to obtain a maximal mannanase activity under the constraint of a certain high level of xylanase activity. The lower limit for the xylanase activities in all experiments was set to 50% of the optimized xylanase. Under this constraint, the level of mannanase activity formed by *T. lanuginosus* and *S. rolfsii* could be at least doubled (Figure 1). In comparison with *S.* rolfsii and S. commune, T. lanuginosus produced only very low mannanase activities.

Table II. Simultaneous production of xylanase and mannanase by *S. rolfsii*, *S. commune* and *T. lanuginosus* using optimized mixtures of carbon sources

Organism	Carbon source*	Concentration [% w/v]	Enzyme activities [µkat/ml] Mannanase	Xylanase
T. lanuginosus:	Xylan (beech) + Konjak meal	1,72 0,73	0,03	20,05
S. commune:	Avicel + Locust bean gum	5,93 1,10	0,54	60,06
S. rolfsii:	α-Cellulose + Konjak meal	3,44 0,38	5,08	5,11

*Carbon sources were added to the basal media. All culture conditions are given in Materials and Methods.

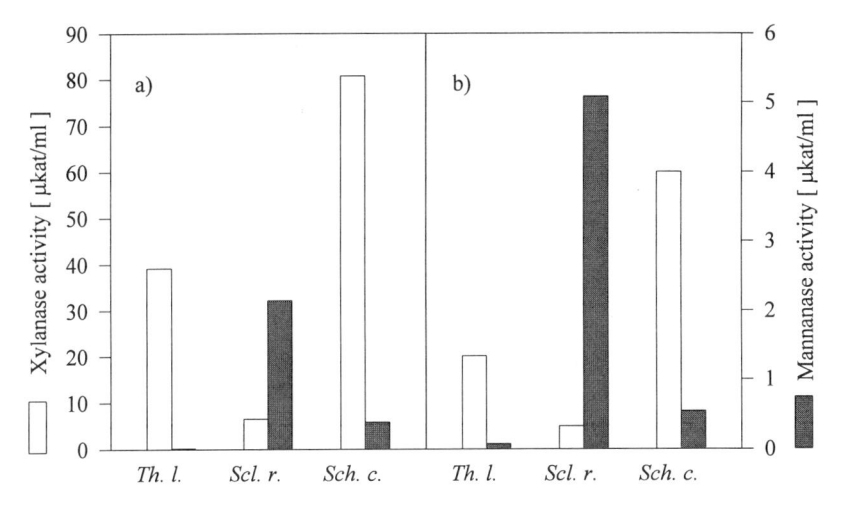

Figure 1. Summary of the enzyme activities produced by *T. lanuginosus (Th.l.), S. rolfsii (Scl.r.)* and *Schizophyllum (Sch.c.)* commune using a) media optimized for xylanase activities and b) media optimized for simultaneous production of xylanase and mannanase.

The time courses of fermentations of the investigated fungi in shake flasks are shown in Figure 2. Compared to the xylanase formation, the production of mannanase by *S. rolfsii* was slower. Xylanase activity reached its maximum after 290 hours and remained unchanged for 52 hours thereafter, while the highest values of mannanase were observed after 336 hours. Mannanase formation compared to xylanase formation was delayed in case of *T. lanuginosus*. Xylanase and mannanase activities reached maximum values after 168 and 172 hours, respectively. Xylanase and mannanase formation by *S. commune* were concurrent, as it is described in the literature for xylanase and cellulase production as well (*15*). Xylanase and mannanase activities reached their maximum value after 230 hours and remained constant for about 40 hours.

The pH optima of the mannanase activities in the culture filtrates of *T. lanuginosus, S. commune* and *S. rolfsii* were all in the acidic region of 4 to 5. The temperature optima of these enzymes were determined to be in a range of 60°C to 70°C after 5 minutes incubation (Table III). Similar results have been reported for the great majority of fungal mannanases (*1,39,40,36,28*). Few mannanases, such as from *Bacillus sp.*, have been reported to perform maximal activity in the alkaline region (pH of 9 and above) (*35*). The pH and temperature stabilities of the examined mannan-degrading systems are shown in Figures 3 and 4.

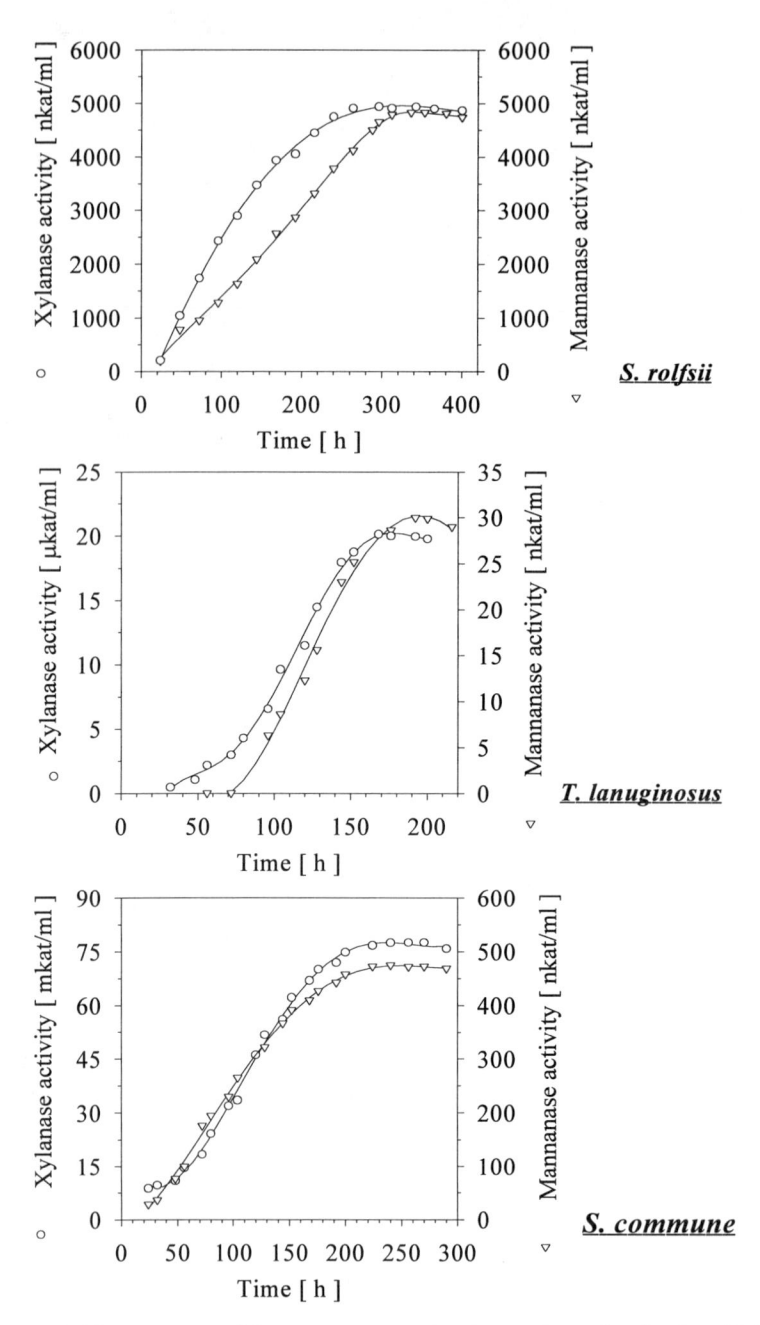

Figure 2. Time courses of the mannanase and xylanase formation by *S. rolfsii*, *T. lanuginosus* and *S. commune*.

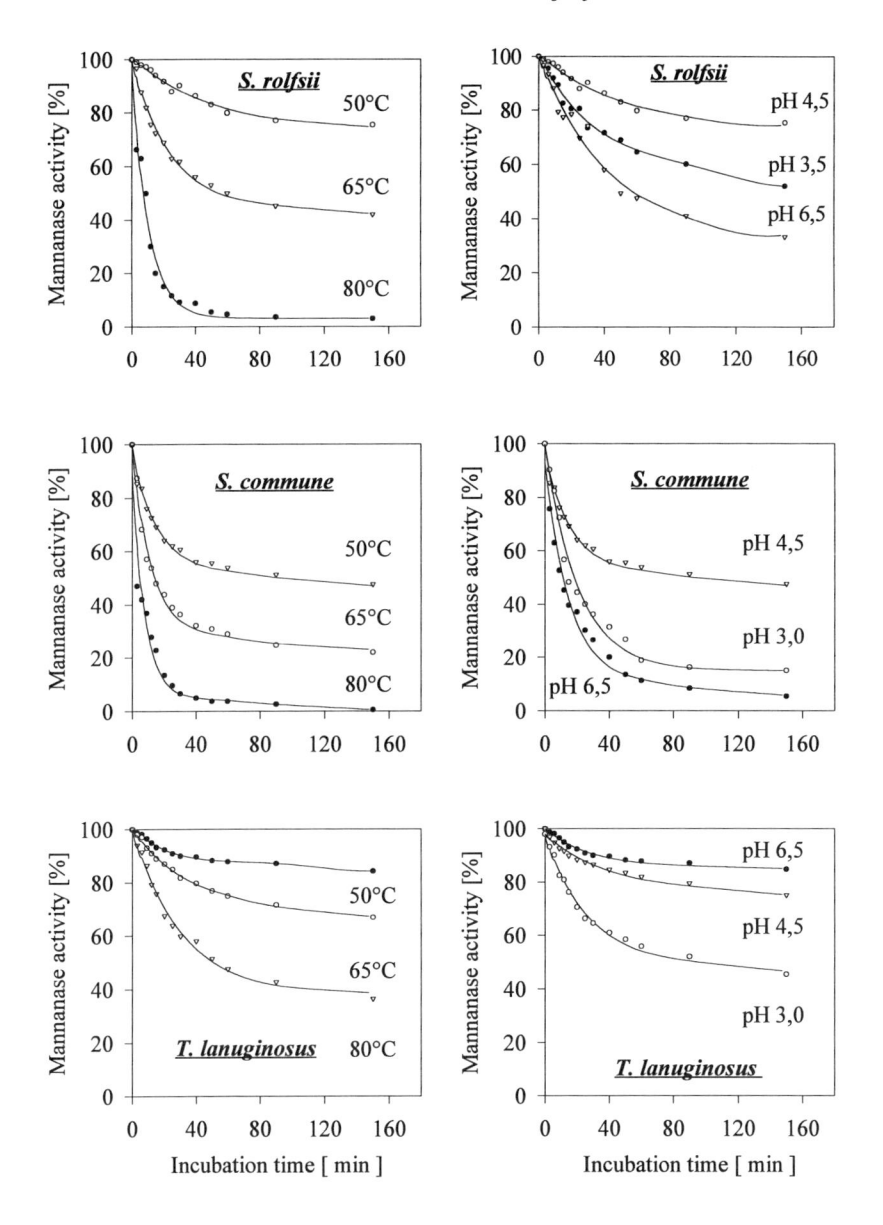

Figure 3. Heat stabilities of *S. rolfsii*, *S. commune* and *T. lanuginosus* mannanases.

Figure 4. pH stabilities of *S. rolfsii*, *S. commune* and *T. lanuginosus* mannanases.

Table III. pH- and temperature optima of the mannanases secreted by *T. lanuginosus*, *S. commune* and *S. rolfsii*

Organism	pH-optimum	T-optimum [°C]
T. lanuginosus	4,7	70
S. commune	4,2	70
S. rolfsii	4,6	60

Enzyme activities were maesured using the standard asssay methods (see Materials and Methods).

The mannanases secreted by *S. rolfsii* and *S. commune* remained most stable around their pH optima at 4,5 whereas the mannanase of *T. lanuginosus* was more stable at higher pH- values up to pH 6,5. Alkaline (pH 7,5 and above) as well as acidic conditions (pH 4,0 and below) caused rapid inactivation of the enzymes. At low pH (3,5) the mannanases of *S. rolfsii* were the most stable enzymes retaining 67% of their activity after 1 hour incubation.

A possible explanation for relatively high stabilities of *S. rolfsii* mannanases at low pH is the ability of this microbial strain to grow at extremely low pH below 2,5. The relationship of production of oxalic acid and hemicellulolytic enzymes by *S. rolfsii* has been described in the literature (*41*). The mannanase of the thermophilic fungus *T. lanuginosus* retained about 50% of its activity after 1 hour of incubation at 80°C, while the mannanases of *S. rolfsii* and *S. commune* were rapidly inactivated at this temperature.

At their pH-optima, the mannanases of both *S. rolfsii* and *S. commune* were more stable than their cellulases (Fig. 5). Fungal xylanases have generally been reported to be more stable than the respective cellulolytic systems (*42*).

■ Mannanase - *S. rolfsii*
○ Mannanase - *S. commune*
◆ Cellulase - *S. rolfsii*
▽ Cellulase - *S. commune*

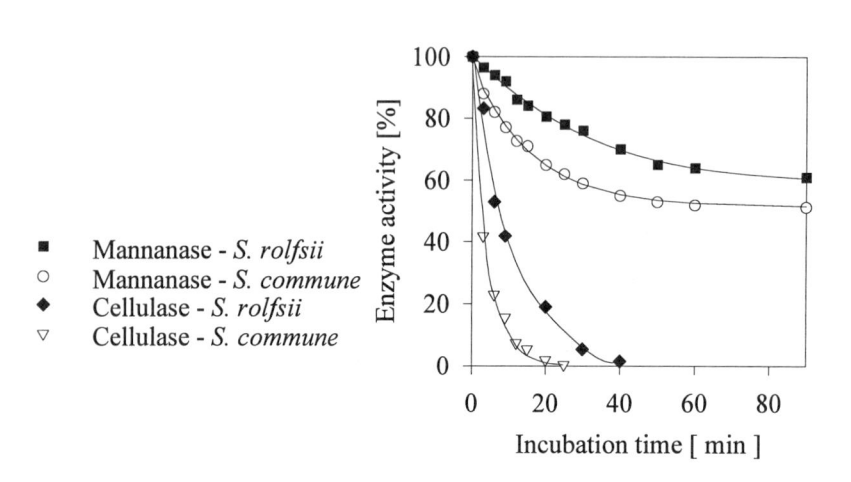

Figure 5. Inactivation of *S. rolfsii* and *S. commune* cellulases and mannanases at pH 4,5 and 60°C.

Using the zymogram technique combined with IEF, mannanase activity in the culture filtrates of *S. commune* was detected in two protein bands. The crude enzymes of *S. commune* had isoelectric points in the acidic region of 3,6 and 4,2. The molecular weights of these mannanases obtained from SDS-PAGE were 49 and 31 kDa. *S. rolfsii* excreted five mannanases with pI values of 3,5; 3,3; 4,1; 4,2 and 4,3 and molecular weights of 71, 22, 57, 37 and 42 kDa, respectively. Only one protein band (pI 4,2) of the culture filtrate of *T. lanuginosus* showed mannanase activity. pI values in the acidic region from 3,5 to 5,5 have been reported for the great majority of fungal mannanases, such as those from *Trichoderma reesei* (*1,39*), *Polyporus versicolor* (*36*), *Trichoderma harzianum* (*8*), *Thielavia terrestris* (*40*) and *Penicillium purpurogenum* (*28*).

Conclusions

The present work shows, that the concurrent occurrence of maximal mannanase and xylanase activities formed by *T. lanuginosus, S. commune* and *S. rolfsii* enables simultaneous production of these enzymes. The mannanase activities in the culture filtrates of *T. lanuginosus, S. commune* and *S. rolfsii* could be significantly increased by media optimization without great loss of xylanase activity. In several biotechnological fields, including bio-bleaching, combinations of hemicellulolytic enzymes are required. These enzymes can be obtained in optimal ratios in simultaneous production. Since these enzymes can be simultaneously obtained in optimal ratios, the downstream processing cost of such applications can be substantially reduced.

Literature Cited

1. Arisan-Atac, I.; Hodits, R.; Kristufek, D.; Kubicek, Ch.P. *Appl. Microbiol. Biotechnol.* **1993**, *39*, 58-62.
2. *Wood chemistry*; Sjöström, E., Ed.; Academic Press, San Diego **1993**
3. Dekker, R.F.H.; Richards, G.N. *Advances in Carbohydrates Chemistry and Biochemistry* **1973**, *32*, 277-352.
4. Komaki, Y.; Oda, T.; Tonomura, K. *J. Fermen. Bioeng.* **1993**, *76*, 14-18.
5. Araujo, A.; Ward, O.P. *J. Ind. Microbiol.* **1991**, *8*, 229-236.
6. Talbot. G.; Sygusch, J. *J. Appl. Environm. Microbiol.* **1990**, *56*, 3505-3510.
7. Yamaura, L.; Matsumoto, T. *Biosci. Biotechnol. Biochem.* **1993**, *57* (8), 1316-1319.
8. Torrie, J.P.; Senior, D.J.; Saddler, J.N. *Appl. Microbiol. Biotechnol.* **1990**, *34* (3), 303-307.
9. Johnson, K.G.; Ross, N.W.; Schneider, II. *World. J. Microbiol. Biotechnol.* **1990**, *6* (3), 245-254.

10. Araujo, A.; Ward, O.P. *J. Ind. Microbiol.* **1990**, *6* (4), 269-274.
11. Civas, A.; Eberhard, R.; Le Dizet, P.; Petek, F. *Biochem. J.* **1984**, *219* (3), 857-863.
12. Ishihara, M.; Shimizu, K. *Mokuza.i Gakkaishi.* **1980**, *26* (12), 811-818.
13. Purkarthofer, H.; Sinner M.; Steiner, W. *Enzyme. Microb. Technol.* **1993**, *15* (8), 677-682.
14. Haltrich, D.; Preiss, M.; Steiner, W. *Enzyme. Microb. Technol.* **1993**, *15* (10), 854-860.
15. Haltrich, D. Doctoral thesis, **1993**, University of Technology Graz.
16. Sadana, J.C.; Lachke, A.H.; Patil, R.V. *Carbohydrate. Research.* **1984**, *133*, 297-312.
17. Sadana, J.C.; Shewale, J.G.; Deshpande, M.V. *Appl. Environm. Microbiol.* **1980**, *39* (4), 935-936.
18. Cole, A.L.J.; Bateman, D.F. *Phytopathology* **1969**, *59*, 1750-1753.
19. Bateman, D.F. *Physiological. Plant Pathology* **1972**, *2*, 175-184.
20. Lachke, A.H.; Desphande, M.V. *FEMS Microbiology reviews* **1988**, *54*, 177-194.
21. Shewale, J.G.; Sadana, J. *Archives of Biochemistry and Biophysics* **1981**, *207* (1), 185-196.
22. Kaji, A.; Yoshihara, O. *Biochim. Biophys. Acta* **1971**, *250* (2), 367-371.
23. Kak, A.; Sato, M.; Shinmyo, N.; Yasuda, M. *Agric. Biol. Chem.* **1972**, *36* (10), 1729-1735.
24. Yoshihara, O.; Matsuo, T.; Kaji, A. *Agric. Biol. Chem.* **1977**, *41* (12), 2335-2341.
25. Steiner, W.; Lafferty, R.M.; Gomes, I.; Esterbauer, H. *Biotechnol. Bioeng.* **1987**, *30*, 169-178.
26. Tenkanen, M.; Puls, J.; Ratto, M.; Viikari, L. *Appl. Microbiol. Biotechnol.* **1993**, *39* (2), 159-165.
27. Biely, P.; MacKenzie, C.R.; Schneider, H. *Can. J. Microbiol.* **1988**, *34* (6), 767-772.
28. Bicho, P.A.; Clark, T.A.; Mackie, K.; Morgan, H.W.; Daniel, R.M. *Appl. Microbiol. Biotechnol.* **1991**, *36*, 337-343.
29. Park, G.G.; Kusakabe, I.; Komatsu, Y. *Agric. Biol. Chem.* **1987**, *51* (10), 2709-2716.
30. Béguin, P. *Anal.Biochem.* **1983**, *108*, 333-336.
31. Bailey, M.J.; Biely, P.; Poutanen, K. *J. Biotechnol.* **1992**, *23*, 257-270.
32. IUPAC, *Pure. Appl. Chem.* **1987**, *59*, 257-268.
33. Arcand, N.; Klueppfel, D.; Paradis, F.W. *Biochem. J.* **1993**, *290* (3), 857-863.
34. Araujo, A.; Ward, P. *J. Ind. Microbiol.* **1990**, *6* (3), 171-178.

35. Araki, T.; Tamaru, Y.; Morishita, T. *J. Gen Appl. Microbiol.* **1992,** *38* (4), 343-351.
36. Akino, T.; Nakamura, N.; Horikoshi, K. *Appl. Microbiol. Biotechnol.* **1987,** *26,* 323-327.
37. Johnson, K.G.; Ross, N.W.; Schneider, H. *World. J. Microbiol. Biotechnol.* **1990,** *6.*(3), 245-254.
38. Lyr, H. *Z. Allg. Mikrobiol.* **1963,** *3,* 25-36.
39. Stalbrand, H.; Siika-aho, M.; Tenkanen, M.; Viikari, L. *J. Biotechnol.* **1993,** *29,* 224-229.
40. Araujo, A.; Ward, O.P. *J. Ind. Microbiol.* **1990,** *6* (4), 269-274.
41. Punja, Z.K.; Huang, J.-S.; Jenkins, S.F. *Can. J. Plant Pathol.* **1985,** *7* (2), 109-117.
42. Biely, P. *Enzymes in Biom. Conversion,* **1991***, ACS-Symp. Ser. 460,* 409-416.

RECEIVED September 11, 1995

Chapter 21

Xylanase Delignification in Traditional and Chlorine-Free Bleaching Sequences in Hardwood Kraft Pulps

Nelson Durán[1], Adriane M. F. Milagres[2], Elisa Esposito[1], and Marcela Haun[3]

[1]Instituto de Quimica, Biological Chemistry Laboratory, Universidade Estadual de Campinas, C.P. 6154, Campinas, CEP 13081–970, Sao Paulo, Brazil
[2]Biotechnology Center, C.P. 16, Lorena, CEP 12600, Sao Paulo, Brazil
[3]Department of Biochemistry, Universidade Estadual de Campinas, C.P. 6154, Campinas, CEP 13081–970, Sao Paulo, Brazil

> In a previous screening of xylanolytic fungi **Penicillium janthinellum** (CRM 87M-115 strain) and **Aspergillus niger** (FTPT-131 strain) were isolated. The xylanase isolated from P. **janthinellum**, Penjanzyme-AB, exhibited similar selectivity than Cartazyme HT. Penjanzyme-AA showed similar reactivity with Cartazyme HS and Novozyme 473. Penjanzyme-AA in a CED bleaching sequence reduced in 25% the chlorine concentration, reaching the same brightness of 80% ISO with a high selectivity than the untreated pulps. In a CEHD sequence compared with XCEHD the most significant results was the economy of 39% of the hypochlorite stage. Similar results with XCEpHD or XCEoHD sequences were obtained. Penjanzyme-AA showed a 12% reduction of the Kappa number when compared with OaP(EOP)DP sequence. Asperzyme 131 from A. **niger** in a XP or XOP short sequence showed an enhancement of the initial pulp viscosity.

Tremendous efforts have been made in the pulp and paper industry to reduce the amount of chlorine used for bleaching. At present, studies have been conducted on the effluent treatment and the use of less toxic bleaching agents such as oxygen and hydrogen peroxide are under study.[1] The residual lignin removal by enzymatic method is actually a very active research area not only at the academic level but also in the R & D laboratories in the industry.[2-4] The use of hemicellulases, particularly the xylanases, for pulp bleaching arises from the idea that they allow a better action of bleaching agents in the subsequent stages. At least three mechanisms are proposed in order to explain the action of these enzymes. The first one proposes that there is xylan hydrolysis inside the cellular wall which results in the formation of micropores. In this way there is an increase of the pulp specific area and consequently a larger accessibility for the bleaching agents.[5] The second one explains its action by

solubilization of the re-precipitated xylan. This redeposition of xylans on the fibres occurs after loss of the side chain in the cooking process.[6] The third one suggests that the prebleaching effect results primarily from depolymerization and not from solubilization of xylan-derived hemicellulose.[7]
Xylanases are produced by different microorganisms[8-10] and in this study, we show that using xylanases from **Penicillium janthinellum** (Penjanzyme-AA and AB) and from **Aspergillus niger** (Asperzyme 131) it is possible to reduce the chlorine charge in **Eucalyptus** pulp bleaching with a simultaneous brightness gain. Also a comparison with some commercial xylanases is attempted.

EXPERIMENTAL CONDITIONS

Tests were carried out using an unbleached Kraft pulp from **Eucalyptus** from São Paulo, Brazil. Commercial Cartazyme HS (5.5×10^4 U/mL, pH 4.0 at 50°C), Cartazyme HT (3.7×10^2 U/mL, pH 9.0, at 65°C)(Repligen-Sandoz) and the Novozyme 473 (1.0×10^4 U/mL, pH 7.0 at 50°C) (Novo-Nordisk) were used. **Penicillium janthinellum** (CRC-87M-115) xylanase (Penjanzyme-AA, 5.2×10^1 U/mL, pH 5.5 at 50°C; Penjanzyme-AB, 1.3×10^1 U/mL , pH 8.0 at 50°C) was obtained as described before.[11,12] Asperzyme 131 was obtained from **Aspergillus niger** FTPT-3132 (3×10^2 U/mL at pH 5.5 at 50°C. The β-xylanase activities were measured by reduced sugar produced from xylan (birch, SIGMA) hydrolysis at 50°C and pH 5.5.[13] Enzyme pre-bleaching (X) was performed at 10% consistency and the conditions are in TABLE 3. For all the bleaching stages, the TAPPI Standards were followed (Kappa number: Tappi T-236 m-60; viscosity: Tappi T-230 su 63, brightness: Tappi T 414-ts). The chlorine (Cl) and residual chlorine, hypochlorite (H), chlorine dioxide (D) and hydrogen peroxide (P) concentrations were measured by published methods.[14] All the results reported are averages of duplicate experiments.

RESULTS AND DISCUSSION

Penjanzyme-AA reduced all the pulp's Kappa numbers with maximum efficiency at 40°C and with 180 min of treatment(TABLE 1).
Also at the same conditions, its selectivity was improved. In alkaline conditions (pH 8.0) the Penjanzyme-AB also exhibited a good capacity for removing lignin at 50°C for 60 min of treatment.
The observed capacities of Penjanzyme-AB (pH 8.0) and Penjanzyme-AA (pH 5.5) were compared with those of commercial xylanases. Table 2 shows that Penjanzyme-AA has a better selectivity than Cartazyme HS and is better than Novozyme 473 in its Kappa number reduction and selectivity while maintaining the same viscosity value. Penjanzyme-AB exhibited similar selectivity as Cartazyme HT, although no significative effects on the pulp as compared with the untreated pulp were observed.
Since Asperzyme-AA presents similar behaviour as Novozyme 473 and Cartazyme HS, a more detailed experiment with a CED bleaching was carried out. TABLE 4

TABLE 1. Bleaching parameters in two stages (X-E) by **P. janthinellum** xylanase (Penjanzyme; 1 U/g of dry pulp) and posterior sodium hydroxide extraction(a-d)

Runs	Temp (°C)	Time (min)	pH	Kappa Numb.	Kappa Reduc. (%)	Visc. (cps)	Delig. Effic. (%)	Decreas. Viscos (%)	Selectivity (d)
Control	40	60	5.5	12.4	-	27.1	18.4	12.9	1.55
+ XYL	40	60	5.5	11.2	9.7	26.7	25.7	14.1	1.82
Control	50	60	5.5	12.8	-	26.2	15.8	15.8	1.00
+ XYL	50	60	5.5	10.9	4.8	26.9	28.3	13.5	2.10
Control	40	180	5.5	13.1	-	29.0	13.8	6.8	2.03
+ XYL	40	180	5.5	10.7	18.3	28.4	29.6	8.6	3.44
Control	50	180	5.5	12.8	-	27.4	15.8	11.9	1.32
+ XYL	50	180	5.5	11.9	7.0	28.6	21.7	8.0	2.71
Control	40	60	8.0	13.9	-	25.3	8.6	18.6	0.46
+ XYL	40	60	8.0	13.2	5.0	25.9	13.2	16.7	0.79
Control	50	60	8.0	13.3	-	25.7	12.5	17.4	0.72
+ XYL	50	60	8.0	11.6	12.8	24.7	23.7	20.6	1.15
Control	40	180	8.0	13.5	-	24.5	11.2	21.2	0.53
+ XYL	40	180	8.0	13.1	2.9	24.5	13.8	21.2	0.65
Control	50	180	8.0	13.5	-	24.2	11.8	22.2	0.53
+ XYL	50	180	8.0	12.5	7.4	23.7	17.8	23.8	0.75

a) Unbleached pulp: Kappa number of 15.2 and viscosity of 31.1 cP. b) 4% NaOH extract. condition: 10% consistency, 70°C and 90 min. c) Pulp consistency in the xylanase treatment was 10%. d) Selec = selectivity= Delignification efficiency (%)/ reduction in viscosity (%). Delignification efficiency (%) =[Kappa number before the treatment-Kappa number after treatment]/[Kappa number before treatment] x 100. d) Penjanzyme AA at pH 5.5 and Penjanzyme at pH 8.0.

shows the Penjanzyme-AA effect on a CED bleaching sequence (conditions are on TABLE 3). In a CED sequence using 0.0-3.0% of chlorine compared to the XCED sequence (TABLE 4) at 2.2% chlorine an increase of 14% selectivity and 11% viscosity was obtained while maintained the same Kappa value. The brightness was improved by 1.9 points under these conditions. In other words, with the xylanase pre-treatment, the same value of brightness (e.g 82.7%) is obtained using a 27% reduction of chlorine, with better pulp quality (viscosity).

TABLE 5 shows the xylanase evaluation in a CEHD bleaching sequence. In a more complex sequence which is used in some pulp and paper industries in Brazil.

The Penjanzyme-AA significantly increased the delignification efficiency. An economy of sodium hypochlorite (38.6%) and of chlorine dioxide (3.6%) with a 9% brightness increase in a XCEHD sequence was found. However, under these conditions, a lower viscosity value than the control was observed. Nevertheless, in spite of a low reduction of the hypochlorite and chlorine dioxide by using Eo (in a XCEoHD sequence) and Ep (in a XCEpHD sequence) stages, a good viscosity protection was obtained.

TABLE 2. Xylanase pre-treatment (1 U/g dry pulp, 10% consistency) in a short bleaching sequence (XE) on **Eucalyptus** pulp(a-c)

	Novozyme	Cartazyme	Cartazyme	Penjanzyme	
	473	HS	HT	AB	AA
pH (Xylanase)	7.0	4.0	9.0	8.0	5.5
Temperature(°C)	50.0	50.0	65.0	50.0(d)	40.0
NaOH (%)	4.0	4.0	4.0	4.0	4.0
Viscosity (cps)	28.0	27.2	24.9	24.7	26.9
	(25.9)(f)	(26.8)	(25.8)	(25.4)	(26.0)
Kappa Number	10.0	9.5	11.0	11.6	9.4
	(10.7)	(10.9)	(10.2)	(10.8)	(10.7)
Selectivity	3.4	3.0	1.4	1.2	2.8
	(1.8)	(2.0)	(1.9)	(1.6)	(1.8)
Selectivity ratio(g)	1.9	1.5	0.7	0.8	1.6

a) Xylanase pretreatment was for 180 min. b) E stage: 4% NaOH concentration at 70°C for 90 min, 10% consistency. c) Unbleached pulp: Kappa number 15.2 and viscosity: 31.1 cP. d) This run was carried out for 60 min. f) No xylanase added. g) Ratio between xylanase pre-treatment and in the absence of the enzyme.

TABLE 3. Bleaching conditions

Stage (a)	Temperature (°C)	Time (min)	Consistency (%)	pH Inicial	Final
C (0.0-3.0%)	50	80	4	-	2.2
E (2.2%)	65	66	12	11.2	11.3
H (1.0%)	45	120	12	11.0	8.2
D (1.0%)	75	180	12	-	4.1
Ep(2.2% NaOH) (0.5% H$_2$O$_2$)	65	66	12	11.4	11.2
Eo(2.2% NaOH) (0.25% O$_2$, 2.0 Kgf/cm^2)	65	16 + 50	12	-	10.9
X (1.3 U/g)	40	180	10	5.5	5.5

a) C: chlorine; E: sodium hydroxide; H: hypochlorite; D: chlorine dioxide; Ep: alkaline/peroxide; Eo: alkaline/oxygen; X: xylanase.

In a similar CEHHH bleaching sequence in India using **Eucalyptus** pulp with Pulzyme HA (Novo Nordisk), the chlorine was reduced by 21.4% in the chlorination stage. The brightness was increased by approximately 3% and no reduction of chlorine in the hypochlorite stage was found.[15]

TABLE 4. Bleaching of Kraft pulp with chlorine reduction pre-treated with Penjanzyme-AA (pH 5.5) (a)

Sequence	Chlorine (Cl$_2$ g/100 g of dry pulp)	Brightness ISO(%)	Kappa Number	Viscosity (cps)
CED	3.0	83.6	1.6	15.8
	2.2	80.8	1.5	16.0
	1.5	78.2	2.2	17.7
	0.0	67.5	4.6	18.9
XCED	3.0	84.6	1.3	14.7
(1 U/g)	2.2	82.7	1.4	17.9
	1.5	80.0	1.7	17.2
	0.0	68.5	4.3	18.0

a) Unbleached pulp: Kappa number 15.2 and viscosity of 31.1 cP.

TABLE 5 Bleaching parameter after Penjanzyme-AA treatment

Sequence	Reagent Consumpt. (%)	Reduct. (%)	Brightness ISO(%)
X	-	-	-
C (a)	99.1(99.7) (b)	0.9	39.7 (36.8)
E	-	-	47.6 (41.5)
H	61.4 (99.0)	38.6	81.0 (71.8)
D	96.4 (99.5)	3.6	88.9 (81.3)
X	-	-	-
C	99.1(99.7)	0.9	39.7 (36.8)
Ep	-	-	59.7 (51.2)
H	82.5(87.9)	17.5	77.5 (69.3)
D	97.4(99.5)	2.6	87.3 (81.2)
X	-	-	-
C	99.1(99.7)	0.9	39.7 (36.8)
Eo	-	-	50.0 (49.8)
H	72.7(88.7)	11.0	79.2 (63.9)
D	97.2(97.9)	2.8	87.8 (77.3)

a) 2.5% chlorine. The other reagents see on TABLE 3. b) In parentheisis are the control value in the absence of xylanase.

Due to these results another xylanase from **Aspergillus niger** (Asperxyme 131) with similar characteristic as those that of **Penicillium janthinellum** was used.
Recently, a study of unbleached and oxygen-bleached **Eucalyptus grandis** pulp treated with **Aurobasidium** xylanase was published.[16] In the unbleached pulp, xylanase decreased the Kappa number by 10%, increased the viscosity by 5% and increased the brightness by 7%. In the bleached pulp, a 11% Kappa number reduction, increase of 10% of the viscosity and increase of 4% of brightness were obtained. However, as seen in TABLE 6 when Asperzyme 131 was used there was a reduction of of 9% of Kappa Number and an increase of 8% in the viscosity compared to the control. On contrary, Penjanzyme-AA slightly decreased slightly the viscosity but a 12% Kappa number reduction was observed. These results probably are indicative that in the case of Asperzyme 131, the xylanolytic pool is efficient in degradation of low molecular weight xylans in the Kraft pulp and that the cellulose was not affected.

TABLE 6. Xylanase treatment in **Eucalyptus grandis** Kraft pulp (a)

Kraft Pulp	Kappa Number Penjan.-AA	Asper.131	Viscosity (cps) Penjan.-AA	Asper.131
Control	15.6	15.6	34.8	34.8
X	13.8	14.3	29.1	37.5
O	12.8	12.8	31.9	31.9
OX	11.9	12.2	30.6	30.8

a) X activity was 2 U/g of dried pulp, 10% consistence, pH 5.5, 180 min at 50°C. The other conditions were the same as TABLE 3.

In an oxygen pre-bleached pulp, no large differences with xylanases either with Penjanzyme-AA (7% reduction) or Asperzyme 131 (5% reduction) were observed.

CONCLUSIONS

These results show that the pre-treatment of Kraft pulp from **Eucalyptus** with the Penjanzyme-AA promotes a Kappa number reduction and a brightness gain. All the facts are indicative that the xylanase facilitates the lignin removal with high specificity, since there was no significative reduction of the final viscosity compared to the controls. Penjanzyme-AA and Penjanzyme-AB as Asperzyme 131 have their commercial equivalents (e.g Cartazyme and Novozyme), and they have shown good potential for **Eucalyptus** pulps treatment..

ACKNOWLEDGEMENT. Supported by CNPq/FINEP/PADCT, FAEP and FAPESP.

REFERENCES

1. V.R. Parthasarathy, Entrained black Liquor Solids and Viscosity Selectivity in Oxygen Delignification Reinforced With Hydrogen Peroxide, TAPPI J. 73(9):243(1990).

2. R.W. Allison, T.A. Clark and S.H. Wrathall, Pretreatment of Radiata Pine Kraft Pulp With a Thermophilic Enzyme Part 1. Effect on Conventional Bleaching, APPITA 46: 269 (1993).
3. D.J. Senior and J. Hamilton, Xylanase Treatment for the Bleaching of Softwood Kraft Pulps: The effect of Chlorine Dioxide Substitution, TAPPI J. 76: 200 (1993).
4. R.P. Scott, F. Young and M.G. Paice, Mill-Scale Enzyme Treatment of a Softwood Kraft Pulp Prior to Bleaching, Pulp Paper Can. 94: 57 (1993).
5. P. Noe, J. Chevalier, F. Mora and J. Comtat, Action of Xylanases on Chemical Pulp Fibres. II. Enzymatic Beating, J. Wood Chem. Technol. 6(2): 167 (1986).
6. L. Viikari, J. Sundquist and J. Pittunen, Xylanases Promote Pulp Bleaching, Paperi ja Puu, 73 (5): 384 (1991).
7. M.G. Paice, N. Gurnagel, D.H. Page and L. Jurasek, Mechanism of Hemicellulose-Directed Prebleaching of Kraft Pulps, Enzyme Microb. Technol. 14: 272 (1992).
8. K.K.Y. Wong and J.N. Saddler, **Trichoderma** Xylanases, Their Properties and Application, Crit. Rev. Biotechnol. 12: 413 (1992).
9. E. Curotto, C. Aguirre, M. Concha, A. Nazal, V. Campos, E. Esposito, R. Angelo, A.M.F. Milagres and N. Durán, New Methodology for Fungal Screening: Xylanolytic Enzymes, Biotechnol. Tech. 7: 821 (1993).
10. M.J. Bailey and L. Viikari, Production of Xylanases by **Aspergillus fumigatus** and **Aspergillus oryzae** on Xylan-Based Media, World J. Microbiol.Biotechnol. 9: 80(1993).
11. A.M.F. Milagres and L.S. Lacis, Efficient Screening of Process Variables in **Penicillium janthinellum** Fermentations, Biotechnol. Lett. 13(2):115 (1991).
12. A.M.F. Milagres and N. Durán, Xylanolytic Enzymes from **Penicillium janthinellum** and its Applications in Bleaching of Pulp", Progress Biotechnol. 7: 539 (1992).
13. M.J. Bailey, P. Biely and K. Poutanen, Interlaboratory testing of Methods for Assay of Xylanase Activity, J. Biotechnol. 23: 257 (1992).
14. R.G. McDonald (Ed.), Pulp and Paper Manufacture, Vol.1. 2nd. Ed. McGraw-Hill Book Co. New York, (1967).
15. P.Bajpai, N.K. Bhardwaj, S. Maheshwari and P.K. Bajpai, Use of Xylanase in Bleaching of Eucalyptus Kraft Pulp, APPITA 46: (4) 274 (1993).
16. J.L. Yang, V.M. Sacon, S.E. Law and K. -E.L. Eriksson, Bleaching of Eucalyptus Kraft pulp with the EnZone Process. TAPPI J. 76: (7) 91 (1993).

RECEIVED September 18, 1995

Chapter 22

Paper Biopulping of Agricultural Wastes by *Lentinus edodes*

G. Giovannozzi-Sermanni[1], A. D'Annibale[1], N. Vitale[2], C. Perani[1], A. Porri[1], and V. Minelli[2]

[1]Department of Agrobiology and Agrochemistry, University of Tuscia, Via San Camillo de Lellis snc, 01100 Viterbo, Italy
[2]Agrital Research Consortium, V. le dell'Industria 24, 00057 Maccarese, Italy

Extra-cellular crude enzymic mixtures of the white-rot fungus *Lentinus edodes*, obtained under solid-state and submerged fermentation conditions, were used to preatreat durum wheat straw and corn stalks prior to alkaline pulping. Cell wall composition, determined after the enzymic incubations, showed a preferential degradation of the hemicellulosic fraction of wheat and corn, whereas lignin removal appeared to be significant only in the case of corn. Nevertheless, the biotreatments resulted in a higher lignin extractability, which was evident after alkaline cooking. Furthermore, the freeness and the strength properties of the paper handsheets obtained with corn and wheat were remarkably improved, as compared with the biologically untreated controls.

In recent years, research efforts have been devoted to the biotechnological application of microorganisms in the pulp and paper industry. Two main approaches have been used, namely the colonization of the lignocellulosic material by microorganisms, in particular white-rot fungi (*1*), or the incubation of such materials with crude or purified mycelial enzymes. The first approach implies several bottlenecks due to the solid-state fermentation (SSF) peculiarity, such as physiological and biochemical behaviour of the colonizing mycelium, process duration, physico-chemical characteristics of the substrate and treatment uniformity. These drawbacks can be overcome by a proper bioreactor design (*2*) and by optimizing the fungal strain and fermentation conditions (*3*). The use of isolated enzymes and/or enzymic mixtures, as an alternative approach, including endoxylanase (*4*), ligninase (*5*), laccase (*6*), hemicellulase systems (*7*), as pulping and bleaching reagents, also for recycled paper (*8*), have been suggested. These enzyme-based processes result in a drastic reduction of the time of treatment.

The use of annual plants, such as sorghum and kenaf, for fibre production, could contribute to a more flexible structure of the agricultural systems and therefore it could allow to follow promptly the market requests (*9*). Keeping in mind that the huge waste residues of wheat and corn crops can be considered also polluting materials, we believed of some interest to explore the potentiality of these materials as fibre sources. In the present paper we report the results of enzymic biotreatments performed on wheat straw and corn stalks and the technical characteristics of the paper handsheets obtained.

Materials and Methods

Lignocellulose Material. Durum wheat straw (*Triticum durum L.*) and corn stalks (*Zea mais L.*), cut into 5-cm pieces, were used as cellulose source, without any further mechanical handling.

Microorganism and Inoculum Production. *Lentinus edodes* (strain SC-465), was kindly provided by Prof. T. H. Quimio (Los Banos, Philippine). The mycelium was maintained and routinely subcultured on potato-dextrose agar slants. Inocula were prepared as previously reported (*10*).

Solid-state Fermentation (SSF) Conditions. A liquid salt medium (*11*), supplemented with 2% of malt extract, was added to wheat straw to obtain a final moisture content of 75%. After steam-sterilization (120° C for one hour), the solid substrate was inoculated aseptically with a 72-hours mycelial suspension (10% v/w). The SSF was carried out in a 3.5 cubic meters tumbling drum bioreactor at 30°C (*12*), and the uniform colonization of lignocellulosic material was improved by a slow rotation rate (4 rev·day^{-1}). After 7 days the colonized substrate was squeezed by hydraulic pressing at 350 atm, filtered through glass wool, centrifuged (11,000 g X 30 min.) and the resulting supernatant was used in the enzymic incubations, denominated solid-state liquid biotreatment (SSLB).

Submerged Fermentation (SM) Conditions. SM was carried out at 30°C in a 25-l stirred tank bioreactor flushed with sterile air (air flow 1 L/min), in the same liquid medium described above; the production of lignolytic exo-enzymes was enhanced by the addition of 0.2% (w/v) soluble lignocellulose (SLC) (*13*). After seven days from the inoculation the culture filtrate was centrifuged at 11,000 g for 30 min. and used in the enzymic incubations, denominated submerged liquid biotreatment (SMLB)

Analytical Methods. Neutral detergent fiber (NDF), acid detergent fiber (ADF) and acid detergent lignin (ADL) were determined by the Van Soest method (*14*). Hemicellulose and cellulose were calculated from the fractions obtained by this procedure: NDF-ADF and ADF-ADL respectively. The Chlorine number was determined by the ATICELCA MC 207-76 method. The protein content was determined by the Bradford method (*15*) after trichloroacetic precipitation, using bovine serum albumin (BSA) as standard. Hydrogen peroxide concentration in the crude extracts was determined by using a peroxidase-coupled assay (*16*). A calibration curve was prepared with different dilutions of a 30% solution of H_2O_2. The concentration of hydrogen peroxide in the commercial solution was determined from its absorbance at 230 nm ($\varepsilon = 81$ $M^{-1}cm^{-1}$). Appropriate blanks were prepared by incubating the crude extracts with 30 U of catalase (SIGMA), as described elsewhere (17).

Enzyme Assays. Mn-dependent peroxidase and laccase (E.C. 1.10.3.2) were assayed according to the method of Glenn and Gould (18) and of Haars & Hutterman (19), respectively. Endo-cellulase (E.C. 3.2.1.4) and exo-cellulase (E.C. 3.2.1.91) were assayed by the method of Shewale & Sadana (20), and endo-xylanase (E. C. 3.2.1.8) according to the method described by Mc Carthy et al. (21). Boiled enzyme extracts were used as controls. The reported data are the means of three different determinations.

Biotreatment Conditions. The biotreatment, based on the mycelial colonization was not longer used, since a significant loss of organic matter (-25%), due to the mycelial metabolism, was evident. Besides, the time required to obtain an uniform colonization was, usually, more than one week and the pulp yield was lowered by about 30%. Therefore the following biotreatments, based on the enzymic incubation of the material, were emphasized:

Solid-state Liquid Biotreatment (SSLB). The enzymic crude mixture arising from SSF, previously was added to dried lignocellulosic material (5:1 v/w) and incubated for 24 hours in a 25-litres tumbling drum bioreactor (64 rev·day^{-1}) at 40° C. Air was flushed inside the bioreactor at a flow rate of 2 l/min. It was assumed, on the basis of the solubility of oxygen at ordinary pressure conditions (1 atm) and at the tested temperature (40° C)(22-23), that its concentration appeared to be sufficient to allow laccase-catalyzed reactions.

Submerged Culture Supernatant Biotreatment (SMLB). The same conditions of the previous biotreatment were adopted by using the crude enzyme mixture obtained from the SM fermentations.

Pulping Conditions and Technical Analysis of Paper Handsheets. The raw materials were cooked in a laboratory rotating digester (Lorentzen & Wettre, Sweden), for 30' at 120°C in the presence of 12% or 16% NaOH (related to the pulp weight) and then beaten in a conical refiner (Toniolo, Italy) for 10, 20, 30 min. Two different NaOH concentrations, depending on the tested cellulose source, were used in order to obtain the same freeness range for both species. In fact, comparable values of freeness were obtained by cooking corn and wheat with 12% and 16% NaOH, respectively. In the case of corn, a higher alkali concentration than 12%, appeared to affect negatively the paper properties. The paper handsheets were obtained by the ATICELCA MC 217-79 method. The freeness was determined by the ATICELCA MC 201-76 method, the breaking length, the burst index and tear index by the TAPPI standard T 405 OM-87, TAPPI standard T 405

OM-85 and ISO 1974-1990 methods, respectively. The technical data obtained were statistically evaluated by the computer program Sigma-Stat.

Results and Discussion

Mycelial Enzymes. The solid-state fermentation conditions reproduce the natural habitat of white-rot fungi and, therefore, induce the production of the enzymatic machinery involved in the degradation of the plant cell wall constituents, mainly resulting in the release of hydrolases (cellulases, hemicellulases, pectinases) and oxido-reductive enzymes (laccase, Mn-peroxidase etc). Table I shows the exo-enzyme compositions obtained after 7 days from the inoculation in solid-state and in submerged cultures of *Lentinus edodes*. Such mixtures differed particularly for the absence of cellulolytic enzymes in the crude extracts arising from the submerged fermentation. Besides, under submerged fermentation conditions, the specific activity of phenol oxidases and Mn-peroxidases appeared to be much higher than under SSF-conditions, while these differences were not so evident if the enzyme activities were referred to volume unit. These findings suggest that under SM-conditions, a more selective release of extra-cellular protein occurred. Therefore SSF system probably produces a more complete spectrum of enzyme activities related to the degradation of lignocellulosic matter, while the lack of cellulolytic activities, obtainable in SM-fermentation, under the previously described conditions, could result in suitable enzymic mixtures for biopulping applications. Hydrogen peroxide was detected in the crude extracts arising from SSF and SM, its concentration being 84 nmol/ml in the former case and 43 nmol/ml in the latter case. It is worth noting that low levels of hydrogen peroxide (ranging from 3 to 12 nmol/ml) were detected in control cultures of *Lentinus edodes*, carried out in the absence of any phenolic effector (data not shown). This finding is in accordance with other reports, describing an increase in H_2O_2 production (24), when white-rot fungi are grown in the presence of aromatic molecules under submerged fermentation conditions. Furthermore, it has been suggested that hydrogen peroxide and derived radicals could be involved in the degradation of the plant cell wall constituents (25-27).

On the basis of the previously mentioned presence of hydrogen peroxide in the crude extracts, no additions of this compound, aimed at allowing Mn-peroxidase to function, were performed in the enzymic incubations of the material.

Effects on Annual Plants. Few preliminary trials, performed on kenaf (*Hibiscus cannabinus*), a fibre crop, showed that the enzymic biotreatments had a negative effect on pulp freeness, which appeared to be negatively affected, whereas breaking length and burst index were not modified, as compared with the untreated controls (data not shown). On the contrary, the biotreatments, carried out on wheat straw and corn stalks, resulted in several remarkable effects. Fiber analysis, performed after the enzymic incubations, showed that the biotreatments didn't significantly reduce lignin content of wheat straw, whereas the cellulosic and

Table I. Enzyme activities of the extracts employed for SSLB- and SMLB biotreatments

Enzyme	SSLB (U/ml)	SSLB (U/mg)	SMLB (U/ml)	SMLB (U/mg)
Phenoloxidase	9.90	3.30	4.19	46.60
Mn-peroxidase	0.84	0.28	0.990	11.11
Endocellulase	2.25	0.751	-	-
Exo-cellulases	2.25	0.751	-	-
Endoxylanase	2.166	0.722	0.05	0.55

hemicellulosic fraction appeared to be more affected by the SSLB-biotreatment, in accordance with the composition of the enzyme mixture (Figure 1). Nevertheless, both biotreatments resulted in the same pulp yield reduction (- 15-20 %), as compared with the control.

Cell wall composition of corn stalks, determined after the biological treatments (Figure 2), showed that SMLB-incubation resulted in the highest hemicellulose degradation (about 40%), while an appreciable lignin removal of 12% (SMLB) and 18% (SSLB) was evident. Both the biotreatments appeared to negatively affect the pulp yield, as shown in Figure 2. The differences in the cell wall composition of corn and wheat and their morphological and chemical structures can significantly affect the accessibility of the enzymes to the substrates, and therefore the results of the biotreatments.

A further reason for the different results could be due to the different amounts of soluble lignocellulose (SLC) of different molecular weights, as determined by size exclusion chromatography (*11*). In fact SLC content is 0.400% on dry weight base in wheat straw, while in corn is 0.080% (*28*). Since soluble lignocellulose are substrates for the exo-enzymes (10-11), the accessibility of the enzymes to the water insoluble cell wall components could be buffered by the presence of water-soluble copolymers, with subsequent different final attack of the lignocellulose structures. Nevertheless, it is noteworthy that the chlorine number, determined after the alkaline cooking, was halved in the samples of biotreated wheat (Figure 1) and corn (Figure 2) with respect to the biologically untreated controls. These results could be explained in terms of a better accessibility of the cooking liquor, due to an increased porosity of the material, or, to an increase in the degree of lignin functionalization (e.g. introduction of phenolic hydroxyl groups in the lignin macromolecule).

When considering the negative impacts of the biotreatments on the pulp yield, one must take into account that the materials were used as such, and, therefore, the observed reduction of pulp yield include also some constituents, such as parenchimatic tissues and epithelial cells, which are not suitable for paper production. These results suggest that appropriate previous handlings of the raw materials could be desirable, in order to minimize the consumption of chemical and biochemical reagents due to these fiber-poor constituents, whose removal after alkaline pulping determines a fictious reduction of pulp yield.

Paper Characteristics after the Biotreatments. Figure 3 illustrates the changes of freeness at different times of beating. The biological pretreatments of corn determined appreciable reduction of freeness values, even if, as expected, this parameter tended to increase with longer times of refining. Such positive effects were comparable in both biotreatments. In the wheat pulp, freeness appeared to be less dependent on the beating times in untreated and in biotreated samples. Generally this parameter was lowered by the biotreatments, but its reduction was less evident for the wheat. The freeness is a very important parameter for paper manufacturing and its reduction could allow a substantial gain in the paper machine

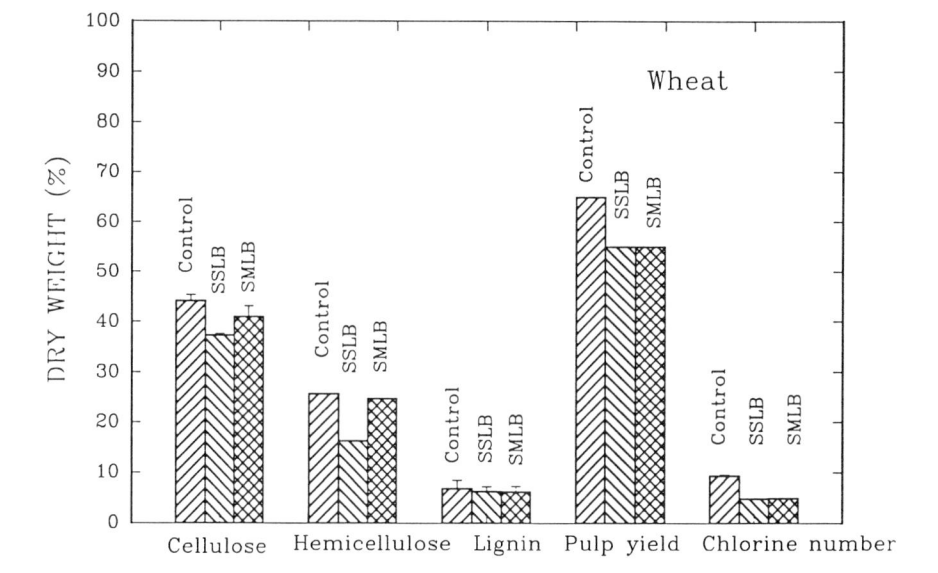

Figure 1. Cell wall constituents (cellulose, hemicellulose and lignin), pulp yield and chlorine number of control and biotreated wheat straw. Cellulose, hemicellulose and lignin were determined after the biological treatments, whereas pulp yield and chlorine number after alkaline cooking. The data are referred to 100 g dry weight.

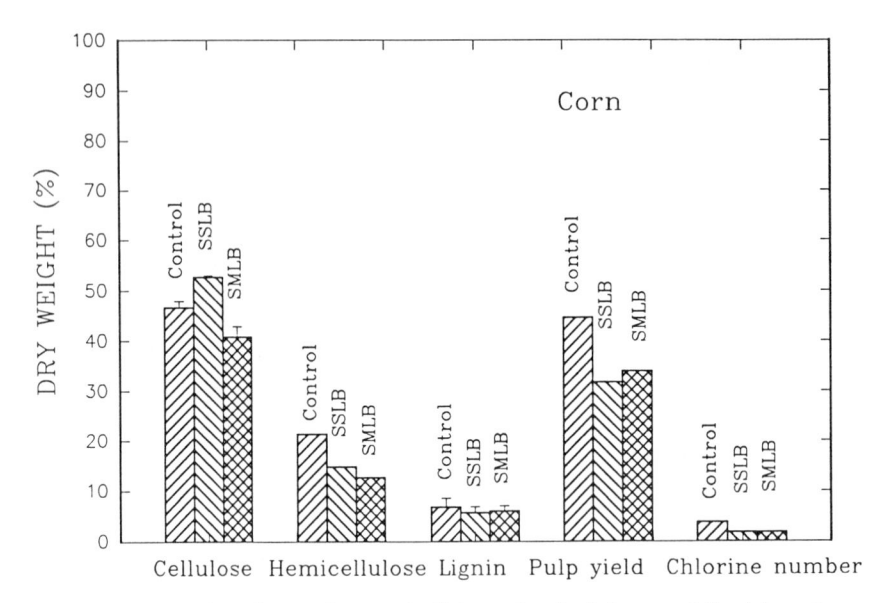

Figure 2. Cell wall constituents (cellulose, hemicellulose and lignin), pulp yield and chlorine number of control and biotreated corn stalks. Cellulose, hemicellulose and lignin were determined after the biological treatments, whereas pulp yield and chlorine number after alkaline cooking. The data are referred to 100 g dry weight.

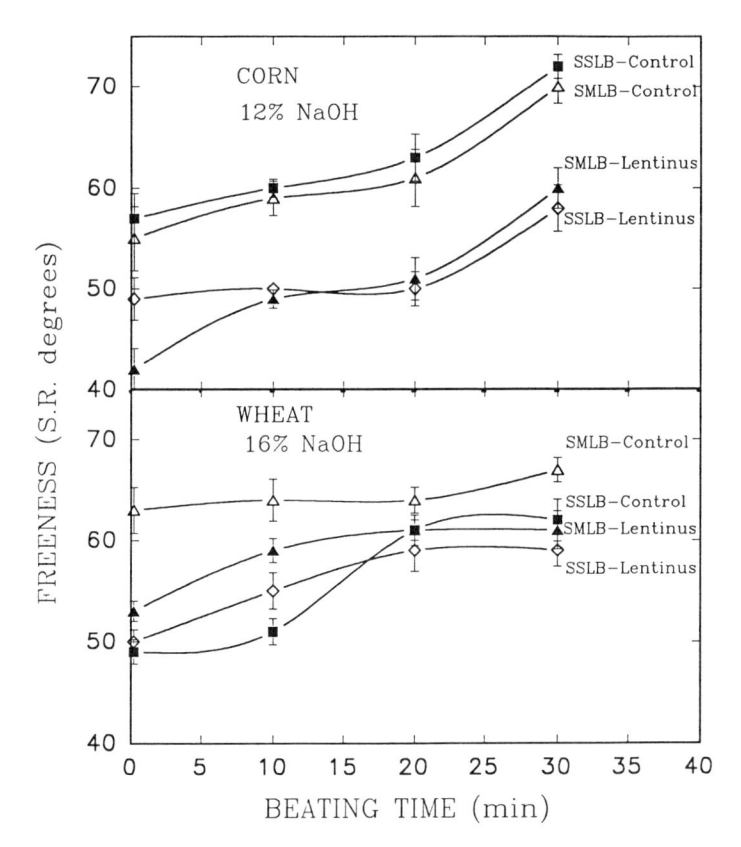

Figure 3. Relationships between freeness (S.R. degrees) and beating time (0, 10, 20, 30 min.).

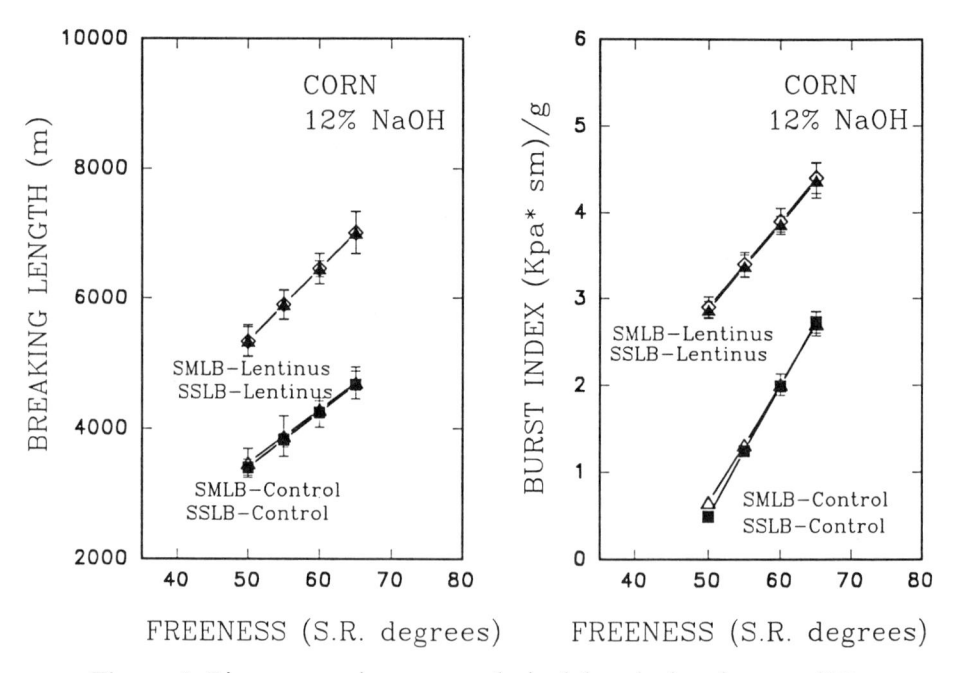

Figure 4. Linear regression curves obtained by plotting freeness (S.R. degrees) versus breaking length (m) and burst index (KPa·m^2·g^{-1}) of the paper handsheets of untreated and biotreated corn stalks (SSLB and SMLB).

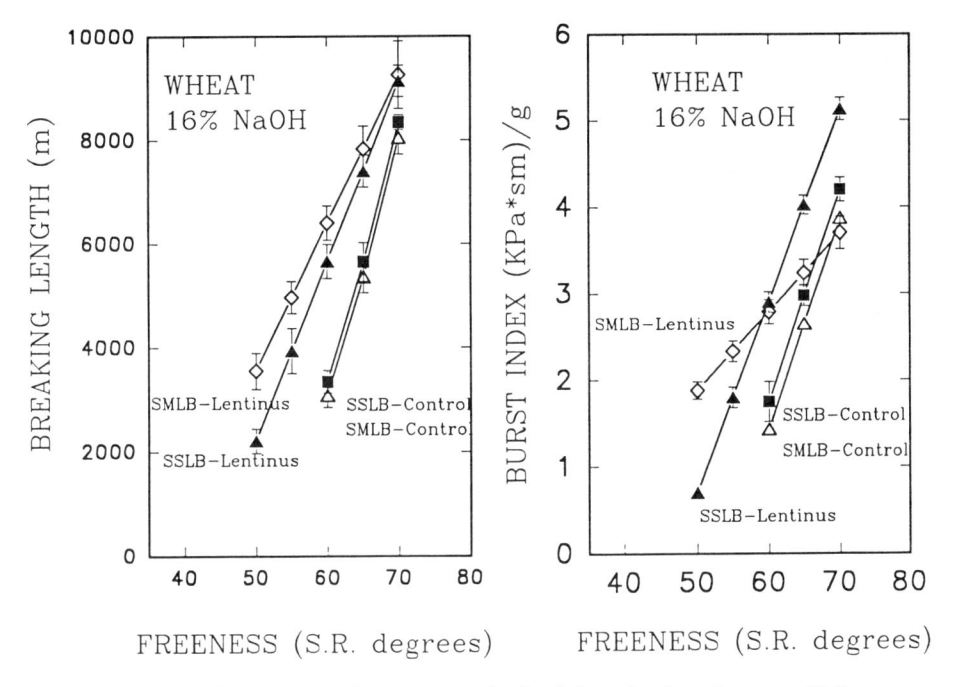

Figure 5. Linear regression curves obtained by plotting freeness (S.R. degrees) versus breaking length (m) and burst index (KPa·m^2·g^{-1}) of the paper handsheets of untreated and biotreated wheat straw (SSLB and SMLB).

speed. To evaluate statistically the technical results due to the biotreatments, freeness data, related to breaking length and burst index values, were submitted to the regression analysis. A high correlation between freeness and burst index or breaking lenght were obtained in all conditions (R squared > 0.9) with confidence levels lower than 5%. Figure 4 shows that, in the case of corn, the biotreatments resulted in a net improvement of the strength properties, even if these positive effects were less remarkable in wheat paper, as shown in Figure 5. Furthermore the values of breaking length and burst index referred to different S.R. degrees in the case of wheat had a wider range than in the case of corn. Tear index, which was not routinely determined, appeared to be also positively affected by the biotreatments. In fact, the untreated controls of durum wheat straw and corn stalks showed similar values of such technical parameter (3.5-3.6 mN/m^2·g), that apppeared to be improved by 45-50% in the case of wheat and by about 20-25% in the case of corn, regardless the enzyme treatment (SSLB and SMLB).

Conclusions

These preliminary results, obtained with annual plants, suggest the possibility of a successful combination of a biological pretreatment with conventional pulping techniques. The optimization of such a process must take into account several factors including the characteristics of the cellulose source, the composition of the exo-enzyme mixtures, the biotreatment and pulping conditions. Until now, the scientific data available on the physiology and biochemistry of lignolytic mycelia, as well as the lignocellulose chemical structure of the raw materials are not sufficient to forecast the results of a biopulping process. These biotechnological applications, based on the use of a single enzyme (lignin peroxidase, phenol-oxidases or xylanase) seem to be less adequate than those based on enzymic mixtures, since the cellulose pulp is obtained from a substrate where cellulose is covalently bonded to a wide variety of aliphatic and aromatic macromolecules. Our results show that the most efficient biotreatment for the improvement of the paper characteristics, as well as for a better bioconversion of the lignocellulose, was obtained by using the enzyme mixture derived from SSF, being the cellulase activity of scarce effect on the yield. Furthermore it's worth noting that the tested agricultural residues (wheat straw and corn stalks) can be successfully used as fibre source to obtain papers with good mechanical characteristics and therefore it could be possible to envisage positive economic results by using such waste materials.

Acknowledgements

The authors wish to express their gratitude to Mr. G. Falesiedi for his skilful technical assistance.

Literature Cited

(1) Trotter, P. C. *TAPPI J.* **1990**, *40*, 198-204.
(2) Giovannozzi-Sermanni, G; Perani, C. *Chimicaoggi* **1987**, *3*, 55-59.
(3) Giovannozzi-Sermanni, G; Porri, A.; Perani, C.; Badalucco, L.; Garzillo, A. M. In *The role of soluble lignocellulose produced during solid-state conversion of plant material by white-rot fungi*, Coughlan, M. P.; Amaral Collaco, M. T., Ed., Advances in biological treatment of lignocellulosic materials, Elsevier, London, 1990, pp. 59-70.
(4) Yang, J. L.; Lou, G.; Eriksson, K. E. L. *TAPPI J.*, **1992**, *42*, 95-101.
(5) Arbeloa, M.; de Leseluec, J.; Goma, G.; Pommier, J.C. *TAPPI J.* **1992**, *42*, 215-221.
(6) Bourbonnais, R.; Paice, M.G. *Appl. Microbiol. Biotechnol.* **1992**, *36*, 823-827.
(7) Kantelinen, A.; Hortling, B.; Ranua, M; Viikari, L. *Holzforschung* **1993**, *47*, 29-35.
(8) Pommier, J. C.; Fuentes, J. C.; Goma, G. *TAPPI J.* **1990**, *40*, 197-202.
(9) Fuwape, J. A. (1993). *Biores. Technol.*, *43:*113-115.
(10) Giovannozzi-Sermanni, G.; D'Annibale, A.; Porri, A.; Perani, C. *AgroFood Ind. Hi-Tech.* **1992**, *3*, 39-42.
(11) Giovannozzi-Sermanni, G ; D'Annibale, A.; Perani, C.; Porri, A. *Ag. Med.* **1993**, *123*,191-199.
(12) Giovannozzi Sermanni, G.; D'Annibale, A.; Perani, C.; Porri, A.; Pastina, F.; Minelli, V.; Vitale, N.; Gelsomino, A. *TAPPI J.*, **1994**, *77*, 151-157.

(13) Giovannozzi-Sermanni, G; Perani, C.; Porri, A.; De Angelis, F.; Barbarulo, M.V.; Mendola, D.; Nicoletti, R. *Agrochimica* **1991**, *35*, 174-189.
(14) Van Soest, P. J. *J. Assoc. Off. Anal. Chem.*, **1963**, *46*, 829-835.
(15) Bradford, M. M. *Anal. Chem.* **1976**, *72*, 248-254.
(16) Pick, E., Keisari, Y . *J. Immunol. Methods* **1980**, *38*, 161-170.
(17) Guillen, F.; Martinez, A.T.; Martinez, M.J.; Evans, C.S. Appl. Microbiol. Biotechnol. **1994**, *41*, 465-470.
(18) Glenn, J. K.; Gould, M. H. *Arch. Biochem. Biophys.* **1985**, *242*, 329-341.
(19) Haars, A., Chet, I., Huttermann, A. *Eur. J. For. Pathol.* **1981**, *11*, 67-76.
(20) Shewale, J. G., Sadana, J. C. *Can. J. Microbiol.* **1978**, *24*, 1204-1208.
(21) Mc Carthy, A. J.; Peace, E.; Broda, P. *Appl. Microbiol. Technol.* **1985**, *21*, 238-244.
(22) Perry, R.H.; Green, D.W.; Maloney, J. O. In *Perry's Chemical Engineer Handbook*. 6th Ed. Mc Graw Hill, New York, U.S.A.,1984, vol. 1, pp. 1-22.
(23) Finn, R.K.. In *Agitation and aeration*. Biochemical and Biological Engineering Science, Blakebrough Ed., Academic Press, New York, U.S.A., 1967, vol.1, pp. 23.
(24) Forney, L.J.; Reddy, C.A.; Pankratz, H.S. *Appl. Environ. Microbiol.* **1982a**, *44*, 732-736.
(25) Forney, L.J.; Reddy, C.A.; Tien, M.; Aust, S.D. *J. Biol. Chem.* **1982b**, *257*, 1455-1462.
(26) Reddy, C.A.; Forney, L.J.; Kelley, R.L. In *Involvement of hydrogen peroxide-derived hydroxyl radical in lignin degradation by the white-rot fungus Phanerochaete chrysosporium*. Higuchi, T.; Chang, H.M. Ed., Recent Advances in Biodegradation Research, , Uni Publishers Co. LTD, Tokyo, Japan, 1983, pp. 153-163.
(27) Venees, R.G.; Evans, C.S. *J. Gen. Microbiol.* **1989**, *135*, 2799-2806.
(28) Giovannozzi-Sermanni, G.; Porri, A. *Chimicaoggi* **1989**, *7*, 15-19.

RECEIVED September 11, 1995

Chapter 23

Possible Roles of Xylan-Derived Chromophores in Xylanase Prebleaching of Softwood Kraft Pulp

Ken K. Y. Wong, Patricia Clarke[1], and Sandra L. Nelson

Chair of Forest Products Biotechnology, Department of Wood Science, Faculty of Forestry, University of British Columbia, 270–2357 Main Mall, Vancouver, British Columbia V6T 1Z4, Canada

> Xylanase treatment can directly brighten partially bleached kraft pulps but not brownstocks. It usually leads to a direct decrease in the kappa number of the pulps. The degree of direct brightening seems to be correlated with the amount of UV-absorbing material solubilized from the pulps, rather than the amount of sugars. Previous results also showed that it did not correspond to the brightness gain after subsequent peroxide bleaching. There is therefore no evidence indicating that xylan-derived chromophores are the main target substrates during xylanase prebleaching of kraft pulp.

Xylanase prebleaching of kraft pulp has been implemented in several kraft mills in North America and Europe in order to reduce the environmental impact of their bleach plant (1). This process enhances subsequent chemical bleaching, thereby reducing the chemical loading required to reach target pulp brightness (2). The mechanism by which xylanase enhances pulp bleaching remains unclear. Kantelinen et al. (3) have proposed that the enzyme attacks xylan redeposited on the surface of pulp fibers at the end of the kraft cook. However, the removal of the surface xylan which is extractable with dimethyl sulfoxide does not enhance bleaching (4). Other studies indicate that xylanase treatment of kraft pulp also leads to extensive depolymerization of xylan (5). To-date, four hypotheses have been made concerning the nature of the target xylan (6):
- xylan in lignin–carbohydrate complexes;
- xylan that physically entraps residual lignin;
- xylan that modifies fiber porosity;
- xylan that is chromophoric.

[1]Current address: Pulp and Paper Research Institute of Canada, 3800 Wesbrook Mall, Vancouver, British Columbia V6S 2L9, Canada

An objective of our work was to eliminate one of these hypotheses, namely the role of xylan–derived chromophores in xylanase prebleaching. It is hypothesized that chromophoric units derived solely from xylan are still carried on xylan polymers. These hypothetical polymers may be, at least in part, hydrolyzed by xylanase and they may contribute to the adsorptive and light–scattering properties of pulp, and thus its reflectance. Lignin–carbohydrate complexes differ in that their chromophoric units are assumed to be derived from the lignin component.

The hypothesis that xylan may contribute directly to the color of kraft pulp is based on the observations of Hartler and Norrström (7). These authors found that holocellulose develops color after kraft cooking, and thus estimated that about 10 % of the light absorption properties of kraft pulp could be attributed to carbohydrates. In addition, the reflectance of α-cellulose was reported to decrease after cooking with xylan (8). When cellulose was cooked with sugars, the resultant decrease in pulp brightness could be easily reversed by bleaching (9) and this decrease was more substantial if tannins were also present (8). Recent studies have also indicated that the kraft or alkaline cooking of monosaccharides produces aromatic compounds such as dihydroxybenzaldehyde, compounds based on benzendiols and acetyophenones (10), and high molecular mass material that yields cresol, ethylphenols and phenol upon pyrolysis (11).

Although xylanases are not known to directly modify lignin, which is considered to be the major contributor to pulp color (7, 8, 12, 13), these enzymes might enhance pulp bleaching by removing xylan–derived chromophores. In the present work, this hypothesis on the mechanism of xylanase prebleaching is evaluated by examining the direct effects that xylanase has on different kraft pulps.

Materials and Methods

Xylanases. One purified xylanase and two commercial xylanases were used in the work. Enzyme A was Irgazyme 40, a crude *Trichoderma* xylanase commercially produced by Genencor (San Francisco, USA). Enzyme B was Pulpzyme HB, a bacterial xylanase commercially produced by Novo Nordisk (Bagsvaerd, Denmark). Enzyme E, a 22 kDa xylanase belonging to Family 11 of glycanases, was purified from *Trichoderma harzianum* E58 as reported by Maringer et al. (14).

Kraft Pulps. Samples of brownstock and oxygen delignified kraft pulp derived from western hemlock (*Tsuga heterophylla*) were kindly provided by Howe Sound Pulp and Paper (Port Mellon, Canada). The kraft pulps derived from Douglas–fir (*Pseudotsuga menzeisii*) and western red cedar (*Thuja plicata*) were provided by Fletcher Challenge (Crofton, Canada).

Pulp Bleaching. Xylanase treatment was carried out at 10 % pulp consistency and 50 °C for 1 h. Its start pH, adjusted using H_2SO_4, was 7 for Enzymes A and B, and 4.8 for Enzyme E. The enzymes were each loaded at

200 nkat xylanase/g pulp, with the activity assayed at the start pH. The dinitrosalicylic acid–based assay of Bailey et al. (15) was used with citrate buffer at pH 4.8 and phosphate buffer at pH 7. The enzyme was replaced with water, or boiled enzyme when specified, in the control treatment. Most of the data presented for Enzymes A and B were collected as parts of a survey on the effects that different xylanases have on the bleaching of different kraft pulps (16).

Chemical bleaching was carried out using peroxide (2 % H_2O_2; 2 % NaOH; 0.05 % $MgSO_4 \cdot 7H_2O$) at 10 % pulp consistency and 80 °C for 3 h. A 30 min. chelation stage was always carried out before peroxide bleaching using 1 % $Na_2 \cdot EDTA \cdot 2H_2O$ at a start pH of 5.5, 3 % consistency and 50 °C, and it was followed by pulp washing. Oxygen delignification of the brownstock derived from Douglas–fir was carried out in a Mark IV reactor (Quantum Technol. Inc., Twinsburg, USA) using 1.5 % NaOH and 0.3 % $MgSO_4 \cdot 7H_2O$ at 70 psi oxygen, 10 % consistency and 90 °C for 1 h. All chemical loadings were based on the dry weight of the pulp. An example of a bleaching sequence is OXQP, where O = oxygen delignification, X = xylanase treatment, Q = chelation, and P = peroxide bleaching.

Analytical Techniques. The brightness of pulp handsheets (CPPA Standard C.5) was measured using a Technibrite TB-1c instrument (Technidyne, New Albany, USA), and the lignin content was quantified by determining the microkappa number (TAPPI Useful Method UM 250, 1991). Filtrates from the enzyme treatment were analyzed for the solubilization of UV–absorbing material and sugars. Absorption at 280 nm was measured after appropriate dilution of the filtrate to give an absorbance ranging from 0.3–0.7, and the relative absorption was then calculated by multiplying the absorbance with the dilution factor and subtracting the enzyme blank. Total sugars were quantified using the phenol–sulfuric acid method (17), with the filtrate diluted eight folds and results verified using samples spiked with a known quantity of xylose.

Results and Discussion

Our study examined the direct effects that three xylanases have on brownstock, oxygen delignified and peroxide bleached kraft pulps derived from three softwood species. Pulp handsheets were made immediately after the enzyme stage and after subsequent peroxide bleaching. They were compared with pulp that had undergone a control treatment, carried out under the same conditions. To simplify the discussion, differences in pulp brightness and kappa number were considered to be substantial when they were greater than 0.5 % ISO and 0.5 pt., respectively.

For brownstocks derived from Douglas–fir and hemlock, treatment with xylanase did not lead to a direct increase in pulp brightness as compared to the control treatment (Figure 1). In most of these trials, a subsequent peroxide stage indicated that bleaching had been enhanced by the enzyme treatment. The exceptions were the pretreatment of the brownstock derived from Douglas–fir with Enzymes A and B, although the former enzyme was found to enhance pulp brightness when a second peroxide stage was used (16). There is therefore no evidence that xylanase directly brightens brownstocks when used under conditions in which it enhances subsequent peroxide bleaching. However, a direct decrease in kappa number was observed in most cases (Figure 1), the exceptions being the pretreatment of hemlock brownstock with Enzymes B and E. Other trials with Enzyme E indicated that it can directly decrease the kappa number, but not increase the brightness, of the brownstock derived from hemlock as well as that from red cedar (Table I, trials 1a, 2a & 3). Boiled enzyme controls were also performed for the experiments carried out with Enzyme E, and the results were essentially indistinguishable from those obtained with the water controls (data not shown).

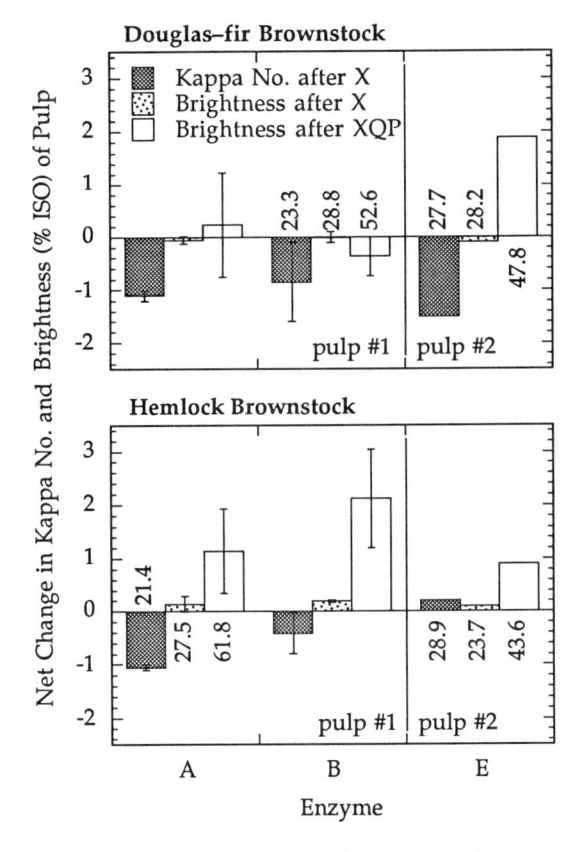

Figure 1. Direct effects of three xylanases on brownstock kraft pulps derived from Douglas–fir and hemlock. Pulp sample #1 was tested using Enzymes A and B while pulp sample #2 was tested with Enzyme E. The histograms show the net change from the controls, the values of which are reported at the base of the histograms. Error bars, when present, show the range for two replicate experiments.

Table I. The direct effects that Enzyme E has on various unbleached and partially bleached kraft pulps

Trial Number	Sample Bleaching Sequence	Wood Species	Directly after Enzyme Treatment Control Kappa No.	Directly after Enzyme Treatment Control Brightness (% ISO)	Test ΔKappa No.	Test ΔBrightness (% ISO)	Brightness after Bleaching (%ISO) Control	Brightness after Bleaching (%ISO) Test (Δ)
1a	XQPP	hemlock (#2)	26.6	24.7	−1.1	0	57.2	+0.7
1b	QPPX			57.1		+0.1		
2a	XQPP	hemlock (#2)	27.6	23.6	−0.6	+0.2	57.1	+1.3
2b	QPPX			57.2		+0.2		
3	XQP	cedar	27.0	27.0	−0.7	−0.3	47.2	+0.6
4	QPX	cedar	14.5	47.5		+0.1		
5	QPX	Douglas–fir	15.4	47.4		+0.1		
6	QPX	hemlock (#2)	16.0	43.4		+0.2		
7	OXQP	hemlock	14.8	35.6	−0.4	+0.1	61.8	+0.3

In general, our results agree with most of the other work reported on the direct effects that xylanase has on kraft pulps. Direct decrease in kappa number has been reported by several workers (18–23), but the case for direct brightening remains debatable. Direct brightening by xylanase was not found in a hardwood kraft pulp by Paice et al. (18) nor in an oxygen delignified softwood pulp by Pedersen et al. (24). However, it was reported in an oxygen delignified hardwood pulp by Pedersen et al. (24), and in both hardwood and softwood pulps by Yang and Eriksson (19). In our work with softwood brownstocks, it is not clear why the direct decrease in kappa number was not associated with direct pulp brightening. More work is therefore required to characterize the materials solubilized from the pulp.

Our results on partially bleached kraft pulps suggest that they can be directly brightened by xylanase (Figure 2). Both Enzymes A and B directly brightened two peroxide bleached pulps, whereas only Enzyme A brightened the oxygen delignified pulp. For Enzyme A, a direct decrease in the kappa number of the pulp was also observed. The scatter plot summarizing the results clearly shows that partially bleached pulps were directly brightened but not brownstocks. Although the degree of direct brightening did not correspond to the brightness gain after subsequent peroxide bleaching (16), the direct effects that xylanase has on more fully bleached pulp warrant further investigation in order to determine whether the enzyme can further brighten the pulp.

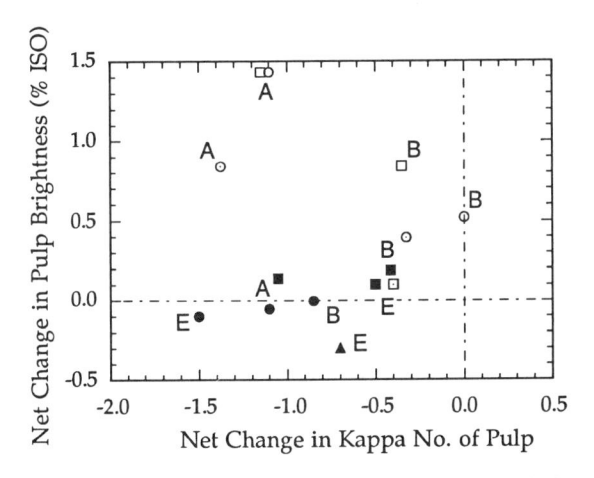

Figure 2. Relationship between direct brightening and direct decrease in the kappa number of kraft pulps by xylanases. Samples include those shown on Figure 1 and Table I, and those previously reported (16). Symbols that are black, that carry a dot at their center and that are white correspond to brownstocks, oxygen delignified pulps and peroxide bleached pulps, respectively. Triangles, circles and squares correspond to cedar, Douglas–fir and hemlock, respectively. The enzyme used in the treatment is labelled with its corresponding letter.

Trials have been carried out on QPP-bleached pulp derived from hemlock using the purified xylanase, Enzyme E (Table I, trials 1b & 2b). The results indicate that Enzyme E cannot directly brighten bleached kraft pulp derived from hemlock, even though it can enhance the bleaching of the brownstock (Table I, trials 1a & 2a). Indeed, this enzyme was found to be unable to directly brighten peroxide bleached pulps from three wood species and oxygen delignified pulp derived from hemlock (Table I, trials 4–7). The absence of direct brightening in these cases could be due to several factors, such as the difference in the enzyme, pulp or start pH used. On the other hand, it is also possible that the removal of metal ions during bleaching has reduced the hydrolytic efficiency of this enzyme, as has been reported for a similar xylanase purified from *T. reesei* (25).

An unexpected correlation was found between the amounts of UV-absorbing material solubilized from pulp and the degree of direct brightening (Figure 3), but these two variables were not correlated with the direct decrease in kappa number (Figures 1 & 3). It would appears that UV absorbance might be more indicative of lignin than kappa no., where a reaction with permanganate is the basis for measuring lignin. It is therefore possible that some of the UV-absorbing material present in pulp do not react with permanganate or that some of the material reacting with permanaganate do not absorb in the UV range. For example, compounds other than lignin might be oxidized by permanganate. These oxidizable compounds, along with certain lignin structures, might have little absorbance in the visible range and thus further complicate the relationship between pulp brightness and kappa no. These considerations could help explain how Enzyme A can decrease the kappa no. of different pulps to the same extent while solubilizing different amounts of UV-absorbing material, and how Enzymes B and E can decrease the kappa no. in three cases without solubilizing large amounts of UV-absorbing material.

Our results indicate that xylanases may vary in their ability to decrease the kappa number of kraft pulps, as has been reported previously (21). Xylanases may also vary in their ability to solubilize sugars from pulp (23, 26, 27). In our work, Enzyme A was found to be consistently more effective than Enzyme B or E (Figure 4). The relationship between the solubilization of sugars or UV-absorbing material and the change in the brightness or lignin content of pulp may be further clarified if these variables are examined in a specific pulp by changing enzyme loadings or reaction times. However, it may be difficult to resolve the relatively small changes in pulp brightness and lignin content.

Although our work does not eliminate the possibility that xylan-derived chromophores occur in kraft pulp, it does suggest that they are not the sole explanation for the mechanism of xylanase prebleaching of kraft pulps. One reason is that when direct pulp brightening was observed in partially bleached pulps, the increase in pulp brightness was not the same as that achieved after additional peroxide bleaching (16). Furthermore, since direct brightening is generally associated with some decrease in kappa number, it would seem that the solubilization of xylan-derived chromophores always leads to some solubilization of residual lignin or lignin–carbohydrate complexes. An alternative explanation is that oxidation by permanganate does not distinguish xylan–derived chromophores from lignin. This possibility reflects the limitations of the approach used in the present study as well as the techniques available for differentiating lignin–containing components from xylan–derived chromophores. There is also a lack of methods to examine the effects that 'non-chromophoric' xylan has on the light–scattering properties of pulp. It appears that further investigation of the other three hypotheses concerning the mechanism of xylanase prebleaching may be more fruitful. Recent reports have indicated that larger lignin molecules can be extracted from kraft pulp with alkali after a treatment with xylanase (28, 29) and that xylanase can reduce the molecular mass of the UV–absorbing material in xylan–containing extracts from kraft pulps (30).

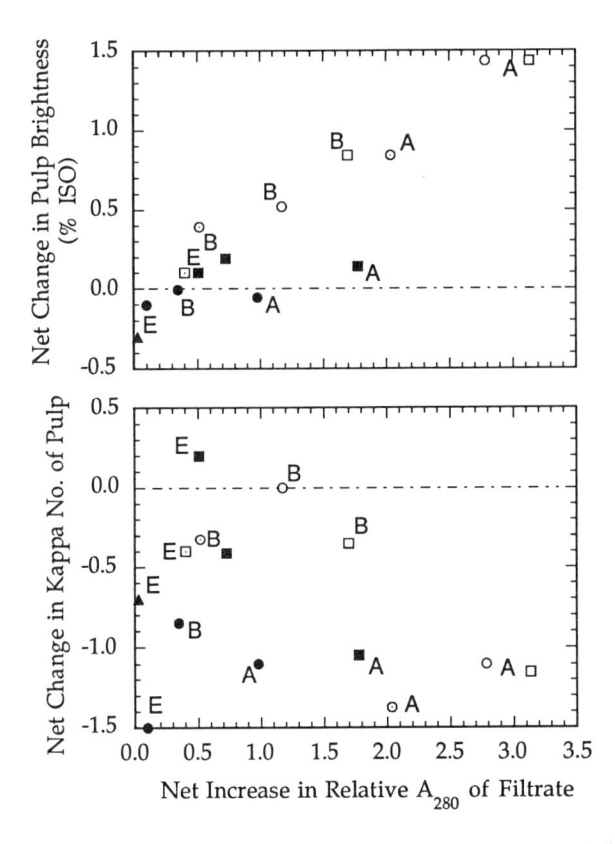

Figure 3. The solubilization of UV–absorbing material from kraft pulps and its relationship with direct brightening and decrease in the kappa number of pulp. Symbols and lettering follow the convention indicated in the legend of Figure 2.

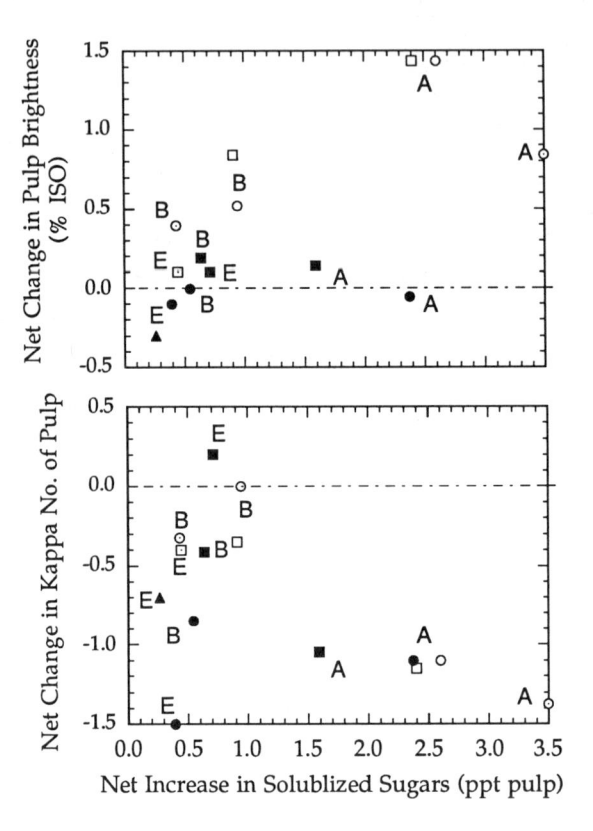

Figure 4. The solubilization of sugars from kraft pulps and its relationship with direct brightening and decrease in the kappa number of pulp. Symbols and lettering follow the convention indicated in the legend of Figure 2.

Conclusions

- Xylanases varied in their ability to directly decrease the kappa number or increase the brightness of kraft pulp
- Direct decrease in kappa number was observed in most cases
- Direct brightening was only observed in partially bleached kraft pulp
- Direct brightening seemed to be correlated with the solubilization of UV-absorbing materials rather than the solubilization of sugars
- Evidence did not indicate that xylan-derived chromophores constitute the sole target substrate during xylanase prebleaching

Acknowledgments

We thank the Pulp and Paper Research Institute (Vancouver, Canada) for the use of their equipment for handsheet making, optical testing and oxygen delignification. We also thank Drs. R.P. Beatson (Canfor R&D Centre, Vancouver) and H. Worster (Fletcher Challenge, Vancouver) for arranging the supply of pulp samples, and Drs. E. de Jong and J.N. Saddler for their critique of this manuscript. This work was partly financed by a grant to J.N. Saddler and K.K.Y. Wong from the Science Council of British Columbia and a grant to J.N. Saddler and C. Breuil from the Natural Sciences and Engineering Research Council of Canada.

Literature Cited

1. Jurasek, L.; Paice, M.G. *Proc.: Int. Symp. Pollution Prevention Manuf. Pulp Pap.* **1992**, pp 105–107.
2. Viikari, L.; Kantelinen, A.; Sundquist, J.; Linko, M., *FEMS Microbiol. Rev.* **1994**, *13*, 335–350.
3. Kantelinen, A.; Hortling, B.; Sundquist, J.; Linko, M.; Viikari, L. *Holzforschung* **1993**, *47*, 318–324.
4. Holm, H.C.; Skjold-Jørgensen, S.; Munk, N.; Pedersen, L.S. *Proc. Pan-Pac. Pulp Pap. Technol. Conf.* **1992**, *A*, 53–57.
5. Paice, M.G.; Gurnagul, N.; Page, D.H.; Jurasek, L. *Enzyme Microb. Technol.* **1992**, *14*, 272–276.
6. Wong, K.K.Y.; Saddler, J.N. *Crit. Rev. Biotechnol.* **1992**, *12*, 413–435.
7. Hartler, N.; Norrström, H. *Tappi* **1969**, *52*(9), 1712–1715.
8. Bard, J.W. *Pap. Trade J.* **1941**, *113*(12), 29–34.
9. Pigman, W.W.; Csellak, W.R. *Techn. Assoc. Pap.* **1948**, *31*, 393–399.
10. Forsskåhl, I.; Popoff, T.; Theander, O. *Carbohydr. Res.* **1976**, *48*, 13–21.
11. Ziobro, G.C. *J. Wood Chem. Technol.* **1990**, *10*, 133–149.
12. Schwartz, H.; McCarthy, J.L.; Hibbert, H. *Pap. Trade J.* **1940**, *111*(18), 30–34.
13. Kimble, G.C. *Pap. Trade J.* **1942**, *115*(3), 37–46.
14. Maringer, U.; Wong, K.K.Y.; Saddler, J.N.; Kubicek, C.P. *Biotechnol. Appl. Biochem.* **1995**, *21*, 49–65.
15. Bailey, M.J.; Biely, P.; Poutanen, K. *J. Biotechnol.* **1992**, *23*, 257–270.
16. Nelson, S.L.; Wong, K.K.Y.; Saddler, J.N.; Beatson, R.P. *Pulp Pap. Can.* **1995**, *96*, in press.
17. Dubois, M.; Gilles, K.A.; Hamilton, J.K.; Rebers, P.A.; Smith, F. *Anal. Chem.* **1956**, *28*, 350–356.
18. Paice, M.G.; Bernier, R., Jr.; Jurasek, L. *Biotechnol. Bioeng.* **1988**, *32*, 235–239.
19. Yang, Y.,L.; Eriksson, K.-E.L. *Holzforschung* **1992**, *46*, 481–488.
20. Allison, R.W.; Clark, T.A.; Wrathall, S.H. *Appita* **1993**, *46*, 269–273,281.
21. Patel, R.N.; Grabski, A.C.; Jeffries, T.W. *Appl. Microbiol. Biotechnol.* **1993**, *39*, 405–412.
22. Pekarovicová, A.; Rybáriková, D.; Kosík, M.; Fiserová, M. *Tappi J.* **1993**, *76*(11), 127–130.
23. Saake, B.; Clark, T.; Puls, J. *Holzforschung* **1995**, *49*, 60–68.
24. Pedersen, L.S.; Kihlgren, P.; Nissen, A.M.; Munk, N.; Holm, H.C.; Choma, P.P. *TAPPI Proc. – Pulp. Conf.* **1992**, *1*, 31–37.

25. Buchert, J.; Viikari, L. *Int. Pulp Bleaching Conf.* – *Preprints* **1994**, pp 59–62.
26. Buchert, J.; Ranua, M.; Kantelinen, A.; Viikari, L. *Appl. Microbiol. Biotechnol.* **1992**, *37*, 825–829.
27. Gilbert, M.; Yaguchi, M.; Watson, D.C.; Wong, K.K.Y.; Breuil, C.; Saddler, J.N. *Appl. Microbiol. Biotechnol.* **1993**, *40*, 508–514.
28. Hortling, B.; Korhonen, M.; Buchert, J.; Sundquist, J.; Viikari, L. *Holzforschung* **1994**, *48*, 441–446.
29. Suurnäkki, A.; Kantelinen, A.; Buchert, J.; Viikari, L. *Tappi J.* **1994**, *77*(11), 111–116.
30. Yokota, S.; Wong, K.K.Y.; Saddler, J.N.; Reid, I.D. *Pulp Pap. Can.* **1995**, *96*(4), 39–41.

RECEIVED July 13, 1995

Author Index

Adney, William S., 113,208,256
Baker, John O., 113,208
Bothast, Rodney J., 197
Brown, R. M., 13
Chou, Yat-Chen, 113,208
Claeyssens, Marc, 90,164
Clarke, Patricia, 352
Cousins, S. K., 13
D'Annibale, A., 339
Damude, Howard G., 174
Domínguez, Juan Manuel, 164
Dowd, M. K., 13
Durán, Nelson, 332
Esposito, Elisa, 332
Freer, Shelby N., 197
French, A. D., 13
Fukui, Sakuzo, 79
Gilkes, Neil R., 142,174
Giovannozzi-Sermanni, G., 339
Gübitz, G. M., 319
Haltrich, Dietmar, 305
Hart, John, 59
Haun, Marcela, 332
Hehre, Edward J., 66
Hendrickson, R., 237
Henrissat, Bernard, 164
Henson, C. A., 51
Himmel, Michael E., 113,208,256
Honda, Koichi, 79
Irwin, Diana, 1
Ishikawa, Kazuhoko, 79
Jacobson, R. H., 38
Karplus, Andrew, 1
Kilburn, Douglas G., 142,174
Kohlmann, K. L., 237
Kuroki, R., 38
Kwan, Emily, 174
Ladisch, M. R., 237
Laymon, Robert A., 113,208
Macarrón, Ricardo, 164
Matsui, Eriko, 79

Matsui, Ikuo, 79
Matthews, B. W., 38
Meinke, Andreas, 174
Mielenz, J. R., 208
Milagres, Adriane M. F., 332
Miller, D. P., 13
Miller, Robert C., Jr., 142,174
Minelli, V., 339
Miyairi, Sachio, 79
Mohagheghi, A., 256
Nelson, Sandra L., 352
Nidetzky, Bernd, 90
Nieves, Rafael A., 113,208
Odt, Christine L., 291
Perani, C., 339
Porri, A., 339
Robertus, Jon D., 59
Saha, Badal C., 197
Sarikaya, A., 237
Sebesta, Brigitte, 305
Shen, Hua, 174
Spezio, Mike, 1
Steiner, Walter, 90,305,319
Sun, Z., 51
Taylor, Jeff, 1
Thomas, Steven R., 113,208,256
Tomme, Peter, 142,174
Tucker, Melvin P., 113,208
van Beeuman, Jozef, 164
Velayudhan, A., 237
Vinzant, Todd B., 113,208
Vitale, N., 339
Warren, R. Antony J., 142,174
Weaver, L. H., 38
Weil, J., 237
Weimer, Paul J., 291
Westgate, P. J., 237
Wilson, David B., 1
Wong, Ken K. Y., 352
Wyman, Charles E., 272
Zhang, X-J., 38

Affiliation Index

Agricultural Research Service,
   13,51,197,291
Agrital Research Consortium, 339
Albert Einstein College of Medicine, 66
Centre National de la Recherche
   Scientifique, 164
Cornell University, 1
Fukuyama University, 79
National Institute of Bioscience and
   Human Technology, 79
National Renewable Energy Laboratory,
   113,208,256,272
Oregon State University, 237
Purdue University, 237
Technical University Graz, 90,319
Universidad Complutense, 164
Universidade Estadual de Campinas, 332
University of Agriculture–Austria,
   90,305
University of British Columbia,
   142,174,352
University of Ghent, 90,164
University of Oregon, 38
University of Technology Graz, 305
University of Texas, 13,59
University of Tuscia, 339
University of Wisconsin, 51,291
U.S. Department of Agriculture,
   13,51,197,291
W. R. Grace and Company, 237

Subject Index

A

Ab initio quantum mechanics calculations,
   modeling of structures and energies, 13
Acidothermus cellulolyticus E1
   expression, 117
Agricultural wastes, paper biopulping by
   Lentinus edodes, 339–350
Air pollution, contribution of
   transportation fuels, 272–273
Amorphous cellulose, description, 187
Amylase(s)
   comparison to other enzymes catalyzing
      hydrolysis of insoluble
      polysaccharides, 1–11
   mutant, with enhanced activity specific
      for short substrates, 79–88
   role in starch degradation, 52–55
   that binds maltooligosaccharide, 79
Arginine, molecular structures, 81
Artificial cellulase systems for biomass
   conversion to ethanol
   advantages, 209
   future work, 231–232
   progress toward development, 220–231
Artificial cellulase systems for biomass
   conversion to ethanol—*Continued*
   review of cellulase literature, 210–217
   strategy, 217–220
Asparagine, molecular structures, 81

B

Bacteria, ruminal, cellulose degradation,
   291–301
Bacterial cellulase(s)
   similarities with fungal cellulase
      systems, 189,191–193
   synergism with fungal cellulases, 113–138
Barley endochitinase
   amino acid sequence, 62
   future work, 64
   isolation, 62
   ribbon drawing, 62,63f
   structure vs. that of hen egg white
      lysozyme, 62–64
Barley seed α-glucosidases, 51–57
   characteristics, 55–57
   experimental description, 51–52
   role in starch degradation, 52–55

INDEX

Bicinchroninic acid method, protein concentration determination, 257,259
Binding domains, occurrence, 2
Biofuels, search for lower cost raw production materials, 197
Biomass
 conversion to ethanol using artificial cellulase systems, 208–232
 economics of ethanol production, 272–289
Biopulping of agricultural wastes by *Lentinus edodes*, 339–350
Biotechnological application of microorganism, pulp and paper industry, 339
Bleaching sequences in hardwood kraft pulps, traditional and chlorine-free, xylanase delignification, 332–337
Bradford dye binding method for protein determination, procedure, 257,259
Brassica napus
 advantages for controlled ecological life support systems, 237
 composition, 237–238
 enhanced enzymatic activities on hydrated lignocellulosic substrates, 238–252
N-Bromosuccinimide oxidation, identification of two tryptophan residues in *Trichoderma reesei* endoglucanase III essential for cellulose binding and catalytic activity, 164–172
Brownstock, role in xylan-derived chromophore role in xylanase prebleaching of softwood kraft pulp, 354–357

C

Carboxyl side chains, role in enzyme activity, 2–3
Cartazyme HS, delignification in traditional and chlorine free bleaching sequences in hardwood kraft pulps, 332–337
Catalytic abilities of hydrolases, role of glycosyl fluorides, 67–69
Catalytic activity
 Cellulomonas fimi cellobiohydrolases, 184–189
 identification of essential tryptophan residues in *Trichoderma reesei* endoglucanase III, 164–172
Catalytic domain, three-dimensional structure, 6–11
Catalytic properties, thermostable β-glucosidases, 200,202–204
Catalysis, glycosylase, existence of separately controlled plastic and conserved phases, 66–77
Cellobiases, definition, 209
Cellobiohydrolases from *Cellulomonas fimi*,
 catalytic activities of CbhA and CbhB, 184–189
 general structural and functional organization of CbhA and CbhB, 175–177
 isolation of CbhA and CbhB, 175
 relationahip of CbhA and CbhB to other β-1,4-glucanases, 178–184
 similarities between bacterial and fungal cellulase systems, 189–193
Cellobiose
 induction of xylanase and cellulase in *Schizophyllum commune*, 305
 molecular deformations and lattice energies, 19–21,33,34t
 xylanase activity induction, 316
Cellooligosaccharides, hydrolysis by *Cellulomonas fimi* cellobiohydrolases, 187–190f
Cellulase(s)
 artificial, for conversion of biomass to ethanol, 208–235
 bacterial and fungal, synergism, 113–140
 comparison to other enzymes catalyzing hydrolysis of insoluble polysaccharides, 1–11
 effect on hydrolysis rate of cellulose, 238
 family classification, 143
 from *Trichoderma reesei* during cellulose degradation, 90–109
 hydrolysis mechanism, 3–5
 induction in *Schizophyllum commune*, 305–317

Cellulase(s)—*Continued*
  protein contents in worldwide
    round-robin assay, 256–270
Cellulolytic bacteria, ruminal, physiological
    and hydrolytic diversity, 291–303
Cellulolytic enzyme(s), categories for
    activity, 113–114
Cellulolytic enzyme production, *Sclerotium
    rolfsii*, 320
Cellulomonas fimi cellobiohydrolases
  catalytic activities
    endoglucanase activities, 184,187–190*f*
    exoglucanase activities, 184,186*f*
    molar specific activity, 184,185*f*
  experimental description, 175
  functional organization, 175,177–178
  isolation, 175,176*f*
  relationship
    to fungal family B cellobiohydrolases,
      178–182
    to β-1,4-glucanases, 178
    to *Trichoderma reesei* CBH II, 178–182
  similarities between bacterial and
    fungal cellulase systems, 189,191–193
  structural organization, 175,177–178
Cellulose
  hydrolysis by *Cellulomonas fimi*
    cellobiohydrolases, 187–190*f*
  importance of enzymatic
    depolymerization, 113
  induction of xylanase and cellulase in
    Schizophyllum commune, 305–317
  role of β-glucosidase in enzymatic
    hydrolysis, 198
  source of fermentable sugars in
    lignocellulosic feedstocks, 209
  synergistic hydrolysis, 92
Cellulose allomorph, molecular deformations
    and lattice energies, 21–23,33,34*t*
Cellulose binding, identification of
    essential tryptophan residues in
    Trichoderma reesei endoglucanase III,
    164–172
Cellulose-binding domains
  classification schemes, 143,144*t*
  family classification, 143
  functions, 143

Cellulose-binding domains—*Continued*
  role
    in endoglucanase synergism, 137
    in substrate hydrolysis, 92
  structures, 157,159–160*f*
Cellulose degradation
  ruminal microbes
    diversity among ruminal cellulolytic
      bacteria in adherence to cellulose,
      294–299
    hydrolytic capability diversity, 299–301
    physiological diversity among ruminal
      cellulolytic bacteria, 292–294
  synergistic interaction of cellulases
    from *Trichoderma reesei*, 90–109
Cellulose reduction to cellobiose,
    enzymatic synergism requirement, 209
Cellulose-rich substrates, induction of
    xylanase and cellulase in
    Schizophyllum commune, 305
Cellulosic substrate concentration, role
    in synergistic interaction of cellulases
    from *Trichoderma reesei* during
    cellulose degradation, 97–100
Cellulosic substrate type, role in
    synergistic interaction of cellulases
    from *Trichoderma reesei* during
    cellulose degradation, 95–98*f*
Cellulosomes, description, 210
Cereal seeds, source of calories, 51
Chemical modification, evaluation of
    importance of tryptophan residues in
    biologically active proteins, 164
Chemical pretreatments, enhancement of
    enzymatic conversion of cellulose to
    glucose, 238
Chitin
  annual production, 59
  hydrolysis by exochitinases, 60–64
  molecular structure, 60,61*t*
  occurrence, 59–60
Chitinases, comparison to other enzymes
    catalyzing hydrolysis of insoluble
    polysaccharides, 1–11
Chlorine-free bleaching sequences in
    hardwood kraft pulps, xylanase
    delignification, 332–337

Chromophores, xylan-derived, possible roles in xylanase prebleaching of softwood kraft pulp, 352–360
Classification, cellulose binding domains, 143
Computerized molecular modeling of carbohydrates, importance, 13
Conserved phase, existence in glycosylase catalysis, 66–77
Controlled ecological life support systems, waste, 237
Coomassie Brilliant Blue G-250 dye binding method, determination of protein concentration, 257
Crystalline cellulose
hydrolysis by fungal cellulase systems, 90
hydrolysis mechanism, 174–175
Crystallography, indication of means of structural control of steric outcome for glycosylases, 74,76

D

Debranching enzyme, role in starch degradation, 52–55
Deformation energy caused by crystal field, types, 14
Degradation, cellulose, synergistic interaction of cellulases from *Trichoderma reesei*, 90–109
Degree of synergistic effect, 114
Delignification by xylanase in traditional and chlorine-free bleaching sequences in hardwood kraft pulps, *See* Xylanase delignification in traditional and chlorine-free bleaching sequences in hardwood kraft pulps
Differing specificity model, endoglucanase, 136
Direct brightening, effect of xylanase treatment, 352–360
Domain that binds tightly to insoluble polymer substrate, occurrence, 2

E

Economics, ethanol production from lignocellulosic biomass, 272–289

Endo–endo synergism, purified bacterial and fungal cellulases, 121,129–138
Endo–exo synergism
purified bacterial and fungal cellulases, 120–127
role of enzymatic pretreatment of filter paper, 108
Endochitinase from barley, three-dimensional structure, 59–64
Endoglucanase
cellulose binding domain effect on synergism, 137
differing specificity model, 136
function, 114
grouping based on synergism, 137–138
self-synergism as model, 129–135
Endoglucanase III from *Trichoderma reesei* identification of two tryptophan residues, 164–172
Endoglucanase activities, *Cellulomonas fimi* cellobiohydrolases, 184,187–190*f*
Endoglucanase E2, three-dimensional structure of catalytic domain, 6–11
Endo-1,4-β-xylanases, occurrence, 320
Energies
modeling methods, 13
source of petroleum, 272
Enhanced enzyme activities on hydrated lignocellulosic substrates
biological pretreatment, 245,247*f*,248*f*,251,252*f*
compositional analysis, 241,242*t*
enzymatic hydrolysis, 241–242,245*t*
experimental procedure, 239–241
future work, 251
water pretreatment, 242–251
Enzymatic conversion of lignocellulosic materials to sugar, interest for space research, 237
Enzymatic degradation of biomass, 142
Enzymatic depolymerization of cellulose, synergism, 114–115
Enzymatic hydrolysis
cellulose, role of β-glucosidase, 198
glucosyl substrate with inversion or retention, transition states, 76

Enzymatic pretreatment of filter paper, role in synergistic interaction of cellulases from *Trichoderma reesei* during cellulose degradation, 104,106–109
Enzymatic protonation of glycal, determination of direction, 71,72f
Enzyme(s)
  enhanced activities on hydrated lignocellulosic substrates, 237–252
  hydrolysis catalyzing, comparison, 1–11
  roles in different organisms, 1–2
Enzyme families, number, 3,4t
Escherichia coli β-galactosidase, 39–43
Esterases, production in *Schizophyllum commune*, 306
Ethanol
  advantages as transportation fuel, 273
  economics of production from lignocellulosic biomass, 274–289
  from biomass using artificial cellulase systems, 208–232
  sources, 273
Exo cyclic enolic substrates, catalysis of glycosylation reactions, 71–73
Exo–exo synergism
  mechanistic concepts, 91–92
  purified bacterial and fungal cellulases, 127,128f
  role of enzymatic pretreatment of filter paper, 108–109
Exoglucanase activities, *Cellulomonas fimi* cellobiohydrolases, 184,186f
Exo-1,4-β-glucosidases, function, 114
Experimental design, role in synergism between purified bacterial and fungal cellulases, 119–120

F

Families of enzymes, number, 3,4t
Family I of cellulose binding domains, description, 143,145–146t,150–151
Family II of cellulose binding domains, description, 146–147t,151–153,155t
Family III of cellulose binding domains, description, 147–148t,153–155f
Family IV of cellulose binding domains, description, 148t,153,156f
Family VI of cellulose binding domains, description, 148–149t,153,156–157
Family IX of cellulose binding domains, description, 149t,157,158f
Family X of cellulose binding domains, description, 149t,157,158f
Fermentable sugars in lignocellulosic feedstocks, sources, 209
Ferulic acid esterase, production by *Schizophyllum commune*, 316
Filamentous fungi, cost-effective resource for industrial cellulases, 209
Fuel ethanol industry, expansion opportunities, 197
Functions, cellulose binding domains, 143
Fungal cellulase(s)
  hydrolysis of crystalline cellulose, 90
  similarities with bacterial cellulase systems, 189,191–193
  synergism with bacterial cellulases, 113–138
Fungal family B cellobiohydrolases, relationship to *Cellulomonas fimi* cellobiohydrolases, 178–182
Fungi, hemicellulolytic, simultaneous production of xylanase and mannanase, 319–328

G

4-*O*-β-Galactopyranosyl-D-mannopyranose, induction of xylanase and cellulase in *Schizophyllum commune*, 305–317
β-Galactosidase
  active site, 41–43
  comparison to T4 lysozyme, 46,49f
  three-dimensional structure, 39–41
β-1,4-Glucanases
  Cellulomonas fimi cellobiohydrolases, 182–185f
  classification, 178
α-D-Glucose, conversion of β-D-glucosyl fluoride by α-glucosidases, 73–75f

INDEX

Glucose inhibition, thermostable
  β-glucosidases, 204
Glucose release, synergism between
  purified bacterial and fungal
  cellulases, 113–138
α-Glucosidases
  barley seed, 51–57
  conversion of β-D-glucosyl fluoride to
    α-D-glucose, 73–75f
β-Glucosidases
  function, 114
  role in enzymatic hydrolysis of
    cellulose, 198
  thermostable, 197–206
β-D-Glucosyl fluoride, conversion of
  α-D-glucose by α-glucosidases, 73–75f
Glucosyl substrate, transition states in
  enzymic hydrolysis with inversion or
  retention, 76
Glycal, determination of enzymic
  protonation direction, 71,72f
Glycosidases
  comparison of β-galactosidase and T4
    lysozyme, 46,49f
  Escherichia coli β-galactosidase, 39–43
  β-galactosidase, 39–43
  reactions with small nonglycosidic
    substrates, 66–67
  T4 lysozyme, 43–46
Glycosyl fluorides, role in catalytic
  abilities of hydrolases, 67–69
Glycosyl substrates, prochiral, *See*
  Prochiral glycosyl substrates
Glycosylase catalysis, existence of
  separately controlled plastic and
  conserved phases, 66–77
Glycosylation reactions, catalysis with
  exo cyclic enolic substrates, 71–73
Glycosyltransferases, reactions with small
  nonglycosidic substrates, 66–67

H

Half-lives, thermostable
  β-glucosidases, 199,200t
Hardwood kraft pulps, xylanase
  delignification, 332–337

Hemicellulases
  domains, 142
  family classification, 143
  use for pulp bleaching, 332–333
Hemicellulolytic fungi, simultaneous
  production of xylanase and mannanase,
  319–328
Hemicellulose
  degradation requirements, 319
  examples, 319
  source of fermentable sugars in ligno-
    cellulosic feedstocks, 209
Hen egg white lysozyme
  hydrolysis mechanism, 62
  structure vs. that of barley chitinase,
    62–64
Hg^{2+}, enzyme inhibition, 3
Hinge region, alteration effect on enzyme
  activity, 2
Hybrid cellulase systems, advantages, 114
Hydrated lignocellulosic substrates,
  enhanced enzyme activities, 237–252
Hydrolases
  catalytic group functional flexibility, 69
  role of glycosyl fluorides in catalytic
    abilities, 67–69
Hydrolysis
  cellulase, 3–5
  crystalline cellulose by fungal cellulase
    systems, 90
  insoluble polysaccharides, comparison of
    enzymes catalyzing, 1–11
Hydrolytic capability, cellulose
  degradation by ruminal microbes,
  299–302

I

Induction of xylanase and cellulase in
  Schizophyllum commune, 305–317
  common regulatory control, 316–317
  compound effect on induction
    cellulase activity, 310,311f
    xylanase activity, 308–310
  experimental procedure, 306–308
  mannanase activity formation, 317

Induction of xylanase and cellulase in
  Schizophyllum commune—Continued
  time course
    batch cultivation, 312–315
    effect of inducer concentration on
      xylanase induction, 310,312
    xylanase synthesis, 310,311*f*
Insoluble polysaccharides, hydrolysis,
  comparison of enzymes catalyzing, 1–11
Isolated disaccharides, molecular
  deformations and lattice energies,
  16–19,33,34*t*
Isolated monosaccharides, molecular
  deformations and lattice energies,
  16,33,34*t*

K

κ number of kraft pulps, xylanase
  treatment effect, 352–360
Kjeldahl nitrogen analysis, 259
Kraft pulps
  xylanase delignification, 332–337
  xylanase prebleaching, role of
    xylan-derived chromophores, 352–360

L

Lactose, induction of xylanase and
  cellulase in *Schizophyllum commune*, 305
Lattice energies, solid saccharide models,
  13–36
Lentinus edodes, paper biopulping of
  agricultural wastes, 339–350
Lignin, degradation, 239
Lignin removal by enzymatic method,
  research interest, 332
Lignocellulosic biomass
  availability, 97–198
  economics of ethanol production, 272–289
  ethanol production with artificial
    cellulase systems, 208–235
Lignocellulosic materials, pretreatments
  to enhance enzymatic conversion of
  cellulose to glucose, 238
Lignocellulosic substrates, hydrated,
  enhanced enzyme substrates, 237–252

Linker region, *See* Hinge region
Lowry method, estimation of protein
  concentrations in complex biological
  samples, 256
Lysozymes
  comparison to other enzymes
    catalyzing hydrolysis of insoluble
    polysaccharides, 1–11
  hydrolysis mechanism, 62

M

Maltodisaccharides, molecular deformations
  and lattice energies, 23–30
β-Maltose monohydrate, molecular
  deformations and lattice energies, 24–28
Mannanases
  production in *Schizophyllum
    commune*, 306
  simultaneous production with xylanase by
    hemicellulolytic fungi, 325,327–328
Mannase activity formation, induction of
  xylanase and cellulase in *Schizophyllum
  commune*, 305–317
Mathematical analysis, synergistic
  hydrolysis of cellulose, 92
Methyl β-maltoside monohydrate,
  molecular deformations and lattice
  energies, 28–30,33,34*t*
Methyl α-maltotrioside tetrahydrate,
  molecular deformations and lattice
  energies, 30,33,34*t*
Microbes, ruminal, cellulose degradation,
  291–301
Miniature carbohydrate crystals, molecular
  deformations and lattice energies,
  19,33,34*t*
MM3, molecular deformations and lattice
  energies of solid saccharide models,
  14–34
Models of solid saccharides, molecular
  deformations and lattice energies, 13–36
Modified Lowry method for protein
  determination, procedure, 259
Molar specific activity, *Cellulomonas fimi*
  cellobiohydrolases, 184,185*f*

Molecular deformations and lattice
energies of solid saccharide models,
13–36
  cellobiose conformations, 19–21
  cellulose allomorph conformations, 21–23
  comparison to other studies, 33–34
  experimental procedure, 13–16
  isolated disaccharides, 16–19
  isolated monosaccharides, 16
  maltodisaccharide conformations
  distortions, 23–24
    β-maltose monohydrate, 24–28
    methyl β-maltoside monohydrate, 28–30
  miniature carbohydrate crystal, 19
  model quality test, 14–15
  trisaccharide conformations
    methyl α-maltotrioside tetrahydrate, 30
    panose, 30–33
Molecular mechanics, modeling of
structures and energies, 13–14
Mushroom, enhanced enzyme activities,
237–252
Mutant α-amylase with enhanced activity
specific for short substrates, 79–88

N

National Renewable Energy Laboratory
  chromatographic protein estimation
  method, procedure, 259–260
  protein contents of cellulase preparations
  in worldwide round-robin assay,
  256–270
Novozyme 473, delignification in traditional
and chlorine-free bleaching sequences
in hardwood kraft pulps, 332–337

P

Paper biopulping of agricultural wastes by
Lentinus edodes, 339–350
  biotreatment
    vs. enzyme composition, 342,344,345*f*
    vs. pulp yield, 344,346*f*
  effect of hydrogen peroxide, 342

Paper biopulping of agricultural wastes by
Lentinus edodes—Continued
  experimental procedure, 339–342
  mycelial enzyme activities, 342,343*t*
  paper characteristics after biotreatments,
  344,347–350
Partially bleached pulp, role in
xylan-derived chromophore role in
xylanase prebleaching of softwood kraft
pulp, 357
Peroxide-bleached pulp, role in
xylan-derived chromophore role in
xylanase prebleaching of softwood kraft
pulp, 358,359*f*
Petroleum, cost of imports and source
of energy, 272
pH, role in simultaneous production
of xylanase and mannanase by
hemicellulolytic fungi, 325,327–328
Physical pretreatments, enhancement of
enzymatic conversion of cellulose to
glucose, 238
Physicochemical characteristics,
thermostable β-glucosidases, 199
Physiological diversity, ruminal
cellulolytic bacteria, 292–294
Plant(s), defense against pathogens, 60
Plant defense proteins, structure studies,
60–61
Plastic phases, existence in glycosylase
catalysis, 66–77
Pleurotus ostreatus, enhanced enzyme
activities, 237–252
Polyporus adustus, induction of cellulase,
xylanase, and mannanase, 316
Polysaccharide(s), insoluble, hydrolysis,
comparison of enzymes catalyzing, 1–11
Potential energy surfaces for isolated
model molecules, construction, 14
Prochiral glycosyl substrates, stereochemistry of catalyzed reactions, 69–73
Production, ethanol production from
lignocellulosic biomass, 275–277
Protein
  concentrations in complex biological
  samples, 256–258

Protein—*Continued*
  contents of cellulase preparations in worldwide round-robin assay, 256–270
    comparison of protein methods, 263,267–270
    experimental procedure, 257,259–260
    laboratories, 257
    Lowry protein determinations, 260–265f,268
    solids content analysis, 263,266t
Pulp and paper industry
  biotechnological application of microorganisms, 339
  reduction of chlorine used for bleaching, 332
Pulps, kraft
  xylanase delignification, 332–337
  xylanase prebleaching, role of xylan-derived chromophores, 352–360

R

Rapeseed, *See Brassica napus*
Reactions catalyzed with prochiral glycosyl substrates, stereochemistry, 69–73
Rumen, description and function, 291
Ruminal cellulolytic bacteria
  diversity in adherence to cellulose, 292,294–299
  physiological diversity, 292–294
Ruminal fermentation, improvements, 291
Ruminal microbes, cellulose degradation, 291–301
Ruminal volume, value, 291

S

Saccharides, solid, molecular deformations and lattice energies, 13–34
Saccharomycopsis fibuligera α-amylase, mutation with enhanced activity specific for short substrates, 79–88
Schizophyllum commune
  induction of xylanase and cellulase, 305–317
  lignocellulolytic enzyme production, 306
  simultaneous production of xylanase and mannanase, 319–328

Sclerotium rolfsii
  cellulolytic enzyme production, 320
  simultaneous production of xylanase and mannanase, 319–328
Self-synergism, endoglucanase, 129–135
Semiempirical quantum mechanical models, modeling of structures and energies, 13
Separately controlled plastic phase, existence in glycosylase catalysis, 73
Short substrates, mutant α-amylase with enhanced activity, 79–88
Simultaneous production of xylanase and mannanase by hemicellulolytic fungi
  effect of carbon source on mannanase production, 322–324
  experimental procedure, 320–322
  mannanase activity vs. high level of xylanase activity, 324–325
  pH, 325,337–328
  stability vs. that of cellulolytic systems, 328
  temperature, 325,327–328
  time courses of formation, 325,326f
Soft rot fungi, cellulase systems, 175
Softwood(s), composition, 320
Softwood kraft pulp, xylanase prebleaching, role of xylan-derived chromophores, 352–360
Solid saccharides, molecular deformations and lattice energies, 13–34
Starch degradation, roles of barley seed α-glucosidases, 51–57
Steam explosion, description, 238–239
Stereochemistry, reactions catalyzed with prochiral glycosyl substrates, 69–73
Structure(s)
  cellulose binding domains, 157,159–160f
  modeling methods, 13
  three-dimensional, barley endochitinase, 59–64
Subsite affinity evaluation method, mutant α-amylase with enhanced activity specific for short substrates, 80
Substrate inhibition, thermostable β-glucosidases, 204
Substrate nature, role in synergism between purified bacterial and fungal cellulases, 118–119

Synergism
  between purified bacterial and fungal cellulases, 113–140
    cellulose digestion procedure, 118
    endo–endo synergism experiments, 121,129–138
    endo–exo synergism experiments, 120–127
    enzyme purification procedure, 115,117
    exo–exo synergism experiments, 127,128f
    experimental approach, 115,116f
    experimental design choice, 119–120
    physical nature of substrate, 118–119
    synergism calculation methods, 119–120
  calculation method for synergism between purified bacterial and fungal cellulases, 119–120
  definition, 6
  in enzymic depolymerization of cellulose, history, 114–115
  in hydrolysis of crystalline cellulose, cellulases, 3,6
  with cellulases, thermostable β-glucosidases, 205
Synergistic hydrolysis of cellulose, mathematical analysis, 92
Synergistic interaction
  cellulases from *Trichoderma reesei* during cellulose degradation, 90–111
    cellulosic substrate concentration, 97–100
    cellulosic substrate type, 95–97,98f
    enzymatic pretreatment of filter paper, 104,106–109
    experimental procedure, 92–94
    individual cellulolytic components, 100–103t
    previous studies, 92
    synergistic combination, 101–102,104,105f
  fungal celluloses, mechanistic concepts, 90–91

T

T4 lysozyme
  comparison to β-galactosidase, 46,49f
  three-dimensional structure, 43–48
Temperature, role in simultaneous production of xylanase and mannanase by hemicellulolytic fungi, 325,327–328
Thermal β-glucosidases, production, 198–199
Thermoactivity, thermostable β-glucosidases, 199–200
Thermomonospora fusca cellulases, increase in insoluble cellulose degrading activity, 6–11
Thermomonospora fusca endocellulase E2, alteration of hinge region, 2
Thermomyces lanuginosus, simultaneous production of xylanase and mannanase, 319–328
Thermostable β-glucosidases, 197–206
  catalytic properties, 200,202–204
  glucose inhibition, 204
  half-lives, 199,200t
  physicochemical characteristics, 199
  substrate inhibition, 204
  synergism with cellulases, 205
  thermoactivity, 199–200
  thermostability, 199–202t
Three-dimensional structure
  barley endochitinase, 59–64
  Escherichia coli β-galactosidase, 39–43
  T4 lysozyme, 43–48
Transportation fuels, contribution to urban air pollution, 272–273
Trichoderma reesei
  description, 210
  mechanism of crystalline cellulose hydrolysis, 174–175
  synergistic interaction of cellulases during cellulose degradation, 90–109
Trichoderma reesei CBH II, relationship to *Cellulomonas fimi* cellobiohydrolases, 178–182
Trichoderma reesei CBH II and EG II, purification, 117

Trichoderma reesei EG I and CBH I, purification, 117
Trichoderma reesei endoglucanase III, identification of two tryptophan residues, 164–172
Trisaccharides, molecular deformations and lattice energies, panose, 30–34
Tryptophan residues in *Trichoderma reesei* endoglucanase III essential for cellulose binding and catalytic activity, 164–172

U

U.S. Department of Energy program for conversion of lignocellulosic biomass to ethanol transportation fuel, generations of technology, 208–209

V

van der Waals contacts and hydrogen bonds, carbohydrate-binding proteins, 164

W

Water, pretreatment agent for controlled ecological life support systems, 238–239
White rot fungus
 cellulase systems, 175
 paper biopulping of agricultural wastes, 339–350
Worldwide round-robin assay, protein contents of cellulase preparation, 256–270

X

D-Xylan(s), factors affecting side-group composition, 305
Xylan-derived chromophores, role in xylanase prebleaching of softwood kraft pulp, 352–360
 brownstock, 354–357,
 experimental procedure, 353–354
 partially bleached pulp, 357
 peroxide-bleached pulp, 358,359*f*
 xylanase, varying abilities, 358–360
Xylanase(s)
 bleaching of hardwood kraft pulps, 332–333
 classification, 178
 comparison to other enzymes catalyzing hydrolysis of insoluble polysaccharides, 1–11
 delignification in traditional and chlorine-free bleaching sequences in hardwood kraft pulps, 333–336
 induction in *Schizophyllum commune*, 305–317
 prebleaching of softwood kraft pulp, 358–360
 production in *Schizophyllum commune*, 306
 simultaneous production with mannanase by hemicellulolytic fungi, 325,327–328
Xylanolytic enzyme systems, examples, 306